Behavior Genetics
and Evolution

Behavior Genetics and Evolution

Lee Ehrman
State University of New York, Purchase

Peter A. Parsons
LaTrobe University
Bundoora, Australia

McGraw-Hill Book Company

New York   St. Louis   San Francisco   Auckland   Bogotá   Hamburg
Johannesburg   London   Madrid   Mexico   Montreal   New Delhi
Panama   Paris   São Paulo   Singapore   Sydney   Tokyo   Toronto

To our favorite physician and dentist,
Louise and Richie

This book was set in Times Roman by Bi-Comp, Incorporated.
The editors were James E. Vastyan and Scott Amerman;
the production supervisor was Diane Renda.
The drawings were done by VIP Graphics.
The cover was designed by Charles A. Carson.
Fairfield Graphics was printer and binder.

BEHAVIOR GENETICS AND EVOLUTION

1 2 3 4 5 6 7 8 9 0   FGFG   8 9 8 7 6 5 4 3 2 1

Library of Congress Cataloging in Publication Data

Ehrman, Lee.
   Behavior genetics and evolution.

   Originally published in 1976 under title: The
genetics of behavior.
   Includes bibliographies and index.
   1. Behavior genetics.   2. Behavior evolution.
I. Parsons, P. A. (Peter Angus)   II. Title.
QH457.E37   1981   591.51   81-1585
ISBN 0-07-019276-6   AACR2

Contents

Preface

Preface to *The Genetics of Behavior*

"The time seems ripe for a modern statement of the division of knowledge we have called 'behavior genetics' . . . not presented as a definitive work, because that would be impossible in a field of study which is in a dynamic stage of growth." The time was May 1960 and the writers, John L. Fuller and W. Robert Thompson, were coauthors of the very first book devoted to this hybrid subject (*Behavior Genetics,* Wiley, New York). Indeed, behavior genetics may be said to have begun in 1869 with the publication of another book, F. Galton's *Hereditary Genius* (followed by his *English Men of Science: Their Nature and Nurture* [1874] and *Inquiry into Human Faculty* [1883], Macmillan, London).

To us and those who advise us (some listed below), the time now seems ripe, one and a half decades later, for the appearance of an advanced textbook devoted to behavior genetics, although the field continues in its dynamic stage of growth. This being so, the present text is again ". . . not presented as a definitive work," a task that even now would still be impossible. Our textbook is directed to those undergraduate- or graduate-level students who already possess a basic background in general genetics. These could be students in biology or psychology or in what now often appears in colleges and universities as programs and majors in psychobiology, itself an increasingly fertile hybrid as can be seen by the rapid development of areas of studies in the behavioral sciences.

Because behavior genetics taught as a formal course is a recent innovation, it is our hope that this book will be useful for those already trained in a variety of ways for a multitude of careers. Professional geneticists, animal biologists, and psychologists come to mind first, but the aspects of behavior genetics covered here also increasingly enter into the work of physicians, veterinarians, animal breeders, sociologists, and educators generally, as should become clear from the examples and the organisms discussed. Thoughts about some major political controversies of our day may also be clarified by an understanding of behavior genetics of human beings and of other organisms as well.

We shall consider as behavior any and all activities performed by the holistic entity, the organism, in relation to this organism's surroundings, its environment. We do so according to the recommendation of Ethel Tobach (1972), but we confine ourselves to those aspects of an organism's muscular, glandular, and neural responses that have been demonstrated, albeit with varying degrees of firmness, to have an underlying hereditary basis—one transmitted via germinal tissues from generation to generation.

The examples we have chosen are necessarily selective, meant to illustrate various aspects of behavior genetics. Omission of some excellent studies is inevitable, just as inclusion of some studies has occurred by virtue of our familiarity with them. Even so, because of the need to be selective, it is our hope that we have managed to provide a relatively advanced and comprehensive text in behavior genetics. We apologize to those who may feel that their work has been neglected, and we would be grateful if our readers draw our attention to matters of this sort that they feel strongly about. Indeed, any comments will be most welcome.

Besides our patient spouses to whom this book is dedicated, we were aided in manifold ways by our students and staff. At the State University of New York at Purchase, they were Geoffrey Ahern, Roslyn Black, Luba Burrows, Dan Cannizzo, Lila Ehrenbard, Toni Faucher, Alena Leff, Max Kirsch, Eileen O'Hara, Dr. Anita Pruzan, Jodi Rucquoi, and Gary Rosenfeld. Bertha Inocencio bore an especially heavy burden. In addition to reading the entire manuscript, she typed parts of it and tended files, phones, and so on, while we wrote and rewrote. Bless her!

At LaTrobe University in Australia, aides were Jeff Cummins, Dr. David Hay, Michele Jones, Lon McCauley, Glenda Wilson, and Cheryl Wynd.

A special mention must be made of Dr. Nikki Erlenmeyer-Kimling, who improved the entire manuscript with her kind and perceptive criticism.

Preface to *Behavior Genetics and Evolution*

Some five years span the appearances of our two books, *The Genetics of Behavior* (1976) and *Behavior Genetics and Evolution* (1981), and they have been hectic ones for behavior geneticists of assorted morphs including ourselves. We coauthors now have so much more primary literature with which to deal that we deem it imperative to state this time that our approach will lean toward the biological side of our topics with special reference to evolutionary matters. The title change for this essentially revised edition follows from this change in emphasis. This simply means we will be assuming some elementary background in genetics and in developmental biology though we review and bolster such preparations using behavioral examples in our rewritten, expanded initial five chapters as well as elsewhere. Even so, as of this writing, more of the active members of the Behavior Genetics Association are trained in psychology and in closely related areas than in genetics specifically such as ourselves. But most important are those few recent graduates trained as behavior geneticists as a result of courses being taught and interdisciplinary graduate programs being newly developed. It is our belief that at various levels graduates with such training will increase rapidly in number; it is our hope that this new book will assist them.

This book incorporates more mechanical details of genetic concerns, such as those of transmission and enzymes, together with the expected incorporation of an updated and enhanced body of literature. Topics previously hardly mentioned now given greater consideration include behavioral changes associated with domestication, cultural versus biological inheritance, plus a discussion of evolutionary strategies involved in habitat selection. Additionally, by way of stressing the evolutionary bias of our book we discuss the behavior genetics of a far greater diversity of organisms than in our antecedent book.

Most importantly, we put even more stress upon what we regard as the unique emphasis that the behavior geneticist must appreciate: In studying the behavioral phenotype of any but the simplest of characters, the precise study of environmental determinants is just as important as the study of genetic determinants. In many ways, this is the justification for writing a book specifically devoted to behavior genetics.

In the last chapter of our initial effort we made predictions concerning future trends in behavior genetics. Some still remain future trends, but information on all has accumulated over the past 5 years. Our previous edition was completed before the widespread discussions of sociobiology, at a time when the heredity-IQ debate was omnipresent. So just as we then attempted to position our discussion of this heredity-IQ debate into perspective within a text on behavior genetics, in this edition we try the same for sociobiology in very brief terms, simply because we regard behavior genetics as the major scientific discipline underlying sociobiology. Finally, we try to develop the role of behavior genetics as a discipline of vital importance in the study of evolutionary biology, including speculation itself.

This book is organized into four main sections:

• Chapters 2 through 5 provide an introduction to genetics as applied to behavior, proceeding from behaviors under the control of single genes and chromosomes to those controlled by many genes. Chapter 2 is a brief introduction to genetics using behavioral examples. Its object is to show that the principles of genetics can be studied while relying upon behavioral examples. Anyone with a scant knowledge of genetics should read Chapter 2 in conjunction with an elementary genetics text, a selection of which is listed at the end of that chapter.

• Chapters 6 and 7 provide the theoretical bases of analyses of traits controlled by many genes in experimental animals and in human beings.

• Chapters 8 through 12 look at behavior phylogenetically, considering bacteria, protozoa, invertebrates (especially *Drosophila*), rodents, *Homo sapiens,* and various other animals on which behavior-genetics studies have been or may be undertaken. Discussion of the heredity-IQ controversy occurs in Chapter 12.

• Chapter 13 discusses the role of behavior in evolution. In this sense it stresses the integration of the material of the preceding chapters. Chapter 14, the concluding chapter, presents a final discussion of the place of behavior genetics in evolutionary biology. Some specific areas where behavior genetics has been uniquely successful are discussed, with comments on likely future trends. In some cases behavior genetics studies are beginning to contribute to other areas of genetics. The point is made that the behavior geneticist of the future must look beyond questions of how heredity and environment, considered discretely, control behavior. At this stage it should have become clear why progress in our understanding of sociobiology will occur, but not rapidly.

Once again we invite corrections, suggestions, reprints, and preprints. We do so as part of the affirmation that scientific advances will make another edition imperative. We also wish to thank all those persons, many unmentioned, who have presented us with helpful comments and indicated errors in our first edition, and we welcome openly a similar response for this edition.

We specifically thank Drs. John McKenzie and Neville White for access to unpublished data and for helpful discussions and Dr. David Hay and Joan Probber for help with literature, interpretations, and helpful discussions. Mrs. Marlene Forrester typed most of a draft and so with Bertha Inocencio simply made this second book possible. And we copiously thank our patient editor, Toni Faucher.

Lee Ehrman
Peter A. Parsons

Introduction

The literature of the earlier part of this century shows clearly that the study of behavior and the study of genetics proceeded independently of each other, with few exceptions. The geneticist, preoccupied with the study of easily defined, mainly morphological or anatomical genetic types, tended to ignore possible genetic components of behavioral traits. No doubt one reason for this was the greater difficulty of measuring behavioral traits as compared with morphological ones; a second reason was that few geneticists had any training in psychology. And when one looks at the psychological literature of the period, it is apparent that experimental and certainly clinical psychologists took little note of genetic components of behavior. Beach (1950), in a rather lighthearted but scientifically serious article, "The Snark Was a Boojum," discussed why genetic variability was largely ignored by psychologists. Nonhuman behavioral work was and still is largely conducted on the Norway rat, *Rattus norvegicus*. A relatively constant genetic makeup was and still is sometimes assumed. This one type is then surveyed for a series of behaviors, so that the behaviors themselves are the variables of study. A geneticist, on the other hand, primarily manipulates genetic types, genotypes, in order to see how traits vary according to the genetic type.

The theoretical and empirical observations necessary for combining the genetic and psychological approaches have been embedded in the literature for a long time. For example, in *Drosophila melanogaster,* the fruit fly commonly used in genetic experiments, differences in male sexual vigor in different strains were reported by Sturtevant as early as 1915. This is all the more remarkable since research on this species began only about five years earlier (by T. H. Morgan and his colleagues in the famous Columbia University *Drosophila* room). These early experiments in *Drosophila* behavior, however, were mainly byproducts of genetic or evolutionary investigations having other objectives. The 1940s did produce a number of pertinent investigations, principally by Dobzhansky, Mayr, and their associates, into sexual isolation among many of the then newly discovered races and species of *Drosophila* (for references see Parsons, 1973). Similarly, during this early period there were reports of behavioral differences among different genetic types in some rodents, principally in house mice and to a lesser extent in rats. These are ably summarized in Fuller and Thompson's (1960) classic, *Behavior Genetics,* a thorough account of behavior genetics literature up to the end of the 1950s. In human beings, despite a few early reports mainly concerning twins (e.g., Newman, Freeman, and Holzinger, 1937), the development of a recognizable behavior-genetics approach is relatively recent. Studies carried out by psychologists dealt mainly with traits of social significance in which measurements are difficult, as are precise genetic interpretations.

In the vast majority of organisms, however, the study of behavior genetics is very recent. Much of the work follows the approach of identifying and studying mutations which alter the nervous systems of protozoans, nematodes, crickets, and other organisms, in addition to those mentioned above—a new field known as *neurogenetics* (Ward, 1977; Quinn and Gould, 1979). Just as recent is the start of a behavior genetics of bacteria, mainly utilizing mutants that manifest different levels of attraction to chemicals (Adler, 1976). The behaviors studied are therefore many and varied, but they are constrained by the organism under study. This facet of behavior genetics is attracting considerable attention, with an ever-increasing literature, but as yet little of it relates to evolutionary processes so it will not be considered in detail here.

With this restriction in mind, what then are the factors that differentiate behavior from other traits, such as morphological ones, that a geneticist may use? Although this question cannot be answered in any absolute sense, the study of behavior genetics does have emphases differing from those of other areas of genetics. As such it must be regarded as a true discipline, but one certainly interacting with other subdivisions of genetics such as developmental, population, and evolutionary genetics, and with other subdivisions of behavioral studies. Three main factors suggest themselves as being of greater concern to the behavior geneticist than to other geneticists; the third is essentially unique to behavior genetics:

1 *Difficulty of environmental control.* In unicellular organisms and invertebrates such as *D. melanogaster,* the environment can be controlled relatively precisely. This means that the effects of environmental variations can be assessed and quantified successfully, given appropriate experimental designs. In rodents this is normally possible, especially if electronic devices are used to monitor behavior. However, with vertebrates complications begin to appear, since variations in early experiences may affect later behavior, an observation true even for *Drosophila* under certain definable circumstances. For example, whether mice are brought up together or separately may influence fighting behavior within a given strain. Often these environmental influences on behavior (by no means restricted to work with rodents) are difficult to assess, or worse, may occur without our being aware of them; differences in results between laboratories could be caused by factors of this sort. With *Homo sapiens* we are dealing with a species in which there is great difficulty in defining early experiences or in using controlled environments. This stress on the need for environmental control and its study was not always considered important by classic geneticists, but it is no less than imperative for the behavior geneticist.

2 *Difficulty of objective measurement.* For an accurate assessment of the relative importance of genetic influences, environmental influences, and interactions between them, a trait must by definition be measured completely objectively, that is, without any bias from the person carrying out the measurements. Clearly, in *Drosophila,* objectivity is normally possible for traits such as mating speed (the time elapsed from meeting to mating), duration of copulation, or phototaxis as measured in a maze. In rodents, objective measurement may be somewhat more difficult. However, for traits such as activity, measured by using automatic counters in activity wheels or photoelectric cells that count the number of times the animal passes a certain defined spot, high objectivity is possible. Objective measurements of mating rituals, social behavior, and territoriality present greater difficulties, though such measurements have indeed been achieved in well-designed experiments. In human beings, except for relatively simple sensory perception traits such as color blindness, objective measurement is a difficult problem. For traits such as intelligence and personality, which are so frequently assessed, it is difficult to avoid the conclusion that some subjectivity is likely to occur in measurement. The problem is that once an element of subjectivity appears, it becomes difficult to assess the relative importance of heredity and environment. In our own species we must cope with the greatest difficulties of all. This element of subjectivity, which should be minimal for biochemical, physiological, or morphological traits, is therefore a factor that partly differentiates the work of the behavior geneticist from the work of other geneticists.

3 *Learning and reasoning.* Behavior geneticists are concerned with learning and reasoning; other geneticists generally are not. These concerns should be regarded as essentially unique to behavior genetics, when viewed as a branch of genetics. Learning may well be of minor significance in *Drosophila,* since most behaviors surveyed are apparently innate (i.e., a direct property of the nervous system) as opposed to acquired behaviors including learning. As an evolutionary biologist, Mayr (1974) finds difficulties in using the terminological dichotomy of innate versus acquired. *Innate* refers to the genotype, and indeed the term has been restricted to functions at the reflex level and to lower animals. *Acquired* refers to the phenotype, so that neither term is the opposite of the other. Mayr essentially has resolved this problem by relating behavior to the concept of a *genetic program*—a concept derived from an interaction of

molecular biology and information theory. Those behaviors based on genetic programs not allowing for appreciable modifications during the process of translation into the phenotype are called *closed programs*. Other genetic programs are modified during translation into the phenotype—by input occurring during the life span of the owner. They thus have an acquired component and are referred to as *open programs*. Closed programs are widespread in organisms with a short life span, which to date must include *Drosophila*, while open programs are more likely in organisms with longer life spans which include parental care. Even so, learned behavior in *Drosophila* occurs in species-recognition patterns, but few other reports of learning exist, and they need further substantiation at this stage. In rodents, there is ample evidence that early experience affects later behavior patterns (see Erlenmeyer-Kimling, 1972, for review). Patterns and rates of learning also are found to vary among different strains. Therefore, both heredity and environment are involved in learning, as are interactions between heredity and environment. In human beings, in whom learning and reasoning are developed to the highest level, we have less hope of environmental control and generally we do not have known behavioral phenotypes. It is here that genetic programs are likely to be the most open of all.

GENERAL READINGS

Ehrman, L., G. S. Omenn, and E. Caspari (eds.). 1972. *Genetics, Environment, and Behavior: Implications for Educational Policy.* New York: Academic Press. The proceedings of a research workshop on the genetics of behavior, human and animal, at molecular, cellular, individual, population, and evolutionary levels, with the aim of seeking possible applications in research of interest to education.

Fuller, J. L., and W. R. Thompson. 1960. *Behavior Genetics.* New York: Wiley. The classic text in the field, ably summarizing it to the end of the 1950s.

Fuller, J. L., and W. R. Thompson. 1978. *Foundations of Behavior Genetics.* St. Louis: Mosby. An updated version of the 1960 book stressing rodents and human beings in particular.

Hirsch, J. (ed.). 1967a. *Behavior-Genetic Analysis.* New York: McGraw-Hill. An overview of much of behavior genetics that developed in the early 1960s.

McClearn, G. E., and J. C. DeFries. 1973. *Introduction to Behavior Genetics.* San Francisco: Freeman. A recent representative account of the field at a relatively elementary level assuming no previous knowledge of genetics.

Manosevitz, M., G. Lindzey, and D. D. Thiessen. 1969. *Behavioral Genetics: Method and Research.* New York: Appleton. A comprehensive collection of important original articles contributing to the development of the field.

Parsons, P. A. 1967a. *The Genetic Analysis of Behaviour.* London: Methuen. An account of how behavior can be analyzed genetically, with specific emphasis on *Drosophila,* rodents, and human beings. A discussion of evolutionary implications is included.

Spuhler, J. N. (ed.). 1967. *Genetic Diversity and Human Behavior.* Chicago: Aldine. The proceedings of a conference on the behavioral consequences of genetic differences in human beings.

Thiessen, D. D. 1972. *Gene Organization and Behavior.* New York: Random House. A brief account of behavior genetics with some stress on evolutionary aspects.

Van Abeelen, J. H. F. (ed.). 1974. *The Genetics of Behaviour.* Amsterdam: North-Holland. A collection of important original articles.

Requisite Genetics

The object of this chapter is to review such basic principles of genetics as are necessary for an understanding of the chapters to follow. A single chapter provides too little space to propound the complete principles of genetics. Any reader who finds this brief survey insufficient should employ a general genetics text to augment the information given here. A list of appropriate texts is given at the end of this chapter.

2-1 MENDELIAN GENETICS

If we observe differences in eye or hair color and find that these differences tend to run in families, it is clearly insufficient to say that such traits are inherited. We wish to find out *how* traits are inherited. This is one of the main objects of study in the science of genetics. Thus we must turn to the transmission of observed traits from one generation to the next to see what rules can be formulated.

The appearance of an organism is referred to as its *phenotype*. Although the phenotype conventionally refers to the outward appearance of an individual, its definition can be extended to include the totality of physiological, anatomical, and behavioral components of that individual. In this book we concentrate on the behavioral components of the total phenotype. The phenotype depends upon the sum of the genes an organism possesses (its *genotype*) and upon any effects of the environment in which the organism lives. As will rapidly become apparent, the environment is particularly significant in the study of behavior, since behavioral patterns are capable of large environmentally influenced changes even when morphological changes are small. *Phenylketonuria* in human beings is a phenotype in which there occurs a genetically controlled disorder arising from an upset in phenylalanine metabolism. Phenylalanine is an essential amino acid that is found in toxic quantities in phenylketonurics. One effect of this metabolic upset is a lowered intelligence quotient (IQ), a commonly used assessor of cognitive abilities. In addition, phenylketonurics have a slightly smaller head size and lighter hair color than do comparable normals. However, if the metabolic error is "corrected" by a special diet low in phenylalanine, IQ is "improved"—that is, not so depressed as it would be without treatment. Already we can see the complexity of a phenotype having physiological and behavioral components.

The major theme of the brief survey contained in this chapter is the nature of the genotype. The effect of variations in environment is not extensively considered; however, this issue is taken up in later chapters because it is important to understand genetic principles as such before introducing complications due to environments. The units of inheritance, *genes,* are carried on *chromosomes.* Chromosomes can be observed during cell division. In human beings there are 46 chromosomes arranged in 23 pairs. Because the chromosome pairs differ in length and in appearance, some of these pairs can be recognized (Figure 2-1). In females both members of the 23 pairs are the same size and are referred to as *homologous.* In males there are 22 homologous pairs plus a nonhomologous pair having different lengths (Figure 2-1). The nonhomologous chromosomes in males are referred to as X and Y. Correspondingly in females there is a homologous pair of X chromosomes. Intuitively these chromosomes appear relevant in determining sex. During sperm and ovum formation, referred to collectively as *gamete* formation, the number of chromosomes is halved. All the chromosomes in a single gamete differ; that is, each gamete has one member of each chromosome pair. This means of course that a male gamete has an X or a Y chromosome but not both (regarding the X and the Y as a "pair"). At fertilization, two gametes unite, each containing a set of 23 chromosomes to form the fertilized cell (*zygote*) with 23 pairs or 46 chromosomes once again. This process is shown diagrammatically in Figure 2-2. The chromosome number of 23 for gametes is referred to as the *haploid* number, and that for zygotes ($2 \times 23 = 46$) as the *diploid* number. More generally, we can write n as the haploid number and $2n$ as the diploid number since chromosome numbers vary among species.

Genes occupy various positions on chromosomes referred to as *loci* (singular, *locus*). In the mouse, which has a diploid number of 40, there is a gene on chromosome V at a locus referred to as *fidget*. When present on both members of chromosome V, the fidget gene leads to a behavioral alteration consisting of a

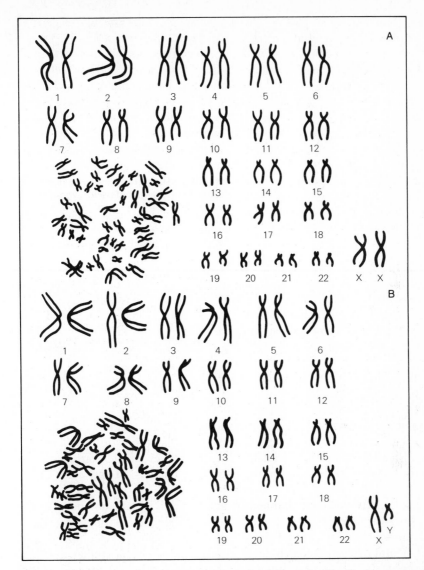

Figure 2-1 Human chromosomes. *A.* Normal female cell with 46 chromosomes and the normal female karyotype (XX). *B.* Normal male cell with 46 chromosomes and the normal male karyotype (XY). (*Courtesy of Professor Raymond Turpin.*)

phenotype in which the mouse's head moves from side to side continuously. The gene can be written *fi* for short. Such a mouse is said to have a *fifi* genotype (the mouse is a diploid individual having chromosomes in pairs). In most mice the fidget gene is not present at this locus, but in that place is its normal alternative gene, which can be written as +. By normal we mean the gene that is usually present at a locus. Two possibilities then occur if the mouse appears

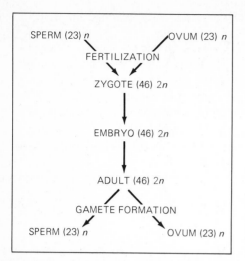

SPERM (23) *n*   OVUM (23) *n*

FERTILIZATION

ZYGOTE (46) 2*n*

EMBRYO (46) 2*n*

ADULT (46) 2*n*

GAMETE FORMATION

SPERM (23) *n*   OVUM (23) *n*

Figure 2-2 Human chromosome number changes during gamete and zygote formation. This can be generalized for sexually reproducing organisms, which will be designated in this book by *n*, where *n* is the haploid number and 2*n* the diploid number.

phenotypically normal. The mouse is genotypically either *fi*+ or ++. In neither case is any behavioral alteration seen. The fidget phenotype is seen only if two *fi* genes are present, so gene *fi* is said to be *recessive* to +. On the other hand, the normal nonfidget phenotype appears if one or two + genes are present, so the gene + is said to be *dominant* to *fi*. While introducing terms, it should be noted that alternative forms of genes at a given locus, in this case *fi* and +, are referred to as *alleles*. Individuals with identical alleles at a given locus on both chromosomes (*fifi* or ++) are referred to as *homozygotes*; individuals with nonidentical alleles such as *fi*+ are referred to as *heterozygotes* (mixed zygotes). These simple terms, introduced in rapid succession, are necessary for understanding what happens in simple crosses. If the explanations of terms here or subsequently in this chapter, are insufficient, the reader should consult one of the texts in the General Readings at the end of the chapter, in particular Crow (1976).

It must be stressed that dominance and recessivity are not necessarily complete because heterozygotes are often distinguishable from both homozygotes. Furthermore, even if at first sight dominance appears to be complete, say at the behavioral or morphological level, detailed biochemical studies or other physiological tests may reveal differences between the heterozygote and the normal homozygote. Untreated phenylketonuria in people is an example. To a superficial observer it is controlled by a recessive gene *p*, with phenylketonurics being *pp* and normals *p*+ or ++. However, at the biochemical level *p*+ and ++ may be distinguishable since *p*+ individuals tend to have more phenylalanine in their serum than do ++ individuals. Recall that phenylketonurics themselves have detrimentally high levels of phenylalanine. Thus dominance is incomplete, a situation called *semidominance*. Therefore, depending on the level of phenotypic observation, it is possible to obtain somewhat differing conclusions about levels of dominance according to the component of the phenotype being measured. Complete dominance is, however, frequently assumed for simplicity.

Suppose a male mouse, *fifi*, is crossed with a female *fi+* mouse. From the *fifi* mouse only *fi* gametes are possible, while from the *fi+* mouse gametes carrying *fi* or + are produced. In other words, in the female there is *segregation* into gametes carrying one gene or the other but not both. By chance from *fi+* mice we should get about half the gametes containing the *fi* gene and about half with the + gene. Diagrammatically the expected gametes and zygotes, formed by the fertilization of the female gametes with a *fi* gamete from males, are

	gametes	
♀	½+	½*fi*
♂ gametes *fi*	zygotes	
	½ +*fi*	½ *fifi*

Thus in the offspring we expect ½ *fi* + : ½ *fifi* or ½ normal : ½ fidget. A parallel result is expected if the sexes are reversed so that a *fi+* male mouse is crossed with a *fifi* female mouse. In other words, offspring clearly show the effects of segregation during gamete formation. (Often slightly fewer than ½ fidget mice occur in breeding data because fidget mice are less likely to survive than are normal ones.) The principle of *segregation* was first demonstrated by Mendel in 1865 in his classic studies on peas and is in fact frequently referred to as *Mendel's first law*.

Mendel also studied the segregation of two pairs of alternative forms of genes or alleles at two loci on separate chromosomes simultaneously. If at one locus there are two alternative alleles *A* and *a*, and at a second locus *B* and *b*, and a double heterozygote *AaBb* is crossed with a double homozygote *aabb*, what should be expected? (We assume *A* and *B* to be dominant to *a* and *b*, respectively.) From the double recessive homozygote *aabb*, we expect only gametes *ab*. From the double heterozygote *AaBb*, the situation is more complicated and, considering each locus separately, there should be ½*A* : ½*a* and ½*B* : ½*b*. When considering simultaneous segregation at these two loci, the simplest hypothesis is that the segregation of the two pairs of alleles occurs independently of each other. At the cellular level this implies that the chromosomes carrying the alleles assort independently during gamete formation. If this happens, the double heterozygote is expected to give the following gametes in equal proportions:

¼*AB* : ¼*Ab* : ¼*aB* : ¼*ab*

This can easily be seen by multiplying (½*A* + ½*a*) and (½*B* + ½*b*). The constitutions of these gametes are then revealed by being fertilized with *ab* gametes

from the double recessive homozygote *aabb* to give four discrete and recognizable phenotypic classes:

$$\frac{1}{4}AaBb : \frac{1}{4}Aabb : \frac{1}{4}aaBb : \frac{1}{4}aabb$$

Many pairs of gene loci in many organisms give ratios approximating this. This
principle of *independent assortment* is known as *Mendel's second law.*

However, if loci are on the *same* chromosome, assortment is not generally
independent. The closer to each other loci are along a given chromosome, the
more closely *linked* they are in gamete formation. This is because, during gamete formation, genes on the same chromosome may recombine with each
other; the frequency of such *recombination* depends on the spatial distance
between the relevant loci. From these frequencies, *chromosome maps* have
been constructed for individual chromosomes. Genes located on the same
chromosomes are said to belong to the same *linkage group.* In human beings we
can expect 23 linkage groups, although they have not yet all been identified. In
any case the expected number of linkage groups corresponds to the haploid
number. The mouse, which is discussed extensively in this book, has 40
chromosomes or 20 pairs of chromosomes and hence 20 linkage groups. In the
fruit fly, *Drosophila melanogaster,* another organism of major importance in
behavior genetics, the figures are 8 and 4, respectively. A chromosome map of
D. melanogaster (Figure 2-3) consists of 4 linkage groups, as expected. The map
given incorporates loci having mainly behavioral effects in addition to some
other loci commonly used in experimental breeding work. (Note that the number of linkage groups corresponds to the haploid number in Figure 2-3.)

A further complication relates to sex. As we have seen, in the human
female there are 23 homologous pairs of chromosomes totaling 46 (Figure 2-1),
including two X chromosomes. The male has 46 chromosomes made up of 22
homologous pairs plus an X chromosome, which corresponds to the X
chromosomes in the female, and a Y chromosome, which does not correspond
to any chromosome in the female (Figure 2-1). Thus the female can be written
as 22 + XX and the male as 22 + XY, these being the 22 *autosomal* pairs plus
the *sex chromosomes* X and Y. Generally, in the organisms discussed in this
book, the sex chromosomes function as the sex-determining mechanism. We
refer to genes on the X chromosome as being sex-linked. Little genetic activity
has been identified as being located on the Y chromosome in organisms with
this type of sex-determining mechanism. This means that in females the principles of heterozygosity and homozygosity apply for the sex chromosomes just as
for the autosomes already discussed. Because the X is paired with a Y in males,
however, a rare sex-linked recessive gene normally shows up more frequently
in males than in females, since the recessive genes cannot be masked by corresponding dominant genes. This follows from the general observation that loci
on the X chromosome are for the most part not matched by corresponding loci
on the Y. Males with loci only on the X are referred to as *hemizygous* for such

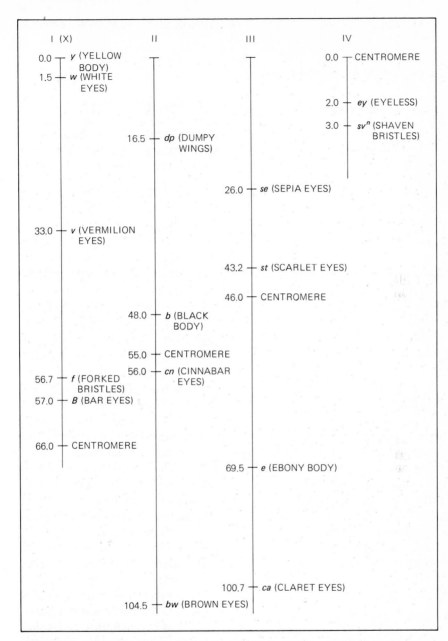

Figure 2-3 Linkage map of *D. melanogaster*. Some of the commonly used genes are included, especially those that have been involved in behavioral work. The *centromere* is the body to which spindle fibers attach during cell division. (*After Bridges and Brehme, 1944, and other sources.*)

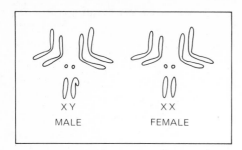

Figure 2-4 Chromosomes of *D. melano-gaster*. Note the X and Y chromosomes of the male and the two X chromosomes of the female.

loci. (Other modes of sex determination exist in other organisms but are of lesser significance for this text.) Of behavioral interest is the fact that in *Homo sapiens* the genes for red-green color blindness (see Section 11-6) and for one form of muscular dystrophy are under the control of sex-linked recessive genes, and, as expected, these conditions occur in males much more frequently than in females. Chromosome 1 in Figure 2-3 is the X chromosome of *D. melanogaster* with the sex-linked genes *yellow* (body color), *white* (eyes), *vermilion* (eyes), and *forked* (bristles), the first three having known behavioral effects involving mating propensity. The microscopic appearance of the X and Y chromosomes of *D. melanogaster*, together with its autosomes, is shown in Figure 2-4. (Note the two dotlike chromosomes that correspond to the very short linkage group IV in Figure 2-3.)

2-2 QUANTITATIVE GENETICS

So far we have discussed genetic variation under the control of specific genes assignable to specific loci on chromosomes, but many behavioral traits are quantitative and do not segregate into discrete classes. Examples in people include height, weight, and IQ within a population. This does not mean that specific genes affecting these traits are not known. Indeed, the phenylketonuria gene has a discrete effect reducing IQ. The frequency distribution of many quantitative traits approximates more or less the continuous *normal distribution* of the statistician (Figure 2-5). The normal distribution can be completely described in terms of two quantities (*parameters*). One is the mean or average

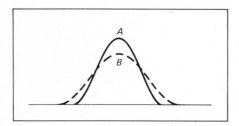

Figure 2-5 Normal distribution. Curves *A* and *B* have the same mean, but the variance of *B* is larger than that of *A*.

value. If x_i is an individual observation, and there are n observations, then the mean $\bar{x}$ is given by

$$\bar{x} = \frac{\Sigma x_i}{n}$$

(Σ is a symbol meaning the sum of, in this case the total of, all the x_i values.) The other parameter is an expression of variability around the mean. In some cases the variability around the mean is small and in other cases large, as shown by curves A and B in Figure 2-5. The term for the parameter measuring variability is the *variance,* which is estimated as

$$\frac{1}{n - 1} \Sigma(x_i - \bar{x})^2$$

The square root of the variance is the *standard deviation.* Much of the theory of quantitative genetics is based on the assumption of a normal distribution. If it is possible to assume a normal distribution, it may be possible to find a suitable algebraic transformation, for example, a transfer to logarithms that will convert the data to an approximately normal distribution.

Assuming that a continuously varying trait is partly under genetic control, it must be asked how the intrinsically discontinuous variation caused by genetic segregation is expressed as continuous variation. Suppose two individuals $A/a \cdot B/b$ are crossed, where A,a and B,b are gene pairs at two unlinked loci, and further suppose that genes A and B act to increase the measurement of a quantitative trait by one unit, and genes a and b act to decrease the trait by one unit. It is perhaps less confusing to write $A/a \cdot B/b$ as $+/- \cdot +/-$, considering A and B genes as $+$ genes and a and b genes as $-$ genes. Counting the number of $+$ and $-$ genes gives a metric or quantitative value for a genotype.

The above cross gives five genotypes distributed as in Figure 2-6, ranging from one with four $-$ genes to one with four $+$ genes. The most common genotype is $+/- \cdot +/-$, having a genotypic value of zero since it has two $+$ and two $-$ genes. This is the mean *genotypic value.* The least common genotypes are the two extremes, $+/+ \cdot +/+$ and $-/- \cdot -/-$, with genotypic values of $+4$ and -4, respectively. If there is a third locus with two "similar" alleles, then for a cross between multiple heterozygotes the number of genotypic classes rises to seven, and with a fourth, to nine, and so on. The differences among the classes become progressively smaller as the number of segregating loci rises. At the stage when the differences among classes become about as small as the error of measurement, the distribution becomes continuous, as in Figure 2-5. In addition, any variation due to nongenetic causes blurs the underlying discontinuity implied by segregation, so that the variation seen may become continuous irrespective of the accuracy of measurement.

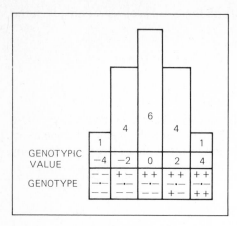

Figure 2-6 Genotypic frequencies from the cross $+/-\cdot+/- \times +/-\cdot+/-$ plotted according to the genotypic value (i.e., the relative number of $+$ and $-$ genes). The frequencies of each genotype are given in the histogram.

Therefore, many genes, each with small effects with respect to the phenotypic trait, superimposed upon variability due to nongenetic or environmental causes lead to the expectation of a continuous distribution similar to that given in Figure 2-5. Genes that contribute to a quantitative trait, but that are not directly identifiable by classic Mendelian segregation (i.e., that cannot be studied individually), are referred to as *polygenes;* genes whose effects can be studied individually are referred to as *major genes.* There is no fundamental biological distinction between major genes and polygenes. The terms are merely a matter of convenience, since the breeding methods used to study the effects of major genes cannot, in general, be used in the study of polygenes. Even so, it is possible under certain circumstances to increase the effects of polygenes by statistical, and perhaps biochemical, techniques to such an extent that, to all intents and purposes, they appear as major genes.

Behavioral traits such as duration of copulation in *Drosophila,* activity scores in mice, and IQ in human beings are essentially quantitative: for their analysis an appreciation of the aims and methods of quantitative genetics is essential. The basic aim is to divide the phenotypic value (P), which we measure, into genotypic (G) and environmental (E) components. This can be written most simply as

$$P = G + E$$

Since continuously varying traits are being considered, we need the phenotypic variance (V_P), which can be split into variance components due to genotype (V_G) and environment (V_E). Assuming no interaction between genotype and environment—which is the simplest assumption possible although it frequently does not hold for behavioral traits—the phenotypic variance is the sum of the genotypic and environmental variances, thus

$$V_P = V_G + V_E$$

It is reasonable to compute the proportion of the total phenotypic variance that is genotypic, thus

$$\frac{V_G}{V_G + V_E} = \frac{V_G}{V_P}$$

This ratio is referred to as the *heritability in the broad sense* or the *degree of genetic determination* of a trait; it is an important entity in the study of quantitative traits including behavioral traits. This and the other concepts mentioned in this section are further developed in Chapters 6 and 7.

Traits controlled by many genes or *polygenes* make up many of the behavioral characters we study, especially in human beings. Except for rare diseases that can be traced in pedigrees and shown to be controlled by a single locus, many deleterious behaviors, including many forms of mental deficiency, are interpreted as being controlled polygenically. It should be clear that a further complicating factor, again especially in human beings, is that such traits are often profoundly labile environmentally when compared with morphological traits such as height and dentition. As already stressed in Chapter 1, this environmental lability is one of the unique difficulties in behavior genetics. Both environmental lability and polygenic control, separately or together, lead to traits showing continuous distributions as in Figure 2-5.

Another type of trait we must consider is the *threshold trait*. These are traits for which organisms can be phenotypically classified into those having a given trait and those not having the trait. (See Section 7-2 for more on threshold traits.) Morphological examples with behavioral consequences in people include gross abnormalities of the nervous system such as anencephaly, hydrocephaly, and spina bifida, probably all originating within the first eight weeks of embryonic life. Polygenic inheritance has often been invoked, even occasionally with major gene effects. In addition, environmental factors may be relevant. This is so because the normal development of complex morphological traits depends on many processes and reactions, offering many possibilities for interference positively or negatively by environmental factors. Information about the heritable component of such traits emerges from comparisons of relatives; the closer the relationship to the *index case* or the *proband,* the higher the expected incidence of the disorder if there is a heritable component. The same occurs for disorders such as epilepsy, schizophrenia, and manic-depressive psychoses (see Chapter 11). Family studies provide evidence for their genetic control, but the problem of variation due to the environment and to interactions between the genotype and the environment makes it difficult to distinguish genetic and environmental components, especially since in some cases family background effects are important (see Chapters 7 and 11).

2-3 POPULATION GENETICS

The first section of this chapter considered segregation in single progeny at the family level. We now extend the unit under consideration to the population, which is made up of a number of individuals with many progenies. In the absence of an elementary behavioral example, the MN blood group system provides a suitable model for the segregation of a pair of alleles in human populations. The blood groups are determined by two alleles, L^M and L^N. The correspondence between genotype and phenotype is direct, since $L^M L^M$ individuals are blood group M, $L^M L^N$ individuals are blood group MN, and $L^N L^N$ individuals are blood group N. The three phenotypes can be detected by serological tests.

Considering a whole population of people, there must be a certain number of L^M alleles and a certain number of L^N alleles. These numbers can be assessed by counting every homozygote ($L^M L^M$) as $2L^M$ alleles, every heterozygote ($L^M L^N$) as $1L^M$ and $1L^N$ allele, and every homozygote ($L^N L^N$) as $2L^N$ alleles.

In a sample of 100 people, let us say there are $40L^M L^M$, $40L^M L^N$, and $20L^N L^N$. We can therefore count the number of L^M and L^N alleles in these individuals.

	Number of L^M alleles	Number of L^N alleles	Total
$40L^M L^M$	80		80
$40L^M L^N$	40	40	80
$20L^N L^N$		40	40
Total	120	80	200

The total number of alleles, of course, adds up to 200, since every diploid individual has two alleles.

The ratio

$$\frac{\text{Number of } L^M \text{ alleles}}{\text{Total number of alleles}} = \frac{120}{200} = 0.60$$

is called the *gene (allele) frequency* of gene L^M, and the ratio

$$\frac{\text{Number of } L^N \text{ alleles}}{\text{Total number of alleles}} = \frac{80}{200} = 0.40$$

is called the *gene (allele) frequency* of gene L^N.

The two ratios add up to unity, as might be expected, as only two alleles are being considered.

Thus in the whole population in the first generation, the eggs and sperm each have a gene frequency of $L^M = 0.6$ and $L^N = 0.4$. What happens if these gametes unite at random? We get

Female gametes	Male gametes	
	$0.6L^M$	$0.4L^N$
$0.6L^M$	$0.36L^M L^M$	$0.24L^M L^N$
$0.4L^N$	$0.24L^M L^N$	$0.16L^N L^N$

In other words, the genotype frequencies are

$$L^M L^M = 0.6^2 = 0.36$$
$$L^M L^N = 2 \times 0.4 \times 0.6 = 0.48$$
$$L^N L^N = 0.4^2 = 0.16$$

which add up to unity.

We then wish to find out what happens in the next generation. From $L^M L^M$ only L^M gametes are produced; from $L^M L^N$, $\frac{1}{2} L^M$ and $\frac{1}{2} L^N$ gametes; and from $L^N L^N$, only L^N gametes. Therefore, the gene frequency of $L^M = 0.36$ from $L^M L^M$ genotypes $+ \frac{1}{2} \times 0.48$ from $L^M L^N$ genotypes $= 0.6$, and the gene frequency of $L^N = \frac{1}{2} \times 0.48$ from $L^M L^N$ genotypes $+ 0.16$ from $L^N L^N$ genotypes $= 0.4$.

Thus after one generation of the gametes uniting at random at fertilization the gene frequencies are unchanged. Similarly, the zygotic (genotypic) proportions remain unaltered from generation to generation.

These conclusions can be shown generally. Let the gene frequency of $L^M = p$ and $L^N = q$, such that $p + q = 1$. This gives, assuming random union of gametes, the following zygotes:

Female gametes	Male gametes	
	pL^M	qL^N
pL^M	$p^2 L^M L^M$	$pq L^M L^N$
qL^N	$pq L^M L^N$	$q^2 L^N L^N$

or $p^2 L^M L^M + 2pq L^M L^N + q^2 L^N L^N$, so that the total of the zygotic frequencies comes to $p^2 + 2pq + q^2 = (p + q)^2$.

In the next generation, the gene frequency of L^M is p^2 from $L^M L^M + \frac{1}{2} 2pq$ from $L^M L^N = p^2 + pq = p(p + q) = p$, since $p + q = 1$; and the gene frequency of L^N is $\frac{1}{2} 2pq$ from $L^M L^N + q^2$ from $L^N L^N = q^2 + pq = q(p + q) = q$, which is what we began with. Computing the genotypic frequencies again gives $p^2 L^M L^M + 2pq L^M L^N + q^2 L^N L^N$.

We have therefore established the Hardy-Weinberg law, so named after its codiscoverers. This law states that (1) gene frequencies do not change from generation to generation under random union of gametes; (2) the progeny genotypes are in the proportions $p^2 : 2pq : q^2$; and (3) irrespective of the initial genotypic frequencies, the Hardy-Weinberg proportions $p^2 : 2pq : q^2$ are established in one generation.

So far, we have discussed a situation where the gene frequency at a locus is estimated assuming the heterozygote to be distinguishable from both its corresponding homozygotes. However, this is not always the situation. For example, there is a locus with behavioral consequences that determines if some members of a population can taste phenylthiocarbamide (PTC). Those who do taste it find it bitter and unpleasant. Tasting is controlled by a locus with two alleles, T and t, such that TT and Tt are tasters and tt are nontasters. In different populations between 50 and 95 percent of people can taste PTC. Unlike the MN blood groups, it is not possible to distinguish the heterozygotes Tt from the homozygotes TT owing to the dominance of the T allele over t. Thus gene frequencies of T and t cannot be estimated by direct allele counting as in the MN blood group example. However, if gene frequencies are p of T and q of t, there are $p^2 + 2pq$ tasters ($TT + Tt$) and q^2 nontasters (tt).

Hence $q = \sqrt{\text{proportion of nontasters}}$. For example, if in a sample of 100, there are 91 tasters and 9 nontasters (or, in proportions, 0.91 tasters and 0.09 nontasters), then $q^2 = 0.09$, so that $q = \sqrt{0.09} = 0.3$, and by subtraction $p = 0.7$, since $p + q = 1$. If the heterozygotes are recognizable, the allele counting method described for the MN blood group system must be used for estimating gene frequencies, since otherwise the information given by the heterozygotes is not taken into account.

So far we have considered *random union of gametes,* but what happens under *random mating* at the phenotypic level (also referred to as *panmixia*)? In order to look at this, the various mating types and their frequencies must be considered as presented in the upper part of Table 2-1. There are six of these mating types shown with their frequencies under random mating in the lower half of the table. Taking mating type $Tt \times Tt$, which has a frequency $4p^2q^2$, then offspring genotypes will have frequencies p^2q^2TT, $2p^2q^2Tt$, and p^2q^2tt. Next, offspring genotypic frequencies are worked out for all six mating types and then summed, to give $p^2TT + 2pqTt + q^2tt$. Table 2-1 therefore demonstrates the Hardy-Weinberg law under random mating. Clearly the genotypic frequencies, and hence gene frequencies, do not change from generation to generation just as for random union of gametes.

Table 2-1 Demonstration of Hardy-Weinberg Law in Random Mating Population

Under random mating there are $p^2TT + 2pqTt + q^2tt$ in each sex:

Males	Females		
	p^2TT	$2pqTt$	q^2tt
p^2TT	p^4	$2p^3q$	p^2q^2
$2pqTt$	$2p^3q$	$4p^2q^2$	$2pq^3$
q^2tt	p^2q^2	$2pq^3$	q^4

From this table we can extract mating types and the offspring they give as follows:

Mating type	Frequency	Offspring		
		TT	Tt	tt
$TT \times TT$	p^4	p^4		
$TT \times Tt$	$4p^3q$	$2p^3q$	$2p^3q$	
$TT \times tt$	$2p^2q^2$		$2p^2q^2$	
$Tt \times Tt$	$4p^2q^2$	p^2q^2	$2p^2q^2$	p^2q^2
$Tt \times tt$	$4pq^3$		$2pq^3$	$2pq^3$
$tt \times tt$	q^4			q^4

Frequency $TT = p^2(p^2 + 2pq + q^2) = p^2$
Frequency $Tt = 2pq(p^2 + 2pq + q^2) = 2pq$
Frequency $tt = q^2(p^2 + 2pq + q^2) = q^2$

Much theory in population genetics is based on the assumption of random mating. However, random mating does not necessarily always apply. One important deviation occurs as a result of *inbreeding* (the mating of individuals related to each other by ancestry). Individuals having a common ancestor are more likely to carry replicates of one of the genes present in the ancestor, and if they mate, the replicates of the ancestral genes may be passed on to their offspring. Such a process increases the number of homozygotes compared with strictly random mating.

Phenotypic assortative mating is the mating of individuals based on phenotypic resemblance. Positive assortative mating occurs when like phenotypes mate more frequently than expected under random mating. Since like phenotypes may be under the control of like genotypes, this leads, as does inbreeding, to more homozygotes than would be expected under random mating. Positive assortative mating has been found for many traits in human beings—height, weight, IQ, and various other behavioral traits. Other forms of nonrandom mating will be considered in different chapters of this book.

However, at this stage it must be stressed again that random mating occurs only when there is no tendency for certain kinds of males and females to pair, when considered with respect to a given trait. Although random mating is almost universally assumed in theoretical considerations, behavioral evidence based on the actual scoring of matings between different genotypes shows that it is in fact a special circumstance. It is unfortunate that theoretical considerations usually become extremely complex once the assumption of random mating is abandoned.

Figure 2-7 gives Hardy-Weinberg equilibrium genotype frequencies for various gene frequencies. One interesting case worth consideration is the rare recessive genetic disorder. If the gene frequency of gene a controlling such a character is $q = 0.01$ so that $p = 0.99$, the expected frequency of affected individuals aa is $q^2 = (0.01)^2 = 0.0001$, which is very small. However, the incidence of heterozygotes Aa, referred to as *carriers* because they carry the gene for the disorder, is $2pq = 2 \times 0.99 \times 0.01 = 0.0198$, which is nearly 200 times as common as the affected individuals. An example is phenylketonuria in humans, which has a frequency of the order of 1/40,000 in some populations. Since it is controlled by a recessive gene, we can write $q^2 = 1/40,000$ or $q = 1/200$, and so the proportion of heterozygous carriers is

$$2pq = 2 \times \frac{199}{200} \times \frac{1}{200} \simeq \frac{1}{100}$$

which is about 400 times as common as the recessive homozygotes. Taking into account all the numerous deleterious recessives found in people, we all possess our load of deleterious genes, including those with behavioral consequences. The other point shown by Figure 2-7 is that as a gene becomes more common, the relative excess of heterozygous carriers Aa relative to aa falls. Thus for $q = 0.10$, carrier individuals are only 18 times as common as aa. Many reces-

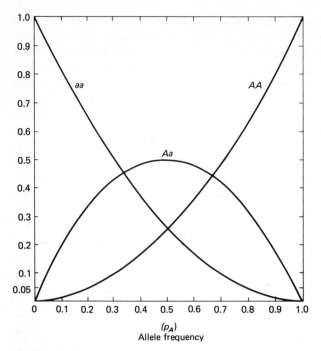

Figure 2-7 Graph of the frequencies of the three genotypes *AA*, *Aa*, and *aa* under random mating with frequencies p^2, $2pq$, and q^2 respectively. (*After Spiess, 1977.*)

sive behavioral disorders under the control of single genetic loci are rare, so in a given population the genes controlling them occur mainly in carriers.

The final situation we must consider are those genes on the X chromosome, the sex-linked genes, described at the end of the first section of this chapter. Since males have only one X, direct gene counting yields an estimate of gene frequency so that, for two alleles, $pA + qa$ would be found in males. Under random mating the familiar Hardy-Weinberg proportions of $p^2AA + 2pqAa + q^2aa$ would be expected in females simply because they have two X chromosomes. For an X-linked recessive gene therefore, the frequency of the trait in females is expected to be the square of that in males. As an example, color blindness is much less common in females than in males, being expected to be 0.64 percent in a population where 8 percent of males are affected. Rare sex-linked recessive genes are thus almost entirely restricted in occurrence to males. Indeed it then follows that extremely rare conditions of this type would almost never be expressed in females.

2-4   CHI-SQUARE TESTS

Segregation and Mating Preference Data

We continue this chapter on requisite genetics with a section on how to assess the meaning of experimental segregation data. Segregation data are often obtained in breeding experiments, and hypotheses can be erected about them. We wish to know whether the data observed fit a hypothesis, since because of random variation, exactly fitting ratios are unlikely. To illustrate this point we consider two crosses in mice: (1) a cross between two heterozygous agouti mice Aa; and (2) a cross between two heterozygous yellow mice A^Ya. A and A^Y are dominant to a. All three are alleles at the same locus, providing our first example of a *multiple allelic series*. Agouti mice have dark fur with yellow tips, yellow mice have yellow fur and are fat and rather sluggish in movement, and non-agouti aa mice have black fur. (Some comments on associated behavioral variants of coat color mutants in mice appear in Chapter 9.)

Considering gametes from Aa mice in each sex, ova and sperm are expected in ratios $\frac{1}{2}A : \frac{1}{2}a$. If these unite at random, the zygotes expected are derived from $(\frac{1}{2}A + \frac{1}{2}a)(\frac{1}{2}A + \frac{1}{2}a)$ yielding $\frac{1}{4}AA + \frac{1}{2}Aa + \frac{1}{4}aa$ genotypically, or $\frac{3}{4}$ agouti : $\frac{1}{4}$ nonagouti phenotypically.

In Table 2-2 some observed data from crosses between heterozygous agouti (Aa) mice are given. Expected numbers based on an expected 3 : 1 ratio are also given. The greater the difference between the observed O and expected E, the greater the deviation of the data from the ratio expected. In this case $O - E$ is small for each class. However, if $O - E$ were, say, 10 times greater, or 112.5, we could well wonder whether the expected ratio of 3 : 1 is realistic. Clearly, some deviation from the exact 3 : 1 is tolerable, just because of chance, but at a certain level of deviation, we begin to suspect the validity of the expected ratio. The testing of deviations from an expected ratio can be done by means of a simple statistical test, which consists of computing $(O - E)^2/E$ for each phenotypic class and adding together. The resultant value is called χ^2 (chi square), which can be written

$$\chi^2 = \sum \frac{(O - E)^2}{E}$$

where Σ means the sum of. For the above data, $\chi^2 = 1.59$, which is small. It can be shown that if $\chi^2 > 3.84$, there is a <5 percent chance of the data fitting a 3 : 1 ratio, and if $\chi^2 > 6.64$, there is a <1 percent chance of the data fitting a 3 : 1 ratio. Values of χ^2 corresponding to various probabilities P are readily available in statistical tables. Table 2-3 provides part of a χ^2 table. Clearly the value of χ^2 increases as P decreases.

Table 2-2   Data for a Cross between Heterozygous Agouti (Aa) Mice

Phenotype	Genotype	Number observed (O)	Number expected (E)	O − E	(O − E)²	(O − E)²/E
Agouti	Aa	306	317.25	−11.25	126.5625	0.3989
Nonagouti	aa	117	105.75	11.25	126.5625	1.1968
Total		423	423.00			1.5957

Table 2-3　Distribution of χ^2

Degrees of freedom n	Probability P						
	0.50	0.30	0.20	0.10	0.05	0.01	0.001
1	0.455	1.074	1.642	2.706	3.841	6.635	10.827
2	1.386	2.408	3.129	4.605	5.991	9.210	13.815
3	2.366	3.665	4.642	6.251	7.815	11.345	16.266

As n increases, χ^2 increases for a given probability, and as the probability decreases χ^2 increases for given n.
Source: Fisher and Yates, 1967.

For the moment, only the top row should be considered; the use of the remaining rows will be explained later. Usually it is judged by convention that if the probability comes out at <5 percent, we begin to suspect that the data do not fit the hypothesis assumed, or in statistical terminology, we say the data *differ significantly* from the hypothesis assumed; that is, the hypothesis can be regarded as not likely to be valid. In the example, χ^2 is extremely small so the data are in agreement with the 3 : 1 ratio as far as they go. However, it can never be absolutely proved that the 3 : 1 ratio is correct, since if more data were collected then ultimately the value of χ^2 might increase and indicate a significant difference from the 3 : 1 ratio. Thus all the χ^2 test can do is show the unlikelihood of a hypothesis at a certain level of probability.

In Table 2-4 are data for the cross between heterozygous yellow mice. Phenotypically we would expect ¾ yellow : ¼ nonagouti just as for the above cross. Considering this simple 3 : 1 hypothesis, $\chi^2 = 16.46$, which is significant at the 0.1 percent level (Table 2-3). If P is the probability of the hypothesis being correct, we can write $P < 0.001$ as the chance of the hypothesis being correct, which is obviously very remote. If we look further at the data in Table 2-4, it is clear that there is a deficiency of yellow mice observed compared with expected. This provides a clue, because a number of situations are known where genotypes do not survive or are *lethal,* so in this case we are suggesting that A^YA^Y homozygotes may be lethal. If this is so, we would then expect ⅔ yellow : ⅓ nonagouti made up of ⅔A^Ya heterozygotes and ⅓ aa homozygotes. Carrying out a χ^2 test on a 2 : 1 ratio gives $\chi^2 = 3.64$ ($P < 0.10$), which does not disagree with the hypothesis. Thus the χ^2 test is useful in determining which of several hypotheses fits best. It is important, however, to proceed from the simplest hypothesis to the most complex, since there is little point in ex-

Table 2-4　Data for a Cross between Yellow Mice A^Ya

Phenotype	Number observed O	Number expected E at		$(O - E)^2/E$	
		3 : 1	2 : 1	3 : 1	2 : 1
Yellow	706	762	677.33	4.12	1.21
Nonagouti	310	254	338.67	12.35	2.43
Total	1016	1016	1016.00	16.46	3.64

plaining data by a complex hypothesis when a simpler one is biologically adequate. Biologically the $2:1$ hypothesis fits, since A^YA^Y genotypes have been shown to die in utero, leaving only A^Ya and aa as the two viable genotypes.

In many cases significant χ^2 values may appear in data because of poor viability of genotypes rather than because of direct lethality. In mice, fidget homozygotes tend to be nonviable; therefore, in a cross $fi+ \times fi+$, which would have an expectation of 3 nonfidget : 1 fidget, there is frequently a deficiency of fidget mice, which would, if enough progeny were obtained, lead to a significant χ^2 value. The same point would apply to a number of the neurological mutants in the mouse discussed in Chapter 9.

What happens if there are more than two classes? In the cross $AaBb \times aabb$, the recognizable phenotypic classes are, genotypically, expected to be $\frac{1}{4}AaBb : \frac{1}{4}Aabb : \frac{1}{4}aaBb : \frac{1}{4}aabb$, if the genes a and b segregate independently of each other. The procedure adopted is to compute χ^2 over the four classes exactly as described previously, namely $\Sigma(O - E)^2/E$. Now the larger the number of classes, the greater the number of components of χ^2. This means that the value of χ^2 is expected to increase by chance as the number of classes increases, or alternatively, the value of χ^2 at which data are judged as significant at, say, the 5 percent level increases. To deal with this, the concept of the *number of degrees of freedom* (n) must be introduced, which in simple cases is the number of phenotypic classes minus one. In Table 2-3, χ^2 values are given for $n = 1, 2$, and 3. In the simple crosses analyzed in Tables 2-2 and 2-4 we are dealing with only two classes for which $n = 1$, and we use the symbol χ_1^2, where the subscript indicates the number of degrees of freedom. For the above cross with four phenotypic classes, a χ_3^2 would be calculated.

As an example where χ_3^2 values should be calculated, in Table 2-5 mating preference data from five geographically distinct strains of *Drosophila pseudoobscura* are given. In each mating test, 10 virgin males and females were

Table 2-5 Mating Preferences in Crosses between Different Geographical Strains of *D. pseudoobscura*

Cross strain A × strain B	Number of matings	A♀ × A♂	A♀ × B♂	B♀ × A♂	B♀ × B♂	χ_3^2 for random mating
		Number of each type of mating				
Berkeley × Okanagan	222	60	50	72	40	10.14*
Berkeley × Austin	160	37	43	42	38	0.65
Berkeley × Hayden	28	7	7	5	9	1.14
Berkeley × Sonora	103	23	22	28	30	1.74
Okanagan × Austin	125	27	33	33	32	0.79
Okanagan × Hayden	51	14	14	10	13	0.84
Okanagan × Sonora	114	26	29	32	26	0.74
Austin × Hayden	103	21	26	30	26	1.58
Austin × Sonora	113	36	28	27	22	3.57

* Significant at 0.05 level.
Source: Anderson and Ehrman, 1969.

used in specially designed mating chambers. The χ_3^2 tests are given for an expectation of random mating, that is, a $1:1:1:1$ ratio among the four possible mating types between strains A and B, namely $A\female \times A\male$, $A\female \times B\male$, $B\female \times A\male$, and $B\female \times B\male$. The results indicate that only in one case, the Berkeley × Okanagan combination, is there a significant deviation that seems to be due to the higher mating propensity of the Berkeley strain male compared to the Okanagan strain male. Indeed there are 132 matings involving A (Berkeley) males and 90 involving B (Okanagan) males, while there are 110 matings with A females and 112 with B females for a $1:1$ expectation in each case. The reader should carry out χ_1^2 tests on these $1:1$ expectations. As will be seen in the next and later chapters, differences in male mating propensities are quite common. The third way of combining these data for a $1:1$ comparison is the total of 100 like matings ($A\female \times A\male$, $B\female \times B\male$) compared with 112 unlike matings ($A\female \times B\male$, $B\female \times A\male$), which also provide material for a χ_1^2 test. Often, but not here, as will be seen later, the number of like matings exceeds the number of unlike matings. If the reader totals the three separate χ_1^2 values mentioned above, they should add to 10.14 or the χ_3^2 value in Table 2-5. This shows the additive property of χ^2 values in appropriately arranged data as these are. In conclusion, χ^2 tests are useful for any observed frequency data, where such data can be compared with what would be expected if a hypothesis were true.

Contingency χ^2 Test

Occasionally data are presented in the form of 2×2 tables. For example, some female twin pairs are tested for smoking habit and classified according to whether they are identical twins derived from the same zygote (*monozygotic* twins) or nonidentical fraternal twins derived from two different zygotes (*dizygotic* twins). The data from Fisher (1958), tallied for concordances in smoking habit, are

	Concordant (both smoked or both did not)	Discordant (one smoked and one did not)	Total
Identical twins (monozygotic)	44	9	53
Fraternal twins (dizygotic)	9	9	18
Total	53	18	71

The question to be answered is whether smoking habits are more alike in monozygotic twins than in dizygotic twins, since monozygotic twins have the same genotype while dizygotic twins have differing genotypes. If there is no association, we expect the concordances for smoking habit to be similar in monozygotic and dizygotic twins.

Algebraically, the 2×2 table can be written

	Alike	Unlike	Total
Monozygotic	a	b	$a + b$
Dizygotic	c	d	$c + d$
Total	$a + c$	$b + d$	$a + b + c + d = N$

where a, b, c, d are observed totals corresponding to the numbers in the above table. We expect $a : b = c : d$ if there is no association. In other words, if there is no association we expect $ad = bc$ or $ad - bc = 0$. It can be proved that

$$\chi_1^2 = \frac{(ad - bc)^2 N}{(a + c)(b + d)(c + d)(a + b)}$$

is a test for association. Note that if $ad - bc = 0$ or $ad = bc$, $\chi_1^2 = 0$, and if $ad \neq bc$, $\chi_1^2 > 0$. The greater the inequality between ad and bc, the greater is χ_1^2 and the greater the association.

Especially for small expected numbers, as is the case in the data under consideration, Yates's continuity correction is generally applied, as it gives a better theoretical fit to the expected χ^2 distribution. The above χ_1^2 formula, applying Yates's correction for continuity, becomes

$$\chi_1^2 = \frac{(|ad - bc| - \frac{1}{2}N)^2 \, N}{(a + c)(b + d)(c + d)(a + b)}$$

For the data $\chi_1^2 = 6.09$ ($P < 0.05$), showing a significant association between smoking habit and twin type, such that smoking habit is more similar in monozygotic twin partners than in dizygotic twin partners. One may therefore argue for a genotypic component in determining smoking habits.

Now, as with much human data of this type, one can argue that the environment of monozygotic twins may be more similar than that of dizygotic twins, so the above result could be environmental. The only way of dealing with this is to compare monozygotic twins separated at birth and reared apart with those reared together. The subdivision of monozygotic twins so obtained (Fisher, 1958) is

	Concordant	Discordant	Total
Separated	23	4	27
Not separated	21	5	26
Total	44	9	53

for which the χ_1^2 for association $= 0.004$. In other words, the difference in upbringing has no significant effects for these limited data. (When expected values are less than about 3 or 4, then χ^2 tests become rather inaccurate, but we

Table 2-6 χ_1^2 **Tests for Random Mating**

	Number observed			Number expected				
	AA	**Aa**	**aa**	**AA**	**Aa**	**aa**	χ_1^2	**P**
Group I	40	240	120	64	192	144	25.00	<<0.001
Group II	85	150	165	64	192	144	19.14	<<0.001
Pooled	125	390	285	128	384	288	0.20	>0.50

are probably in the safe range here.) See Chapter 7 for detailed discussions of twins in genetic studies of both continuous and discontinuous traits.

χ^2 Test for Random Mating

To test for random mating, one must first carefully define the population and trait being measured. The population must be as uniform as possible. Mixing populations together that do not themselves show random mating for a trait may lead to simulated random mating or panmixia. The most commonly used method of testing for random mating is to see if the distribution of phenotypes is in agreement with Hardy-Weinberg equilibrium. This involves an elementary knowledge of the χ^2 test. We compute the gene frequencies from the observed data. Thus in the hypothetical example presented in Table 2-6, both groups I and II have gene frequencies of A or p 0.4 and of a or q 0.6. Hence the expected genotypic frequencies are ($N = 400$, the population size):

$$AA = p^2N = 0.4^2 \times 400 = 64$$
$$Aa = 2pqN = 2 \times 0.4 \times 0.6 \times 400 = 192$$
$$aa = q^2N = 0.6^2 \times 400 = 144$$

From the observed and expected frequencies so obtained, a χ^2 value can be calculated in the usual way as $\Sigma(O - E)^2/E$.

There is, however, one difference between these data and those considered previously: to obtain the expected frequencies, a parameter, namely the gene frequency (p), is estimated from the observed data. In these circumstances, the rule derived from statistics is that the number of degrees of freedom equals the number of phenotypic classes minus the number of independent parameters estimated from the observed data minus 1. Clearly only one independent parameter is determined from the data, since $p + q = 1$. Therefore, in this case χ_1^2 is being calculated as a test for random mating.

Table 2-6 therefore shows the results of testing each of two groups as well as the two groups combined for Hardy-Weinberg equilibrium. The first group does not fit expectation because the observed number of heterozygotes exceeds the expected number, so that there is a deficiency of homozygotes for an expectation of random mating. An excess of heterozygotes, quite commonly observed in both experimental and natural populations, may be due to natural selection favoring the heterozygotes at the expense of the corresponding homozygotes. At the behavioral level there exists the possibility of greater than expected mating frequencies among unlike types or negative assortative mating, a phenomenon encountered much less frequently than is positive assortative mating. The second group also gives a poor fit, with the two homozygote classes having excess numbers. This could occur as a result of positive assortative mating or inbreeding. If we ignore the differences between the two groups and pool them to test for Hardy-Weinberg equilibrium, the fit is very good. But to conclude that the combined population shows panmixia is false owing to the heterogeneity of the population.

The opposite effect can be obtained if two groups with different gene frequencies for a particular trait are pooled together to test for random mating. The resulting population does not necessarily show evidence of random mating, although within each homogeneous group mating may well be at random. Such a pooled population shows an excess of homozygotes over those expected, an effect first described by Wahlund (1928). An example of this may result if two ethnic groups, although geographically intermingled, continue to be partially isolated with respect to mating patterns. Some characteristics (e.g., blood groups), although not apparently influencing mating choice, may maintain gene frequency differences in the two groups. Other forms of partial mating isolation (e.g., assortative mating for height) may not be directly associated with corresponding discrete gene frequency differences of other traits (such as blood groups, see Falk and Ehrman, 1975), so the criterion of a homogeneous population is fulfilled for these traits. A more complete discussion of the importance of homogeneity of groups when either association or independence is being sought can be found in Li (1976).

It should be emphasized that knowledge of the homogeneity of the population is imperative before tests for random mating can be meaningful, and in many cases it is assumed that no heterogeneity is present when, in fact, it is just not detected. If the homogeneity of a population can be demonstrated satisfactorily, a test for random mating can be carried out, preferably by testing mating classes rather than merely by looking for Hardy-Weinberg equilibria. However, conclusions drawn, either in favor of random mating or against it, should always be assessed carefully, and they should be no stronger than the confidence one can place in the demonstration of an acceptably homogeneous population.

Finally, as a matter of completeness, it is appropriate to list causes, including those already discussed, whereby deviations from panmixia may occur in a statistical sense as described in this section:

• *Selection*. This possibility is considered in the discussion of Table 2-6, group I. Selection occurs when some genotypes leave more offspring in the next generation because of differences in overall viability or *fitness* between different genotypes. The likelihood is that behavioral parameters, especially those associated with mating, are of considerable importance in fitness differences, as discussed in later chapters.

• *Mutation.* Genes may change, say from *A* to *a,* at a low frequency. Over long periods of time, mutations play a role in evolutionary change, but the mutation rate is usually so low that over a few generations it can normally be ignored.

• *Migration.* In some senses this is related to mutation, in that new genes may be introduced into populations, but the influence on the gene pool may be much greater than mutation if there are many immigrants.

• *Inbreeding.* As already mentioned, inbreeding leads to an increased incidence of homozygotes. In people, inbreeding is of importance in isolated populations where high frequencies of marriages between relatives (*consanguineous marriages*) may occur.

• *Assortative mating.* This has been, and will be further, discussed.

• *Random genetic drift.* This is the term used to describe the chance events that may occur generation by generation in the determination of gene frequencies. For example, if the population size is relatively small, it may happen just by chance that a sample of gametes from the population that provides the next generation is not representative. The consequence is a change in gene frequency in the new population merely because of chance phenomena. Clearly the importance of drift progressively diminishes as the population size increases.

2-5 GENE ACTION

So far we have considered genotypes as assessed directly by their phenotypes in families and populations and have briefly considered some of the principles needed to understand their patterns of inheritance from generation to generation. However, as will become apparent in certain parts of this book, we also need to consider the path between the gene and the behavioral phenotype that is observed.

Almost all cells of a given organism, except its gametes, contain the same amount and type of a material known chemically as *deoxyribonucleic acid* (DNA), which is organized in the chromosomes. Experiments in microorganisms show clearly that DNA contains the information necessary to make new cells essentially identical to the parent cells. The essentials of this principle have also been shown in higher organisms. The amount of DNA per cell is not large; in the human somatic or fertilized egg cell the figure is about 6×10^{-12} g. In spite of this very small quantity, the amount of information contained is enormous and is sufficient to direct the synthesis of a human individual.

DNA is made up of chemical units consisting of

• A base belonging to the family of purines or pyrimidines, which are nitrogen containing compounds; the two possible purines are adenine (A) and guanine (G); the two possible pyrimidines are cytosine (C) and thymine (T).

• A pentose (five-carbon sugar)—deoxyribose.

• A phosphate group.

The DNA molecule is made up of *nucleotides,* each comprising one base, one sugar, and one phosphate group. For a given DNA molecule the phosphate and sugar groups are all the same; only the bases vary. Because only the bases A, G, C, and T can vary, apparently the information determining heredity resides in these and in the degree to which they vary.

The quantity of these bases is constant in a given species but varies among species. However, in every species the quantity of A = T, and of G = C. From this it follows that A and T are always paired together and similarly for G and C.

The overall structure of DNA was demonstrated in 1953 by Watson and Crick. The bases are attached to a sugar-phosphate backbone forming a chain of nucleotides

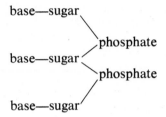

Watson and Crick found DNA to be a double chain of nucleotides

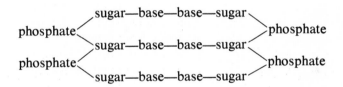

The two chains are joined by hydrogen bonding of the bases and twist around each other, forming a double helix (Figure 2-8). There is a distance of 3.4 Å between the nucleotide pairs. (Å = an angstrom unit or 10^{-7} mm.) The spiral makes a complete revolution in 10 pairs or 34 Å. Because of the pairing rules—A always with T, and G always with C—if we know the sequence of bases on one helix, we also know the sequence on the other.

According to the Watson-Crick theory, the linear sequence of nucleotides is constant for a given species although there are fluctuating, minor heritable variations within species. It is the actual sequence of nucleotides that leads to variations in proteins which have as their primary structure a chain of amino acids. The genetic component of the phenotype observed depends on this linear sequence of nucleotides. Furthermore, variations in the observed phenotype are likely to be due to minor changes in the sequence of nucleotides (if environmental effects can be eliminated). Therefore, the sequence of nucleotides can be regarded as a *genetic code.*

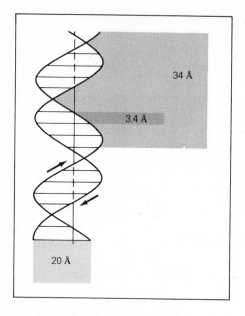

Figure 2-8 The double helix of DNA.

There are 20 essential amino acids that are specified by the genetic code. Since there are four possible bases (A, T, G, and C), one- and two-nucleotide code sequences are inadequate, specifying 4 and $4^2 = 16$ combinations only. The triplet code (a sequence of three nucleotides) specifying $4^3 = 64$ combinations is a necessary (and apparently sufficient) minimum. The nucleotide triplet is referred to as a *codon*. Because a triplet code gives 64 different combinations, or words, of which only 20 are needed for amino acid synthesis, the code is referred to as a *degenerate code*. In fact, some amino acids are coded by more than one triplet (as can be seen in Figure 2-9). The names and abbreviations of the 20 essential amino acids are

Alanine	Ala	Leucine	Leu
Arginine	Arg	Lysine	Lys
Asparagine	Asn	Methionine	Met
Aspartic acid	Asp	Phenylalanine	Phe
Cysteine	Cys	Proline	Pro
Glutamic acid	Glu	Serine	Ser
Glutamine	Gln	Threonine	Thr
Glycine	Gly	Tryptophan	Try
Histidine	His	Tyrosine	Tyr
Isoleucine	Ilu	Valine	Val

Note the inclusion of phenylalanine, which we have seen to occur in elevated toxic quantities in phenylketonuria.

From the DNA code, a linear message of triplets or codons is transcribed to a form of RNA (or ribonucleic acid) called *messenger* RNA (mRNA). RNA is

SECOND LETTER

		U	C	A	G	
	U	UUU⎫ UUC⎭ Phe UUA⎫ UUG⎭ Leu	UCU⎫ UCC⎪ UCA⎬ Ser UCG⎭	UAU⎫ Tyr UAC⎭ UAA Chain End UAG Chain End	UGU⎫ Cys UGC⎭ UGA Chain End UGG Try	U C A G
C	CUU⎫ CUC⎪ Leu CUA⎪ CUG⎭	CCU⎫ CCC⎬ Pro CCA⎪ CCG⎭	CAU⎫ His CAC⎭ CAA⎫ Gln CAG⎭	CGU⎫ CGC⎪ Arg CGA⎪ CGG⎭	U C A G
A	AUU⎫ AUC⎪ Ilu AUA⎭ AUG Met	ACU⎫ ACC⎬ Thr ACA⎪ ACG⎭	AAU⎫ Asn AAC⎭ AAA⎫ Lys AAG⎭	AGU⎫ Ser AGC⎭ AGA⎫ Arg AGG⎭	U C A G
G	GUU⎫ GUC⎪ Val GUA⎪ GUG⎭	GCU⎫ GCC⎬ Ala GCA⎪ GCG⎭	GAU⎫ Asp GAC⎭ GAA⎫ Glu GAG⎭	GGU⎫ GGC⎪ Gly GGA⎪ GGG⎭	U C A G

FIRST LETTER (left side) · THIRD LETTER (right side)

Figure 2-9 The genetic code of RNA. Triplet nucleotides code for the 20 essential amino acids and punctuation.

chemically very similar to DNA except that (1) it has ribose rather than the sugar deoxyribose; (2) it has the base uracil (U) substituting for thymine; and (3) it is single-stranded rather than double-stranded. In transcription from DNA to mRNA, the following pairing rules therefore occur:

Base in DNA	Base in mRNA
A	U
T	A
C	G
G	C

There are three kinds of RNA upon which protein synthesis depends: messenger RNA, transfer RNA, and ribosomal RNA, all carrying codons corresponding to the DNA. Transfer RNA brings amino acids to the cytoplasmic ribosomes where ribosomal RNA arranges them into protein chains according to the instructions of messenger RNA.

Before amino acids located in the cytoplasm are assembled into protein chains there, they must be "activated" by the attachment of a special phosphoric acid group. They are then attached to *transfer* RNA (tRNA). In fact, there are as many tRNA molecules as there are triplets determining amino acids.

Alignment of the tRNA and mRNA so as to assemble protein chains in an orderly way is mediated by special particles in the cytoplasm of the cell called *ribosomes,* which are made from the third form of RNA, *ribosomal* RNA (rRNA).

The process of the formation of protein from the code carried by the mRNA is referred to as *translation,* so we can summarize what happens as

$$\text{DNA} \xrightarrow{\text{transcription}} \text{mRNA} \xrightarrow{\text{translation}} \text{protein}$$

The important point then, is that the amino acid sequence in proteins is directly determined by the genetic code carried by DNA molecules. A great deal of additional information about this process can be found in many texts, but the details of the process, largely established in microorganisms, are of only peripheral interest to the behavior geneticist at the present stage in the development of our field. However, with time there will certainly be a trend toward metabolic explanations of behavioral processes, so an understanding of the biochemical basis of gene action will assume progressively more importance. Even so, it should be clear that the hereditary unit discussed so far in this chapter has a definite structural and functional meaning.

As far as protein synthesis itself is concerned, most proteins are made only when they are useful. In other words, there are various means of regulation present. In fact, *regulator genes* have been delineated, particularly in microorganisms. These regulator genes determine whether the genes that specify types of proteins (*structural genes*) are active or otherwise. The regulator genes themselves are controlled by cytoplasmic events and thus are open to environmental influences. For example, if a certain amino acid is necessary for growth and is present in the environment, then the cells can cease making it and the enzymes (biological catalysts, see Section 11-4) necessary for its synthesis (enzyme repression). Undoubtedly this process of regulation of protein synthesis must be the basis of *differentiation,* which is the development of different cells and tissues. At different stages during development it is clear that various portions of DNA are active in different cells and tissues. Such temporal regulation of gene action must be carefully studied in order to understand behavior. It is clear that genes act in sequence during development so that a given gene may start one event which may lead to a series of others. Gene–hormone interactions, for example, are probably involved in sex differentiation, the onset of puberty, and the development of learning in human beings. A good example of a gene–hormone interaction in experimental animals is provided by molting patterns in Diptera and other insects. These patterns depend upon the production of the molting hormone ecdysone and its effect on certain loci.

The other important consequence of understanding the physiological processes of a gene substitution is that their pattern can be modified by appropriate treatment. In mice there is a condition, *pallid,* due to a single recessive gene in linkage group V. The mutant mouse has an impairment of the calcified otoliths in the inner ear. These otoliths normally move in response to an animal's change in position. In this fashion neural responses are induced relevant to the organism's gravity response (Erway, Hurley, and Fraser, 1966). The pallid gene destroys otoliths in one or both ears, and so behavioral equilibrium is disrupted. Otolith destruction can be produced by withholding manganese from the diet of normal mice; that is, the phenotype can be induced environmentally—a phenomenon referred to as *phenocopy*. Conversely, if gestating females bearing the pallid gene are given large supplements of manganese, the mutant offspring do not show the defect. Therefore, we have a gene-behavior relation capable of environmental control once the condition is understood.

A good example of environmental alteration in human beings is phenylketonuria, which we have already considered in certain other respects. To recapitulate, individuals homozygous for this recessive gene generally have IQs < 30 (occasionally higher) and their skin and hair pigmentation is generally lighter than that of the population from which they arise. Phenylketonuria is due to a deficiency or absence of the enzyme phenylalanine hydroxylase, which is essential in the metabolism of phenylalanine, an amino acid that is an essential dietary constituent. Normally, phenylalanine → tyrosine . . . → various metabolic breakdown products (Figure 2-10). In phenylketonuria, this step is blocked; phenylalanine accumulates to a level 40 to 50 times that found in nonphenylketonurics. This excess leads to mental deficiency. Hence a likely treatment would be to feed a phenylalanine-deficient diet. This poses problems, as no known protein is deficient in phenylalanine. However, such a diet can be synthesized by breaking down protein and reconstituting it without phenylalanine, but still containing other essential amino acids. The diet must begin early in life to retard damage to IQ. It is likely to be proportionately less

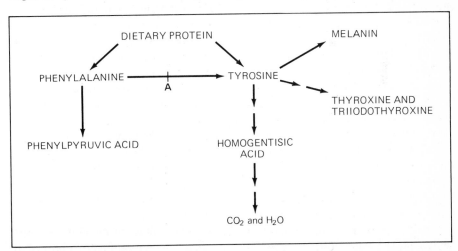

Figure 2-10 Phenylalanine metabolism. Normally, phenylalanine is transformed to tyrosine and various other compounds derived from tyrosine. In phenylketonurics, when block occurs at *A* (i.e., phenylalanine hydroxylase is absent), the alternative pathway through phenylpyruvic acid assumes importance. The altered pigmentation of phenylketonurics is expected, because melanin formation depends partly on tyrosine. (*After Harris, 1959.*)

effective if delayed beyond infancy. The treatment must strike a delicate balance between malnutrition (i.e., inadequacy of the essential amino acid phenylalanine) and intoxication.

Until recently, phenylketonuria was detected by a test that depends for its success on secretion of an abnormal product of phenylalanine, phenylpyruvic acid, in the urine of phenylketonurics (Figure 2-10). However, phenylpyruvic acid secretion may be delayed for up to 5 to 6 weeks after birth. The test itself merely requires adding $FeCl_3$ to urine acidified with $1 N$ HCl; if phenylpyruvic acid is present, the urine turns green. A more recent and more efficient procedure is the Guthrie test, based on the ability of certain strains of bacteria to grow only in a medium containing a high level of phenylalanine. To perform this test, blood is added to a bacterial culture lacking phenylalanine. If the culture grows, this indicates that the phenylalanine level of the blood is high, perhaps indicating phenylketonuria. Most states of the United States now make the Guthrie test compulsory for all newborn babies, and it is commonly carried out in other countries.

Individuals heterozygous for the phenylketonuria gene can be detected by a *phenylalanine tolerance test*. This consists of feeding the fasting individual with phenylalanine and then testing serum levels of phenylalanine at various periods after fasting. Many persons heterozygous for the phenylketonuria gene ($p+$) metabolize phenylalanine more slowly than do normals ($++$). Thus $p+$ individuals can often be distinguished from $++$, showing that at the biochemical level the p gene is not completely recessive. Heterozygote detection is of importance in *genetic counseling* (advice given to people about genetic risks of having abnormal children) and in other problems where genetic advice is given. Thus if two individuals are known to be $p+$, the chance of pp offspring from them is 25 percent.

And what of the phenylketonuric heterozygote gestated by "metabolically cured" homozygous mothers? The mothers are usually uninstitutionalized individuals appropriately treated with phenylalanine-reduced diets after detection via the postnatal tests described just above. Ordinarily such heterozygous fetuses develop normally, but their quantity of the pivotal enzyme, phenylalanine hydroxylase, is relatively low and they cannot cope with the elevated phenylalanine levels to which their genetically phenylketonuric mothers expose them (Sutton, 1975). Consequently they are very often born more defective than wholly untreated homozygotes. (Maternal PKU may also cause abortions and other undesirable congenital effects.) Here we have a cross-generation genotype–environment interaction.

As a final fillip in this complex story, diagnostic errors are possible, especially but not only with treated mothers. If a heterozygote or homozygous normal child is placed on the special PKU diet as a result of diagnostic error, mental deficiencies will be the result of phenylalanine deficiencies. (For attempts at diagnostic improvements, see Paul et al., 1978.)

In conclusion, understanding the mechanisms of gene action supporting a given behavioral phenotype is possible in a few instances, even if incompletely. To find the molecular correlates of behavioral patterns is an exciting possibility that is beginning to be realized especially in unicellular organisms such as bacteria and protozoa (for attraction or repulsion in response to certain chemicals). In human beings this will require more prolonged and intense investigatory efforts. Note though that in a few cases in higher organisms it is possible to assess genetic code changes associated with different alleles at a locus. The future will undoubtedly reveal more examples as our behavioral phenotypes become progressively better understood.

SUMMARY

The phenotype of an organism conventionally refers to its outward appearance. This definition can be extended to include the totality of physiological, anatomical, and behavioral components of that individual. Genotype and environment control behavior as for other traits. Breeding experiments in experimental animals such as *Drosophila* and mice have demonstrated this convincingly. This applies where both genetic variation is under the control of specific genes assignable to specific loci on chromosomes, as well as for quantitative traits which do not segregate into discrete classes.

The principles of genetics reviewed here at the family level can be extrapolated to the population level. This is most easily demonstrated if mating is assumed to be at random; however, the behavior geneticist must always be aware that random mating is rare in the real world. Thus phenotypic assortative mating, which is the mating of individuals based upon phenotypic resemblance, is usual in man for many traits such as height, weight, and IQ.

Another important consideration in this chapter is the recent trend towards metabolic explanations of behavior. An understanding of the biochemical basis of gene action, and hence of behavioral phenotypes, will assume progressively more importance with time. With such knowledge, it is possible in some cases to modify behavioral defects to ameliorate their severity.

GENERAL READINGS

1 Principles of Genetics

Crow, J. F. 1976. *Genetics Notes,* 7th ed. Minneapolis: Burgess. A concise elementary text useful for beginners. A glossary of definitions is provided. Chapters 1 to 3 would be useful for those finding the introduction in the first section of this chapter too rapid.

Goodenough, U., and R. P. Levine. 1974. *Genetics.* New York: Holt. A good general text with a stronger molecular emphasis than either Crow or Strickberger.

Strickberger, M. W. 1976. *Genetics,* 2d ed. New York: Macmillan. An advanced but excellent general text for all sections of Chapter 2.

2 Human Genetics

Bodmer, W. F., and L. L. Cavalli-Sforza. 1976. *Genetics, Evolution and Man.* San Francisco: Freeman. An excellent, very readable book, which considers human behavior genetics in some depth. A useful glossary is provided.
Stern, C. 1973. *Principles of Human Genetics,* 3d ed. San Francisco: Freeman. A comprehensive text in human genetics assuming no prior knowledge of genetics and including a consideration of behavior genetics.

3 Population Genetics

Li, C. C. 1976. *First Course in Population Genetics.* Pacific Grove, Calif.: Boxwood Press. An expanded version of a classic text in this field.
Spiess, E. B. 1977. *Genes in Populations.* New York: Wiley. A comprehensive and well-presented quantitative account, assuming a basic knowledge of genetics.

Single Genes and Behavior

Traits under the control of single genes are the best known simply because they are easiest to trace, as we saw in the last chapter. This applies to all traits, whether morphological, physiological, or behavioral. Many such genes are rare and deleterious and so may not be of great importance in a population, but because their effects are easily traceable, the phenotypes they control may provide information on behavioral variation in the species in question. First, there are genes that produce a visible alteration in appearance with a simultaneous change in behavior. For example, phenylketonurics, discussed in Chapter 2, as well as having low IQs, tend to have lighter hair pigmentation than does the population from which they are derived. In other words, the gene has more than one observable effect, a phenomenon referred to as *pleiotropy*. As shown in this and other chapters, pleiotropic effects involving morphological, physiological, and behavioral traits are common, although all these effects can presumably ultimately be referred back to one particular nucleotide sequence of a DNA molecule. Second, we may ask whether a single gene producing no known morphological effects can, primarily or even exclusively, produce behavioral changes. Outwardly this often appears to be the situation, but in fact detailed research frequently reveals associated physiological or biochemical variables as one might expect.

   This chapter discusses some behavioral traits for which single-gene effects are known, in some cases showing pleiotropic effects and in other cases not (although more detailed studies are likely to reveal such effects in all instances).

3-1 NEST CLEANSING BY HONEYBEES

Rothenbuhler (1964) carried out an elegant analysis of the nest cleansing of honeybee larvae killed by a disease, American foulbrood (pathogen, *Bacillus larvae*). The maintenance of the hygienic environment within a hive requires the opening of combs housing afflicted young and their immediate evacuation. If this is not done, cadavers with their associated spores remain inside the hive as a continuous source of contamination. Genes at two independently segregating loci account for hygienic or nonhygienic behavior; one involves the uncapping of cells and the other the removal of their contents. Where *u* represents the recessive gene for uncapping behavior and *r* the recessive gene for removal, the genetic makeup of the bees of a hygienic colony is *uurr*.

   No physical or physiological differences have been recorded between totally hygienic, partially hygienic, or totally nonhygienic honeybees, although detailed investigations might uncover some. From the genetic point of view, the example is of interest because the fractionation of hygienic behavior into two distinct components leads to the understanding of its genetic basis. Surely both these acts enhance the survival and perpetuation of the reproductive unit built by these social insects—the hive and its inhabitants. We have here, therefore, a remarkable example of behavior controlled by two single-gene loci with obvious fitness effects.

3-2 MATING SUCCESS IN *DROSOPHILA*

Cinnabar and *vermilion* are mutant genes in *Drosophila melanogaster* affecting eye colors, as their names imply. (*Cinnabar* is an autosomal recessive and *vermilion* a sex-linked recessive.) Flies with either mutant gene have bright red eyes, compared with the dull red eyes of the wild type. Bösiger (1957, 1967) compared the speed of mating of *D. melanogaster* with the *vermilion* and *cinnabar* mutants. After 12 days the following percentages of females, each of which was confined with a single male, were found to be gravid:

	Vermilion ♀ × vermilion ♂	Vermilion ♀ × cinnabar ♂	Cinnabar ♀ × vermilion ♂	Cinnabar ♀ × cinnabar ♂
Couples tested	200	302	200	325
Percent fertile	61.0	80.1	54.0	73.8

   In another experiment, groups of females were confined with males, and the percentages of the females that copulated after a lapse of different time intervals were recorded. The results are as follows:

Time min	Vermilion ♀ × vermilion ♂	Vermilion ♀ × cinnabar ♂	Cinnabar ♀ × vermilion ♂	Cinnabar ♀ × cinnabar ♂
0–5	12.9	48.3	0	13.0
5–10	32.3	65.5	21.1	39.1
10–15	35.5	79.3	36.8	43.5
15–20	35.5	82.8	42.1	52.2
20–25	38.7	86.2	47.4	56.5
25–30	38.7	89.7	47.4	56.5

In both experiments, when males had the *vermilion* gene the success rate was lower than when they had the *cinnabar* gene. We can therefore say that males with the *vermilion* mutant are disadvantaged compared with others of the same sex in respect to reproduction. Differential reproductive success rates of this nature are referred to as *sexual selection.* Many examples are discussed in this and in later chapters (see Table 2-5 for an example of sexual selection in natural populations).

Bastock (1956) investigated the effects on mating success of the sex-linked recessive mutant *yellow* compared with the wild type in *D. melanogaster.* In her main experiments, a wild stock was crossed with a yellow stock for seven generations, so that the wild stock was genetically similar to the yellow stock except in the region of the *yellow* locus. Males with the *yellow* mutant are less successful in mating with females of the normal gray body color than are normal males. Bastock found that the courtship pattern of the males had been altered by the mutation from wild-type to yellow body color. Figure 3-1 illustrates this schematically by dividing courtship behavior into three components. *Orientation* normally occurs at the very start of courtship when the male follows the female, circles her, or stands in front of her. *Vibration* means wing movements and display following orientation. This is followed by *licking,* contact between the male proboscis and the female genitalia. All these are preludes to attempted mounts. Note that rows A and B contain the longer bouts of licking, and especially of vibrating, which are characteristic of wild-type males of this species. Rows C and D illustrate shorter intervals for everything but orientation; this is typical of the yellow-bodied *D. melanogaster* males.

Bastock's (1956) data illustrate another point: even in flies phenotypically *yellow,* background genotype may be relevant. Table 3-1 presents a comparison between an ordinary wild stock and the one that was crossed to the yellow stock for seven generations. In this latter wild stock the percentage mating success of the *yellow* × *yellow* flies is far lower than that of the wild × wild. In matings between *yellow* and wild flies the percentage success of the *yellow* male × wild female is lower than that of the wild male × *yellow* female. Thus in the crosses involving males with the *yellow* mutant, the percentage success is much lower than in crosses involving wild-type males; the genotypes of the females have little differential effect. However, before the wild stock was crossed with the *yellow* stock for seven generations, there was a significant difference between females as well as between males. The initial high female receptivity is therefore partly dependent on the genetic background. It can be

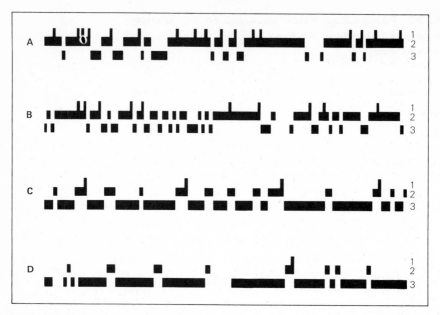

Figure 3-1 Trisected courtship patterns. Each of the four rows (A & B wild type; C & D *yellow*) represents the behavior of a male *D. melanogaster* as time progresses from left to right. 1, licking; 2, vibration; 3, orientation. (*From Bastock, 1967.*)

argued that for matings to occur reasonably frequently between *yellow* flies, there would be selection for *yellow* females with high receptivity in view of the low level of stimulus offered by the *yellow* males; that is, there is likely to be a balance between the level of female receptivity and the mating propensity of the males.

There have been many experiments where males of one or more genotypes are placed with females of one or more genotypes. Variations between genotypes in choice experiments are presumably due to differences in courtship behavior. Sturtevant (1915) carried out experiments, based on direct observation, in which males were offered two types of females (*male-choice* experiments) and others in which females were offered two types of males (*female-*

Table 3-1   Percentage Success in One Hour from Pair Matings Using Yellow-Bodied and Wild-Type *D. melanogaster*

Matings	Before crossing wild stock with *yellow* stock for seven generations	After crossing wild stock with *yellow* stock for seven generations
Wild male × wild female	62	75
Yellow male × wild female	34	47
Wild male × *yellow* female	87	81
Yellow male × *yellow* female	78	59

Source: Bastock, 1956.

Table 3-2 Results from Male-Choice and Female-Choice Experiments between White-Eyed and Wild-Type D. melanogaster

Male choice	Number of females mated	
	Wild-type	White-eyed
Wild-type male	54	82
White-eyed male	40	93
Female choice	**Number of males mated**	
	Wild-type	White-eyed
Wild-type female	53	14
White-eyed female	62	19

Source: Sturtevant, 1915.

choice experiments). Some data for a white-eyed (sex-linked) strain and a wild strain are given in Table 3-2. Clearly, the wild-type males have an advantage in sexual selection over the white-eyed males leading to nonrandom mating. In order to quantify data of this nature, certain indices have been proposed in the literature. These indices give estimates of the strength of sexual selection and of *sexual isolation,* which comes from comparing the proportion of *homogamic* (like-to-like) matings and *heterogamic* (unlike) matings. Under random mating the proportion of homogamic and heterogamic matings is expected to be equal.

For the male-choice situation, let there be n_1 females of type 1 and n_2 of type 2, together with males of type 1. Further, let $x_{1,1}$ and $x_{1,2}$ be the numbers of type 1 and 2 females inseminated, respectively, and let $p_{1,1} = (x_{1,1})/n_1$ and $p_{1,2} = (x_{1,2})/n_2$, so that $p_{1,1}$ and $p_{1,2}$ represent the proportions of each type of female inseminated. An *isolation index* devised by Stalker (1942) is

$$b_{1,2} = \frac{p_{1,1} - p_{1,2}}{p_{1,1} + p_{1,2}}$$

which ranges from $+1$ for 100 percent homogamic matings to -1 for 100 percent heterogamic matings and is zero if mating is at random. Simple χ^2 tests can be used on the raw data to see if deviations from $b_{1,2} = 0$ are significant. If the male is of type 2, a reciprocal index is

$$b_{2,1} = \frac{p_{2,2} - p_{2,1}}{p_{2,2} + p_{2,1}}$$

Joint isolation indices based on combinations of the pairs of experiments with males of types 1 and 2 have been proposed. If there are equal numbers of females or couples of each of the two types, the *joint isolation index* is

$$\frac{x_{1,1} + x_{2,2} - x_{1,2} - x_{2,1}}{N}$$

where $N = x_{1,1} + x_{2,2} + x_{1,2} + x_{2,1}$,   the   total   number   of   matings (Malagolowkin-Cohen, Simmons, and Levene, 1965). If there are not equal numbers of females or couples, the arithmetic mean of the two indices $b_{1,2}$ and $b_{2,1}$ is used

$$\frac{b_{1,2} + b_{2,1}}{2}$$

From female-choice data analogous indices can be computed. Sturtevant's data give a joint isolation index of 0.097 in the male-choice experiment and 0.026 in the female-choice experiment. There is, therefore, little evidence for sexual isolation, since both values are close to zero.

Bateman (1949) proposed an index that measures the relative mating propensity of females:

$$a_{1,2} = \frac{b_{1,2} - b_{2,1}}{2}$$

This is positive if there is an excess of females of type 1 and negative if there is an excess of females of type 2 in a male-choice experiment in which the males are of type 1. A similar index can be derived from the female-choice experiment. These indices, therefore, measure sexual selection. Sturtevant's data show that the relative mating propensity of wild-type females compared with white-eyed females is -0.303 in the male-choice experiments; the relative mating propensity of wild-type males compared with white-eyed males in the female-choice experiments is 0.558. There is, therefore, clear evidence of non-random mating due to differences in the vigor of sexual behavior: there is sexual selection.

In recent years, multiple-choice experiments have become common where males and females of types 1 and 2 are all placed together in an observation chamber. Several designs are available; one of the commonly used chambers, constructed by Elens and Wattiaux (1964), is shown in Figure 3-2. Direct observation is possible, and quite a large number of flies can be introduced—say 60 or more virgin pairs—but this depends on the species. Since copulating pairs do not generally move, they can be localized on the checkered canvas of the chamber. The Elens-Wattiaux technique permits the observation not only of the types of males and females in a mating but also of the time at which a given mating takes place, its sequence among other matings, and the duration of copulation. Furthermore, from this design all the various indices described can be calculated. Remember, though, that the biological situation in a multiple-choice experiment is different from that in a male-choice or female-choice experiment. Table 2-5 gives data on geographically isolated populations of D. *pseudoobscura* using this technique with consideration of the χ^2 tests routinely employed in testing for both sexual selection and sexual isolation.

A number of other genes have been shown to affect mating success in D.

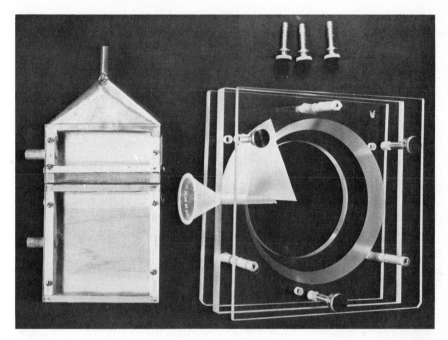

Figure 3-2 Alternate versions of Elens-Wattiaux chamber utilized for the direct scoring of *Drosophila* courtship and copulation. More than one chamber may be linked, either one atop the other or laterally. The older chamber (*left*) is plugged laterally with wooden pegs to prevent loss of moving air, a weak current being either sent from the copper apex or drawn into it. The flies producing the airborne odor are confined in the rectangular chamber nearest the apex, and a screened (on two sides) dead space separates them from the main, square chamber. It is in the square chamber that matings are scored. The dismantled plastic chamber (*right*) contains only a sector of the filter paper constituting the elevated circular center of its floor.

melanogaster, mainly due to variations in sexual selection. Thus in males homozygosity for the autosomal recessive gene *scabrous* (*sca*), resulting in eyes with rough surfaces, leads to low sexual vigor; in females, it produces enhanced receptivity compared with wild-type flies (McKenzie and Parsons, 1971). In other words, there is a threshold switch for mating in both sexes. This cancels out when *scabrous* flies are mated together, since the frequency of *scabrous* female × *scabrous* male matings is similar to matings between wild-type flies. A parallel example of such a balance has already been described for yellow-bodied flies. The low vigor of *scabrous* males with wild-type females is due to the fact that the males are blind (Crossley, unpublished).

Eye color is relevant to mating success in some instances. Normally *D. melanogaster* homozygous for the *v/bw* (*vermilion/brown*) genes have pale sherry-colored eyes and an associated marked attenuation of visual acuity in terms of optomotor response. The mutation at the *vermilion* locus results in a block in brown pigment synthesis, causing the flies to have bright red eyes. However, if the chemical kynurenine is added to the diet, the *vermilion* block is

**Table 3-3   Results of Competition between Kynurenine-Treated *v/bw* Males and
bw Males and between Kynurenine-Treated *v/bw* Males and *v/bw* Males**

Total number of mating competitions	Description of male	Number of males mating	χ_1^2 for 1 : 1 ratio
126	*bw*	52 ⎫	3.5
	Kynurenine-treated *v/bw*	74 ⎭	
83	*v/bw*	15 ⎫	30.12
	Kynurenine-treated *v/bw*	68 ⎭	($P < 0.001$)

Source: Connolly, Burnet, and Sewell, 1969.

bypassed, and brown pigment is formed in the eye. Table 3-3 shows the effect of kynurenine in enhancing mating success of male flies homozygous for the *v/bw* genes compared with those not treated with kynurenine (Connolly, Burnet, and Sewell, 1969). It seems likely that the mating disadvantage shown by flies lacking the eye pigment is due to a sensory defect accompanying the absence of screening pigment in the compound eye. This lack can be alleviated by biochemical supplements. Therefore, Connolly, Burnet, and Sewell (1969) suggest that the role of vision in the courtship of *D. melanogaster* has been underestimated—a result in agreement with work on the stock homozygous for the *scabrous* gene. Comparison of the courtship behavior of males with pigmented and nonpigmented eyes shows that the inferior courtship of *v/bw* males is attributable to difficulties in establishing and maintaining contacts with females. Thus *v/bw* male flies were found to have a significantly shorter bout length (the sum of licking plus vibration) than those to whose diet kynurenine was added. Generally, there was a close correlation between mating success and eye-pigment density. It is also of interest that only the presence of brown pigment is involved in mating success, since the absence of red pigment, as in *bw* flies, does not lead to an attenuation of optomotor response and does not affect courtship latency or duration. In an earlier observation, Parsons and Green (1959) showed a general correlation between brown eye-pigment density and fitness, such that the fitness of *v/bw* flies increased in competition experiments with increasing amounts of kynurenine. Various behavioral variables are therefore associated directly with biochemical changes and fitness changes (pleiotropy).

The final mutants of *D. melanogaster* to be considered are the sex-linked mutant genes that produce *Bar* and white eyes. *Bar* eyes are narrower than normal eyes, and the mode of inheritance is dominant. In a mixture of *Bar* and wild-type flies, the *Bar* males are less successful in mating; their disadvantage is, however, reduced when only a few of them are present; it increases as their frequency in relation to wild-type males becomes greater. With white-eyed males, mating success is greater when white-eyed males are rare or are predominant (Petit, 1958). Ehrman et al. (1965) found a similar situation in experiments with *Drosophila pseudoobscura*. It may indeed be that preferential mating correlated with frequency will prove to be of considerable importance in evolutionary processes, if it is at all widespread. (See Section 8-4 for further discussion of this topic.)

Sexual discrimination appears to be a trial-and-error affair among the drosophilids reported upon here. Males court females (sometimes, even other males) of any species and try to repeat courting and mating. The acceptance is controlled mainly by the female, as are other "turning points" in the courtship-mating-insemination sequence. Even so, as is shown in Sections 4-2 and 13-1, given receptive females, mating speed differences between males rather than between females are common. Presumably, as Bateman (1948) suggested, these differences have evolved, since male reproductive success is determined by the number of matings achieved, while females need only mate once to achieve their reproductive success in a given breeding cycle. Furthermore, duration of copulation is mainly male-determined, at least in *D. melanogaster* (MacBean and Parsons, 1967) and in *D. pseudoobscura* (Kaul and Parsons, 1965). For a description of the courtship and mating behavior of *Drosophila*, see Spieth (1952); for a pictorial description, see Ehrman and Strickberger (1960) and Ehrman (1964).

3-3 SINGLE-GENE EFFECTS IN MICE

Mating Success

Albinism in mice and other animals (e.g., one type in human beings) is frequently controlled by an autosomal recessive gene. Levine (1958) compared the relative mating success of homozygous black agouti male mice and albino mice. Both strains were maintained separately by brother-sister matings and so were inbred. All males were proved fertile both at the beginning and at the end of the experiment. The procedure was first to match at random 10 male albinos with 10 male black agoutis. Each pair of males was then placed in a pen with a single albino female. From each pen 10 litters were obtained, giving a total of 100 litters. The results are shown in Figure 3-3.

Three types of litters were produced: those containing only albino offspring, only black agouti offspring, or mixed offspring (some albino and some black agouti). This last type of litter was the result of double inseminations. There was no statistical difference in litter size among the three types of litters. It was found that 76 percent of the litters were fathered solely by the albino males, 12 percent of the litters were fathered solely by the black agouti males, and 12 percent of the litters were the result of double inseminations. Within the mixed litters, albino offspring were more than twice as common as black agouti offspring. Of a total of 552 mice born in the 10 cages, 458 were fathered by the albino males while only 94 were fathered by the black agouti males.

Interpretation is not easy because selective fertilization favoring the sperm of males belonging to the same strain as the female cannot be ruled out. Levine

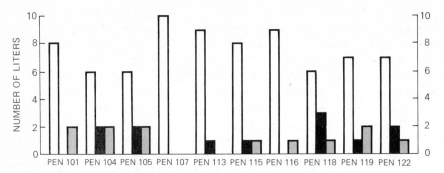

Figure 3-3 Results of competitive matings in mice of different inbred strains. White columns represent number of litters fathered solely by albino males. Black columns represent number of litters fathered solely by black agouti males. Grey columns represent number of litters fathered by both types of males (mixed litters). (*From Levine, 1958.*)

(1958) observed albino versus black agouti fights and noted a predominance of successfully aggressive albino males. Each fight was watched until one opponent assumed a "submission reaction" sitting on his hind legs with forelegs extended in a defensive position. One wonders about the correlation between fighting ability and reproductive success in these mice. This then may represent a true example of sexual selection in mice. Even so, the possibility that albino females may have some preference for albino males (homogamy) cannot be ruled out.

A final point about the mice utilized in Levine's experiment and in many other experiments: unknown to him and ascertained some years later, all the mice he used were homozygous for the *rd* (retinal degeneration) allele and, therefore, blind (Sidman and Green, 1965). Even so, this does not alter the conclusion that sexual selection probably occurs. A further general discussion of behavior in mice is presented in Chapter 9.

Obesity Genes

The mouse mutant genes obese (*ob*) and diabetes (*db*) cause similar obesity-diabetes states in homozygotes. Under standard conditions, these obese mice are less active than and eat and drink more than normal littermates. They are sterile, relatively short-lived, with body weights two to three times normals, and sometimes have very high blood sugar levels; they are therefore referred to as obese-diabetic mice. If food intake is restricted, then life span is increased and male infertility is partially prevented. Much of the syndrome therefore follows from overeating and inactivity.

Metabolically these obesity syndromes are characterized by a more efficient conversion of food to lipid than occurs in normals. Both genes when homozygous, *obob* or *dbdb,* cause the pathology associated with severe diabetes when in the inbred strain C57BL/KsJ, although in the closely related and similar strain C57BL/6S diabetes is less severe and transient (Coleman and

Hummel, 1973). This represents an effect of genetic background as described for the mating behavior of *yellow* flies in the last section.

Coleman (1979) has shown that once food is stored in *obob* or *dbdb* mice, it is released much more slowly than in normals, leading to greatly increased food efficiency and a remarkable ability to withstand fasting up to 40 days. In addition, he has shown that heterozygous mice, either *ob+* or *db+*, survive a prolonged fast significantly longer than normal homozygotes (Table 3-4). This suggests that the heterozygote exhibits greater metabolic efficiency than do the homozygous mutants.

The existence of this "thriftiness" trait, if shown by heterozygotes in natural populations, lends credence to the thrifty gene concept of diabetes in human beings (Neel, 1962). Indeed the relatively common occurrence of diabetes in human beings has been suggested to be the result of a thrifty genotype rendered detrimental by progress in nutrition. In undeveloped countries, that is, in hunter-gatherer societies, people foraged for a limited food supply and were subjected to periods of abundance alternating with periods of food deprivation and even famine. Under these circumstances, thrifty individuals with a predisposition to diabetes could utilize a limited food supply more efficiently and therefore maintain a selective advantage when food was scarce. Increasing affluence would then result in this thrifty genotype becoming a liability, with the development of hyperinsulinemia, obesity, stress of the insulin synthesizing and secreting capacities of the pancreas, and, often, diabetes. It has been suggested that in this way the diabetes genotype has persisted in both animal and human populations, despite strong selection against it. In the human context, this represents an instance where culture is one determining factor.

This fascinating example of two recessive genes in mice illustrates the possibility of using animal models to assist interpretations of human studies, simply because in animals such as mice, breeding experiments can be done

Table 3-4 The Effects of Genotype on Ability to Survive Fasting in Mice
(Data are Means ± Standard Error of Mean)

Strain	Genotype	Starting body weight, grams	Number of animals N	Mean survival time, days
C57BL/6S	+/+	36.7 ± 0.7	32	10.8 ± 0.4
C57BL/6S	ob/+	36.6 ± 0.6	29	12.2 ± 0.4*
C57BL/6S	+/+	33.3 ± 0.3	15	8.6 ± 0.3
C57BL/6S	db/+	33.1 ± 0.4	14	10.6 ± 0.4†
C57BL/KsJ	+/+	29.7 ± 0.3	26	7.2 ± 0.3
C57BL/KsJ	db/+	29.9 ± 0.4	26	10.5 ± 0.3‡

* $P < 0.05$, Student's t-test.
† $P < 0.01$.
‡ $P < 0.001$.
Source: Coleman, 1979.

under controlled environments. However, the gene pools of all species are unique, and so extrapolations across species should only be made with extreme care. Although the search for a clear-cut biochemical explanation for the obesity-diabetes syndrome in mice has not been entirely successful, the search for underlying physiological and biochemical bases of genetically controlled syndromes is an important approach. As will be seen in Section 7-6, diabetes in human beings is extremely difficult to study. Even in mice the complexities extend to effects of genetic background. However, in mice the possibility of ultimately specifying these effects should occur at the biochemical and physiological levels. Then investigations of the human condition will be aided.

3-4 SINGLE-GENE EFFECTS IN HUMAN BEINGS

Huntington's Chorea

What of the simple genetics of behavior in human beings, in whom breeding experiments cannot be carried out? Phenylketonuria was discussed in the last chapter. Another good example is Huntington's chorea, a heritable and invariably fatal disorder. (*Chorea* is a neurological disease marked by muscular twitching, from the Greek word for "dance." Huntington's chorea is named after three generations of physicians who practiced medicine in Connecticut and maintained longitudinal family records.) The onset of the disorder is obscure, its primary metabolic lesion unknown, and its deterioration characteristically progressive from choreic movements to ataxia, dementia, and death. The progressive dementia is characterized by the degeneration of ganglion cells of the forebrain and corpus striatum. As Figure 3-4 and other pedigrees indicate, Huntington's chorea is the result of heterozygosity for an autosomal dominant gene with delayed age of onset. Symptoms usually do not appear until reproductive age has been reached or passed, so that although the condition is fatal, those who carry the gene are in most cases able to produce progeny before they are even aware of their affliction. The average age of onset is 35 years, with a range mainly from 15 to 65 years of age for the initial twitching, although some childhood cases are known. Potegal (1971) has demonstrated spatial-motor defects in patients with Huntington's disease. They are inaccurate in egocentric spatial localization wherein the position of a target in space must be defined in terms of its distance and location from the observer, for example, "straight ahead" or "a yard to my left."

The gene for Huntington's chorea was brought to the United States by three young men who departed from Bures St. Mary, Suffolk, England, with the same ship convoy in 1830. They "left town" because of difficulties incurred by their unusual, even outrageous behavior (Vessie, 1932). All three wed and fathered children in their new country. There are now more than 7,000 people afflicted with Huntington's chorea in the United States; the incidence of this disorder is approximately 1 in 25,000, and occurrences have been reported throughout the world.

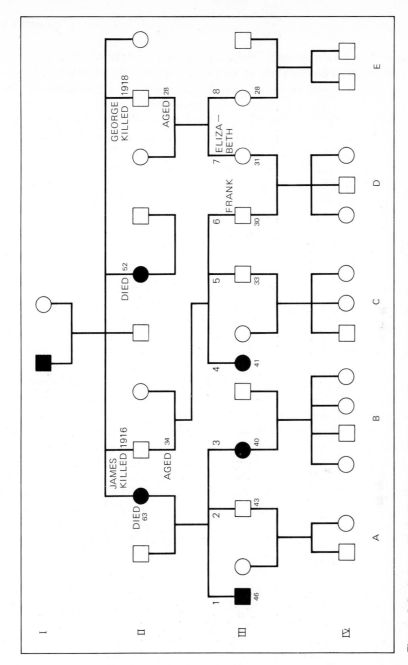

Figure 3-4 Autosomal dominant transmission evidenced in Huntington's chorea traced through four generations. Small numbers represent some ages at the time this pedigree was compiled (when all 14 children in generation IV were under 15 years of age). (*From Roderick, 1968.*)

Consider the pedigree constituting Figure 3-4. The probabilities of inheriting the gene H for Huntington's chorea in generation IV (assuming all unaffected individuals where ages are given to be potential carriers) are

- Either member of family E: $\frac{1}{2} \times \frac{1}{2} \times \frac{1}{2} = \frac{1}{8}$
- Any member of family B: $\frac{1}{2}$
- Any member of family C: $\frac{1}{2} \times \frac{1}{2} = \frac{1}{4}$
- Any member of family D:

  (Assuming two doses of the gene HH to be lethal, three mating types exist. Frank's being Hh has a probability $\frac{1}{2}$; Elizabeth's $\frac{1}{4}$.)

Child

$$\textbf{a} \quad \frac{1}{8}(Hh \times Hh) = \frac{1}{8}(\frac{2}{3}Hh + \frac{1}{3}hh)^* = \frac{2}{24}Hh$$

$$\textbf{b} \quad \frac{4}{8}(Hh \times hh) = \frac{4}{8}(\frac{1}{2}Hh + \frac{1}{2}hh) = \frac{6}{24}Hh$$

$$\textbf{c} \quad \frac{3}{8}(hh \times hh) = \text{unaffected}$$

giving a total of $\frac{1}{3}Hh$

- Either member of family E: $\frac{1}{2} \times \frac{1}{2} \times \frac{1}{2} = \frac{1}{8}$

Consult Stern (1973), Thompson and Thompson (1973), Fuhrmann and Vogel (1969), and especially Porter (1968) for more details and more evidence of the analysis of genetic patterns in nonexperimental subjects.

Before any symptoms develop in a person who had a parent with this disease, the probability that this person has the gene is $\frac{1}{2}$. (This gene is rare enough to allow the assumption that the affected parent was not homozygous or that the homozygosity is lethal.) Therefore, before diagnosis is possible, the probability that the person in question will have an afflicted child is $\frac{1}{2}$ (the probability the person in question has the gene) $\times \frac{1}{2}$ (the probability the child will inherit the gene if the parent has it) = $\frac{1}{4}$. Once diagnosis is certain, the probability of a child manifesting the disease becomes $\frac{1}{2}$ (if a parent has it) or 0 (if the parent is not afflicted). See Falek and Britton (1974) on the psychology of this tense situation. Sometimes gaps occur in pedigrees such that affected persons are derived through unaffected individuals. The likely explanation is the late age of onset in an afflicted parent who died of other causes before initial symptoms of Huntington's chorea appeared.

Lactase Deficiency

By stretching our definition of behavior genetics a bit, we are able to include in this chapter consideration of the role culture has played in the evolution of the three allelic genes controlling lactase deficiency and milk consumption in human populations. The emerging story (McCracken, 1971; Gottesman and Heston, 1972; Kretchmer, 1972) is fascinating. Lactose is the primary sugar

---

* $Hh \times Hh = 1HH$ (dies?) : $2Hh$ (afflicted) : $1hh$ (normal) = $\frac{2}{3}Hh : \frac{1}{3}hh$

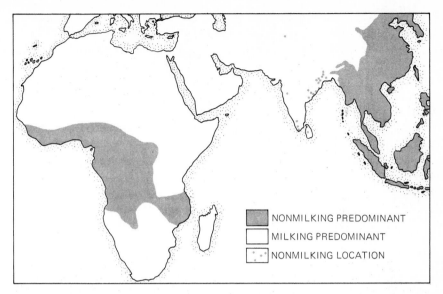

NONMILKING PREDOMINANT

MILKING PREDOMINANT

NONMILKING LOCATION

Figure 3-5 Milking and nonmilking areas. Map shows predominant traditional patterns in Africa and part of Asia. (*From Simoons, 1970.*)

found in milk; it is metabolized by lactase, an enzyme produced in the villi of the small intestine. Simply put, the reaction is

$$\text{Lactose (disaccharide)} \xrightarrow{\text{lactase}} \text{glucose} + \text{galactose (monosaccharides)}$$

The end products can subsequently be absorbed into the human circulatory system, but if lactase is absent, lactose passes through the intestines without providing any nutritional value. This passage sometimes results in bloating, cramps, and diarrhea.

Three alleles (L, l_1, and l_2) seem to occupy the autosomal locus that controls lactase production. Both l_1 and l_2 are recessive to L, the wild-type or normal allele, and l_2 is recessive to l_1. LL, Ll_1, or Ll_2 individuals produce lactase both as children and as adults; l_1l_1 and l_1l_2 individuals do not produce lactase as adults; and l_2l_2 is a very rare combination, usually lethal, because milk cannot be digested even in infancy. In Northern Europe, 80 to 100 percent of all adults are LL, Ll_1, or Ll_2, while just the reverse is true in, for example, Oriental, Amerindian, Southern European, Australian Aborigine, and African populations (l_1l_1 or l_1l_2). Note that these l_1l_1 and l_1l_2 adults can manage the digestion of sour milk products such as sour milk itself, yogurt, and cheese.

McCracken (1971) suggests:

It is hypothesized that prior to the domestication of animals [400 generations ago for the start of the domestication of sheep and goats?] and the development of dairying, the normal condition for all men was adult lactase deficiency, but with the introduction of lactose into the adult diet in certain cultures, new selective pressures were created that favored the genotype for adult lactase production.

Simoons (1970) cautions that it must not be assumed that where milkable animals were kept, people did in fact milk them and, furthermore, consume the milk as adults. Finally, lactase activity may be inducible; that is, lactase may be produced at a rate in parallel with the demands of the diet—the more lactose consistently ingested, the more lactase synthesized. This is known as adaptive enzyme formation, a phenomenon not always observed in response to apparently significant dietary challenges. At any rate, culture is surely a major factor in the evolution of the one human species, and culture implied behavioral adaptations, at least some of which are likely to be under genetic control. Look at Figure 3-5 and judge for yourself the magnitude of the advantage of milk as a supplementary nutrient for adults and whether it seems reasonable that this particular advantage could eventually have altered the appropriate gene frequencies involved.

SUMMARY

Many single genes directly or indirectly influence behavior. Behavioral alterations commonly result from morphological and physiological changes. Even where there are no ostensible effects apart from behavioral ones, detailed studies frequently reveal underlying physiological or biochemical causes.

The analysis of the effects of single genes affecting behavioral traits is relatively simple in experimental animals such as honeybees, *Drosophila,* and mice. In the case of *Homo sapiens* the analysis of pedigrees must be carried out, where variables such as differing ages of onset of a disorder may complicate interpretations.

From the evolutionary point of view, the behavior geneticist is vitally interested in genes affecting mating behavior. Many of these genes lead to variations in sexual vigor of one or both sexes in mating success, a process called *sexual selection.* This phenomenon is considered in numerous places in this book.

Chromosomes and Behavior

In Chapter 3 the effect of single genes on simple behavioral traits was considered. This chapter extends the genetic unit under consideration to the chromosome. Before discussing behavior as such, a short account is given of the various types of gross chromosomal changes commonly found. A good detailed account is in Herskowitz (1973); other presentations are in the general texts listed at the end of Chapter 2.

4-1 CHROMOSOMAL CHANGES

Changes involving unbroken chromosomes are common. Although most sexually reproducing organisms have a diploid chromosome complement, or genome, the occurrence of triploids and tetraploids (three and four complete chromosome sets instead of two) is common in plants. In *Drosophila,* triploid and tetraploid females occur and haploid/diploid somatic mosaics have been found. (A chromosomal *mosaic* is an individual with tissues of differing chromosomal constitutions due to abnormal somatic cell division[s] early in fetal life.) In human beings complete triploidy is lethal, but individuals who are diploid/triploid mosaics may survive, though they are physically and mentally defective. The presence of a complete set ($2n$) of chromosomes is referred to as

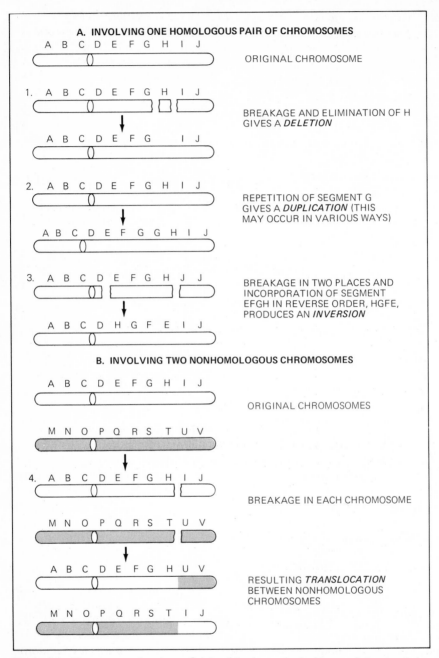

Figure 4-1   Chromosome breakage. Origin of the four principal types of structural change by means of chromosome breakage: deletion, duplication, inversion, and translocation.

euploidy. In contrast, *aneuploidy* is the addition or subtraction of single chromosomes from a chromosome set. Aneuploids may arise as a result of aberrations during the process of cell division at the time of gamete formation. The process of cell division producing haploid gametes from diploid cells is referred to as *meiosis,* while the process of cell division after fertilization during growth and development is referred to as *mitosis;* chromosome number changes do not often occur during mitosis. In *gametogenesis* (the production of mature eggs and sperm) chromosomes normally disjoin in a regular fashion during meiosis to enable one representative from each chromosomal pair to be included in each daughter cell. In contrast, *nondisjunction* of the fourth chromosome of *Drosophila melanogaster* for example yields individuals with either one or three fourth chromosomes, the former individual being *monosomic* and the latter *trisomic.* In human beings, individuals trisomic for one of the smallest chromosomes, number 21, have Down's syndrome, characterized by gross morphological and mental impairment. (For a detailed account of chromosome behavior during meiosis, refer to a general text such as Strickberger, 1976.)

In contrast to the modifications of chromosome number discussed above are changes involving *chromosome breakage* (Figure 4-1), of which there are four possible types: deletion, duplication, inversion, and translocation.

- *Deletion,* or removal, of a gene locus or a group of loci is most often lethal in the homozygous form. From an evolutionary point of view, deletions are relatively insignificant.
- *Duplication* of a gene locus, which may occur in various ways (Strickberger, 1976), may cause an imbalance of gene activity, reducing the viability of an organism. However, since some organisms can tolerate duplications of chromosomal material, duplications may play an evolutionary role. If a particular locus is duplicated, one of the twin loci could mutate to an allele having a different function without reducing fitness. Presumably the unchanged alleles at the other locus could adequately perform the original function of the locus. In this way evolutionary change may occur—and indeed has been postulated to occur in the evolution of the four genes for the hemoglobin molecule in human beings (for details see Herskowitz, 1973). The genetic control of some other complex molecules has evolved in a similar way.
- An *inversion* occurs when a chromosome breaks in two places and the segment between the breaks then rotates 180° leading to a reversal of gene order with respect to that on the unbroken complementary chromosome. Inversions occur spontaneously during various chromosomal movements during cell division. The significant effects of inversions are due to the fact that during meiosis homologous chromosomes pair exactly, gene by gene, which leads to the formation of characteristic loops during meiosis in individuals heterozygous for inversions (Figure 4-2).

During meiosis, chromosome breakage and rejoining of homologous partners routinely occurs and is referred to as *crossingover.* In Figure 4-2 this process is illustrated for a heterozygous inversion. The figure shows the consequence to be two normal chromosomes not showing the effect of crossingover and two abnormal structures showing the effect of crossingover—one (without

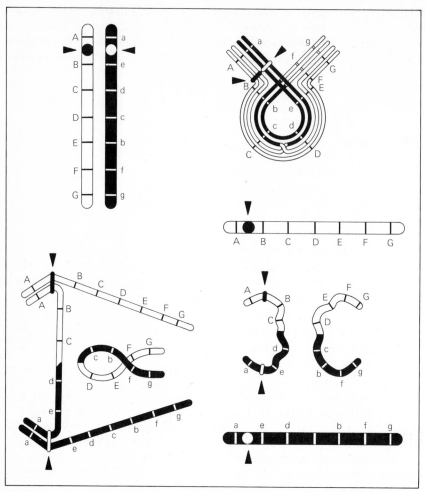

Figure 4-2 Crossingover in an inversion heterozygote. *Upper left:* two chromosomes differing by one (paracentric) inversion. *Upper right:* pairing at the earliest stage of meiosis. *Lower left:* after the first meiotic division a chromatid bridge and acentric piece of chromosome are formed. *Lower right:* two viable and two nonviable chromosomes, the result of crossingover at the completion of meiosis. *Black triangles* indicate centromeres. (Reprinted from T. Dobzhansky: *Genetics and the Origin of Species,* 3d ed., revised. New York: Columbia University Press, 1951, p. 125. By permission of the publisher.)

a centromere) that becomes lost during the meiotic process and the other (with two centromeres) that breaks in cell division and so also is eventually lost. If an organism is heterozygous for an inversion, therefore, crossingover within the inverted segment is generally ineffective in contrast to crossingover exclusive of inversion heterozygosity, where crossingover leads to recombination of genes (see Section 2-1). Therefore, the genes of an inverted segment are transmitted as a single unit in inversion heterozygotes, since only those chromosomes not showing the effect of crossingover remain. This is a point of some considerable evolutionary significance, especially in *Drosophila* (Dobzhansky, 1970).

• A *translocation* occurs when two nonhomologous chromosomes break simultaneously and exchange segments. If an organism becomes homozygous for such a rearrangement, some of its genes have been transferred to a completely different chromosome, and linkage relationships of the genes are considerably changed, as shown in Figure 4-1.

With this brief background we can now turn to an assessment of chromosomal effects on behavior. It should be stressed, however, that, according to our current knowledge, not all the above chromosomal changes are important in this regard. To date, inversions and chromosome number changes appear to form the main categories of importance.

4-2 INVERSIONS IN *DROSOPHILA*

Many species of Diptera have two or more inversions in natural populations in such frequencies that these cannot be explained by recurrent mutation. When a population has two or more hereditary alternatives consistently maintained over generations, it is said to exhibit genetic *polymorphism*. The occurrence of polymorphism indicates a situation of particular genetic and evolutionary interest, since there must be a balance of selective forces maintaining the inversions in the population. Thus any behavioral differences associated with inversions could be of major evolutionary significance. Experimental work, in particular in *D. pseudoobscura,* has shown that inversion heterozygotes (*heterokaryotypes*) often have a fitness superior to that of inversion homozygotes (*homokaryotypes*). (A *karyotype* is the chromosomal constitution of an organism.) The introduction of two sequences, Standard (ST) and Chiricahua (CH), of chromosome III of *D. pseudoobscura* into population cages at 25°C leads to ultimate inversion frequencies at about 0.7 ST and 0.3 CH irrespective of the initial frequencies (Wright and Dobzhansky, 1946). In other words, the inversion frequencies arrive at an *equilibrium*—a result in contrast with the Hardy-Weinberg situation discussed in Sections 2-3 and 2-4. We therefore should ask which of the various factors listed in Section 2-4 is important. It is known from an extensive series of experiments by Dobzhansky and his associates (references in Parsons, 1973, and see Anderson and McGuire, 1978) that various components of fitness, such as innate capacity for increase, population size, productivity, and egg-to-adult viability, show heterokaryotype superiority over the corresponding homokaryotypes. In other words, the fitnesses of the karyotypes differ.

Because ST and CH inversions essentially segregate as single genes, it is important to consider briefly the conditions under which two alleles, *A* and *a*, are polymorphic at a locus. In Section 2-3 the Hardy-Weinberg law under random mating is discussed. The additional complication to be considered now is that the fitnesses of the three genotypes *AA, Aa,* and *aa* are not necessarily equal, as assumed heretofore. We therefore let the fitnesses of genotypes *AA,*

Aa, and *aa* be 1−*s*, 1, and 1−*t* respectively, so that the genotypic proportions before and after selection become:

	AA	Aa	aa	Total
Fitnesses	$1 - s$	1	$1 - t$	
Frequencies before selection	p^2	$2pq$	q^2	1
Frequencies after selection	$p^2(1 - s)$	$2pq$	$q^2(1 - t)$	$\bar{W}$

where $\bar{W}$ is the average fitness of the population after selection. Letting p' and q' represent the gene frequencies of A and a in the next generation, then

$$p' = \frac{p^2 - p^2 s + pq}{\bar{W}} = \frac{p - sp^2}{\bar{W}}$$

and

$$q' = \frac{pq + q^2 - q^2 t}{\bar{W}} = \frac{q - tq^2}{\bar{W}}$$

The reason for dividing by $\bar{W}$ is to ensure that $p' + q' = p + q = 1$. At equilibrium, gene frequencies are fixed from generation to generation. If we write the change in gene frequency from generation to generation as Δp, then at equilibrium it is expected that

$$\Delta p = p' - p = 0$$

or

$$\Delta p = \frac{p - sp^2}{\bar{W}} - p = \frac{pq(tq - sp)}{\bar{W}}$$

It is not difficult to show that when the change in gene frequency $\Delta p = 0$, then $p = 0$, $q = 0$, or $tq = sp$. The first two solutions are trivial when the population is either all AA or aa, which is not a polymorphic situation. In other words, either gene A or a is lost while the other is fixed. It is obvious and can be demonstrated algebraically that if fitnesses are (1) $AA > Aa > aa$, then A will be fixed and a lost (i.e., $p = 1$, $q = 0$); and (2) $aa > Aa > AA$, then a will be fixed and A lost (i.e., $p = 0$, $q = 1$).

The solution $tq = sp$ gives, by algebraic rearrangement, equilibrium gene frequencies $p = t/(s + t)$ and $q = s/(s + t)$, which therefore depend only on the selective values s and t. This means that irrespective of the initial values of p and q the same equilibrium is expected. Clearly, the only circumstances under which these equilibria can exist are for $s,t > 0$ or $s,t < 0$, since otherwise one or the other equilibrium gene frequencies would then be negative, which is impossible. For these two conditions the stability of the equilibria must be examined. A stable equilibrium occurs if after there is a small displacement from the equilibrium gene frequency, as may occur by chance in a finite population, the population tends to return to that gene frequency in subsequent generations. It can be shown that if $s,t > 0$, a stable equilibrium is expected. This corresponds to the situation where $Aa > AA,aa$ in fitness; in other words, there is heterozygote advantage over both homozygotes, called *overdominance*. Conversely, if $s,t < 0$, which means that $Aa < AA,aa$ in fitness, then a small displacement from the equilibrium gene frequency is accentuated, generation by generation, and, eventually, one allele or the other is fixed. This is an unstable equilibrium. All these conclusions can be derived algebraically. The important conclusion is, assuming random mating with unequal genotypic fitnesses, that if a heterozygote is more fit than its corresponding homozygotes, a stable equilibrium is to be expected with an associated polymorphism; then the equilibrium gene frequencies depend only on selective values.

These theoretical calculations show that one situation (but not the only one) where polymorphism is expected is where heterozygotes (heterokaryotypes) are more fit than the corresponding homozygotes (homokaryotypes). For the situation of inversion polymorphisms, we are dealing with the gene complex making up the block of genes locked into the inversion rather than a single locus. Because effective recombination is suppressed in inversion heterozygotes, complexes are built up in which the genes interact to produce *coadapted gene complexes* by natural selection. One may ask whether there is evidence of such coadaptation. The answer comes from comparisons of pairs of chromosomal heterozygotes derived from different ecogeographical regions, often not very distant from each other (Dobzhansky, 1950). On the whole, heterokaryotype superiority breaks down completely in crosses between regions even though it still exists within ecogeographical regions. In other words, the gene arrangements within the inversions are unique to specific ecogeographical populations, owing to the mutual adjustment of gene complexes within these populations to give high heterozygote fitness, which Dobzhansky calls *coadaptation*. Clearly, gene complexes from different populations have not had the same opportunity for mutual adjustment of their genic contents; that is, they have not been selected for coadaptedness, so that heterozygosity for inversion sequences from different localities is not expected to lead to high fitnesses.

Turning now to behavioral examples, Brncic and Koref-Santibañez (1963, 1964) studied the relation between sexual selection and chromosomal inversions in the South American species *D. pavani*, which is found mainly in the southern part of the continent. In most natural populations the proportion of heterokaryotypes exists in fairly uniform frequencies. Mating activity was assessed using virgin females of *D. gaucha*, a sibling species of *D. pavani*. (*Sibling*—sister or brother—*species* are closely related and often morphologically indistinguishable.) Pair matings were observed for 30 minutes. Brncic and Koref-Santibañez scored (1) pairs that mated during the period of observation; (2) those that courted but did not copulate; and (3) those that were sexually inactive during the observation period. The frequency of heterokaryotypes was significantly higher among males that courted and/or mated during the first few minutes of being placed with the females. These results suggest that the heterokaryotypes are superior at least up to the first mating. It can therefore be argued that superiority in mating activity of the heterokaryotypes is likely to be one important factor in the maintenance of this polymorphism in natural populations of *D. pavani*. It is an example of overdominance, as discussed in the theoretical considerations above.

We should also consider the extensive studies of Spiess and associates (Spiess, 1962; Spiess and Langer, 1961, 1964a,b; Spiess, Langer, and Spiess, 1966; Spiess and Spiess, 1967) on homo- and heterokaryotypes in *D. pseudoobscura* and its sibling species *D. persimilis*. In *D. pseudoobscura*, great differences in mating speed were found between homokaryotypes derived from stocks collected at Mather, California. Homokaryotypes for Standard (ST), Chiricahua (CH), Tree Line (TL), Pike's Peak (PP), and Arrowhead (AR) inversions were used (Figure 4-3). The experimental procedure involved the direct

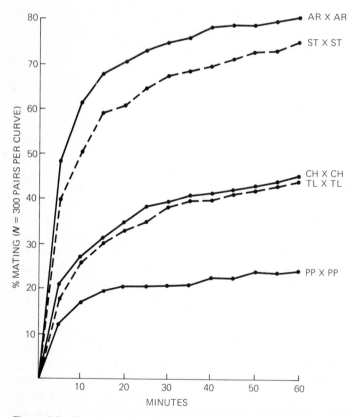

Figure 4-3 Homokaryotype homogamic matings. Cumulative percentage curves for matings during 1 hour's observation. AR, ST, CH, TL, and PP represent different inversions in the third chromosome of *D. pseudoobscura*. (*From Spiess and Langer, 1964a.*)

observation of 10 pairs of 6-day-old flies in mating chambers over a 1-hour period at 25°C. Apart from the pairs AR,ST and CH,TL, all other pairings differed significantly at 60 minutes. For AR and ST rapid matings occurred; CH and TL were intermediate, and PP was slow. The rapidity of acceptance, mounting, and insemination (other things being equal) increases the fitness of the carriers of a given karyotype. With rapid mating, the deposition of fertile eggs by the females is likely to be accomplished without delay, and the females depleting their sperm supply are ready to mate again and receive a fresh supply. This is consistent with the widely accepted definition of natural selection: "Natural selection occurs when the carriers of some genotypes contribute more surviving progeny to the succeeding generations in relation to what the carriers of other genotypes contribute" (Dobzhansky, 1964). Spiess and Langer (1964a) point out that the observed frequencies of the inversions at Mather rank in approximately the same order as the mating frequencies, with AR and ST being the most frequent and PP the least. From this it is tempting to suppose that mating speed is a major factor in maintaining the observed frequencies of the chromosomes in this population and that it is indeed an important component of fitness. Spiess and Langer (1964b) summarize the results of their studies as follows: "If mating speed is constant for each karyotype under 'competitive conditions,' net adaptive values (relative fitnesses) of karyotypes will be frequency dependent [see Sections 3-2 and 8-4]. In natural populations these karyotypes controlling mating behavior must contribute to a major portion of the population's total fitness."

Considering mating speed behavior between karyotypes, for both heterokaryotypes and homokaryotypes, our initial conclusion is that mating speed is almost entirely male-determined (Kaul and Parsons, 1965, 1966; Spiess, Langer, and Spiess, 1966). For example, Kaul and Parsons (1966) showed this to be so in two series of choice experiments consisting of one female with three males and the reverse, three females with one male. The mean period elapsing to the first mating was 0.53 minute in the experiments with the three females and 1.40 minutes in those with the three males (Table 4-1). The likely interpretation is that in experiments with three males, competition that occurs among males delays mating, whereas in the reverse situation with three females, the one male tends to mate more rapidly, having no competition from other males. The second conclusion that emerges is that the male heterokaryotypes consistently have a faster mating speed than do the homokaryotypes. Spiess, Langer, and Spiess (1966) studied mating speeds for a number of combinations employing 10 pairs of flies per mating chamber and found that male heterokaryotypes had a consistently faster mating speed than did the corresponding homokaryotypes. Females displayed no such consistent superiority, which if it did exist would imply variations in receptivity. Clearly then, the overdominance (often called *heterosis*) displayed is due to the greater activity or persistence in courtship of males or to greater female acceptance of heterokaryotype males probably because of the males' increased sexual activity.

Table 4-1   Time Elapsing to First Mating and Matings Occurring in 1 Minute in Male- and Female-Choice Experiments in *D. Pseudoobscura*

	Time elapsing to first mating, min	Number mated*	Number unmated†	χ_1^2
ST/ST				
Male choice	0.56	34	16 ⎫	
Female choice	1.22	22	28 ⎭	4.91‡
ST/CH				
Male choice	0.35	39	11 ⎫	
Female choice	1.08	23	27 ⎭	9.55§
CH/CH				
Male choice	1.00	25	25 ⎫	
Female choice	2.28	14	36 ⎭	4.20‡
Pooled data				
Male choice	0.53	98	52 ⎫	
Female choice	1.40	59	91 ⎭	19.30¶

* Number of replications out of 50 (or out of 150 for pooled data) in which mating occurred in one minute.
† Number of replications out of 50 (or out of 150 for pooled data) in which mating did not take place in one minute.
‡ $P < 0.05$.
§ $P < 0.01$.
¶ $P < 0.001$.
Source: Kaul and Parsons, 1966.

In *D. pseudoobscura,* male mating speed may well be an important component of fitness. It must be considered in relation to all other components of fitness, some of which were cited earlier in this section—innate capacity for increase, population size, productivity, and egg-to-adult viability. The many associations among all these components in a given population have been relatively unexplored but must be of considerable importance in the study of the fitness of organisms. Parsons (1974a) concludes, after a review of available evidence, that male mating behavior is a very important component of fitness, at least in *D. pseudoobscura, D. pavani,* and probably also in *D. melanogaster.* However, in *D. persimilis,* Spiess and Langer (1964b) found a less one-sided situation. Certain females accept males readily, others tend to refuse them, and certain males court more actively than others. In other words, the differences found can be interpreted in terms of the relative intensities of the copulatory tendency of males and the acceptance (or, conversely, the avoidance) tendency of females. Recent evidence for mating speed as a component of fitness comes from observations of large (circa 1,100 individuals) populations of *D. pseudoobscura* incorporating no less than six inversion karyotypes. Anderson and McGuire (1978) have again found significant differences among karyotypes, and between the two sexes within a karyotype, for that component of fitness clearly due to mating success. In order to extrapolate to nature, more such large population cage experiments should be done. The evolutionary significance of these matters is further discussed in Section 13-1.

4-3 KARYOTYPE VARIATIONS IN HUMAN BEINGS

The human chromosome complement is not rigidly uniform. It varies cytologically in normal individuals as regards

1 Lengths—arm ratios, centromere indices
2 Satellites—small appendages with apparently concentrated ribosomal RNA
3 Secondary constrictions—nonstaining or weakly staining regions
4 Minor mosaicism—infrequent and abnormal nondiploid somatic cells
5 Balanced structural rearrangements—reciprocal translocations summing to a typical chromosomal endowment

These variations have little or no established associations with behavior in our species due to an almost complete absence of data because of technical, analytic difficulties. We have no doubt however that in the future such data will accrue. (For one example, see Say et al., 1977.) In contrast, chromosomal aberrations (see Table 4-2) of assorted magnitudes—mostly gross—have been established as underlying either subnormal mental and/or psychomotor performances or atypical behavior (Bergsma, 1979).

Down's Syndrome

One of the best known of the chromosomal aberrations is trisomy for one of the smallest of the human autosomes called trisomy-21, trisomy-G (Figure 4-4), or Down's syndrome (or, by its Victorian name, mongolism, because of what was perceived to be an Oriental appearance about the face and eyes). One in 600 to 700 newborn from all human populations is afflicted with this syndrome. It is characterized by congenitally retarded mental, motor, and sexual development,

Table 4-2 Chromosomal Defects

Autosomal	Sex chromosomes	Others
Chromosome		
Eighteen p syndrome	Gonosomal intersex-	De Lange syndrome-chromo-
Eighteen q syndrome	uality-45,X/46,XY	somal translocations
Eighteen trisomy syndrome		Monosomy-G syndrome type I
Five p syndrome	Klinefelter's syndrome*	
Four p syndrome	True hermaphroditism	Monosomy-G syndrome type II
Thirteen q syndrome		Trisomy-C syndrome
Thirteen trisomy syndrome	Turner's syndrome*	
Twenty-one trisomy syndrome		
Coloboma and anal atresia syn-drome—G group chromosome imbalance due to a small additional chromosome		

* Discussed elsewhere in this chapter and/or see index.
p = short arm of chromosome; q = long arm.
Source: Bergsma, 1979.

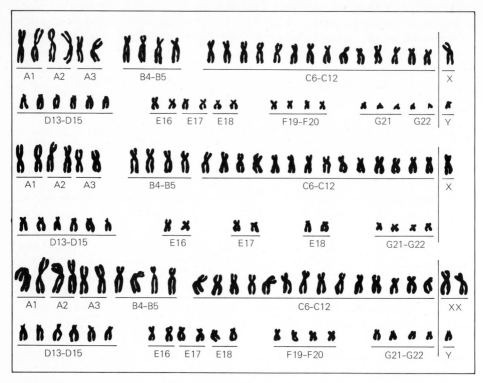

Figure 4-4 Abnormal human karyotypes prepared from chromosomes in human white blood cells. *Top:* male with Down's syndrome. Note the additional chromosomes to set number 21. This chromosome abnormality is phenotypically expressed as a form of idiocy or imbecility. *Middle:* female with Turner's syndrome. Note the single X chromosome. Phenotypic expression of this chromosome abnormality results in retarded sexual development and sterility. *Bottom:* male with Klinefelter's syndrome. Note the XXY genotype expressed phenotypically as lack of male secondary sex characteristics, development of some female secondary sex characteristics, or both. (From *Biology Today,* 1972, pp. 259–260. CRM Inc., Del Mar, California.) Intelligence may range from normal to retarded. With increasing numbers of X chromosomes—3, 4, or even 5—retardation is increasingly severe (see Figure 4-7).

and a number of physical stigmata. Life span is abbreviated. The mental retardation ranges from IQs of less than 20 to less than 65, and therefore represents idiocy or imbecility. Behaviorally, these individuals are often happy and friendly; they often imitate well and are very affectionate. Dingman (1968) studied psychological test patterns in patients with Down's syndrome and noted no systematic behavioral differences between individuals with Down's syndrome and other mentally retarded patients; the differences he recorded are apparently due to degrees of retardation.

Most of the time the presence of the extra chromosome number 21 (or G-group chromosome, as it is also categorized because chromosome pairs 21 and 22 were until recently morphologically indistinguishable) is due to an error or errors in meiosis. (Hungerford, 1971; Hungerford et al., 1971, present compelling evidence that the supernumerary chromosome is actually number 22; this evidence, based on length, comes from testicular biopsy of chromosomes in one of the stages of meiosis called pachytene, when chromosomes are relatively short and thick.) The extra chromosome that characterizes Down's syndrome presumably arises by nondisjunction. This nondisjunction is probably restricted primarily to females becauce the frequency of affected individuals increases rapidly with maternal age. For a woman 45 years old at pregnancy the risk of producing a child with Down's syndrome is approximately 1 : 50 compared to approximately 1 : 3,000 for a woman of 20. Presumably the increase in nondisjunction with age is due to a change in the environment of the oocytes (eggs) induced by increasing age (see Penrose, 1963, for a detailed account of work on this syndrome). About 2 percent of those affected with Down's syndrome may have chromosomal breakages, such as translocations, involving a crucial G chromosome. One example is an individual with this syndrome and only 46 chromosomes—no extra chromosome at all. Such an individual has two G chromosome pairs and an extra long D-group chromosome (chromosome 14 or 15). This suggests a translocation between a third G and a D chromosome leading to a new large chromosome consisting of almost all the material of both (i.e., D and G). Thus the affected individual has the material of three G chromosomes, as occurs through nondisjunction. Such a translocation can be inherited, and for Down's syndrome due to translocation, a familial pattern of inheritance may be expected with carrier individuals having only 45 chromsomes. On the other hand, some one-half of such cases represent new gross mutations.

In passing it is of interest that Down's syndrome may not be restricted to human beings. A syndrome resembling Down's has been reported in *Pan troglodytes,* a chimpanzee (McClure, Belden, and Pieper, 1969). Figure 4-5 depicts the proband's karyotype, and Figure 4-6 shows the results of behavioral tests on this young female indicating retarded growth and retarded neurological development compared with other nursery animals of her species. She exhibited general inactivity. At 40 weeks of age she was still unable to sit up or move about. Such investigations involving animal models of human conditions (see Leader, 1967) are recommended for behavior geneticists as well as for pathologists, who have been fruitfully employing them for some time.

Sex Chromosome Aberrations

The frequency of all types of sex chromosome anomalies is 21 out of 10,000 live births (in males 27 in 10,000; in females 15 in 10,000) (Robinson, Lubs, and Bergsma, 1979). Extrapolating this rate against an estimated world population of 4.5 billion (and assuming no preferential mortality) leads one to expect nearly 9.5 million people worldwide having sex chromosome anomalies. A general conclusion can be made that autosomal aberrations appear to cause more severe effects (morphologically and behaviorally) than do X or Y chromosomal aberrations (but see Figure 4-7). An interpretation is presented later in this section.

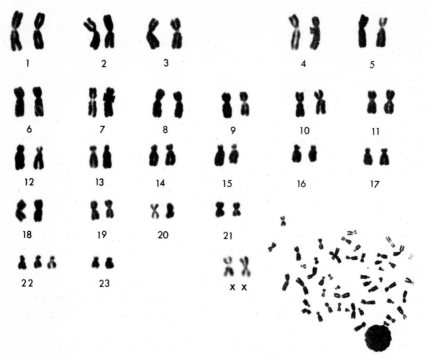

Figure 4-5 Blood cell karyotype of trisomic chimpanzee. This animal, trisomic for chromosome number 22, exhibited mental retardation and other stigmata associated with Down's syndrome in human beings. (*Courtesy of Dr. Harold McClure, Yerkes Primate Research Center, Emory University, Atlanta, Ga.*)

Turner's syndrome, or gonadal dysgenesis (Figures 4-4 *middle* and 4-8), is characterized by having only one sex chromosome, an X chromosome, and therefore a karyotype with 45 chromosomes (written XO). In appearance, individuals with Turner's syndrome are female; in behavior, they are characterized by hypertension, with a normal verbal intelligence but a specific space-form

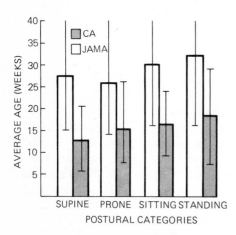

Figure 4-6 Behavioral development of trisomic chimpanzee (Jama) compared with average ages at which 50 percent of 14 tested control chimpanzees (CA) completed the 34 behavioral items here condensed into four categories. (*After McClure et al., 1969.*)

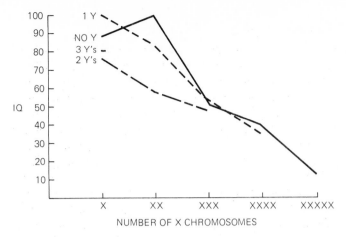

Figure 4-7 Sex chromosome abnormalities and IQ. Effects on mean IQ of abnormal sex chromosome aneuploidies. (*Compiled by Vandenberg, 1972, from Moor, 1967.*)

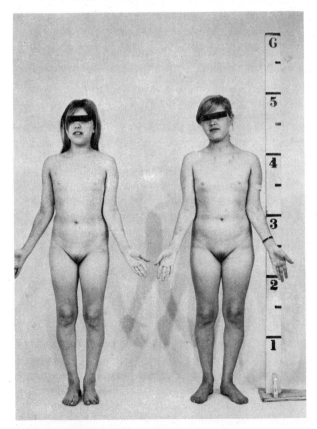

Figure 4-8 XO, or Turner's syndrome. Monozygotic twins with 45,XO karyotype. (From P. L. Riekhof, et al., 1972, Monozygotic twins with the Turner syndrome. *Am. J. Obstet. Gynecol. 112:*59–61.)

defect—what can be called a degree of space-form blindness or, more technically, partial congenital agnosia (Schaffer, 1962; Money, 1970).

Klinefelter's syndrome (Figure 4-4 *bottom*) is characterized by an extra X
and, therefore, a karyotype with 47 chromosomes (written XXY). In appearance these individuals are male, but usually they are sterile and of low libido.
The syndrome is not necessarily marked by mental retardation, but it may be.
XXY males are often socially inadequate, often quitting school and other activities requiring social contacts. Some XXY males are more than socially
inept; they are antisocial and may require institutionalization. Many reports list
passivity, dependency, withdrawal from reality, limited interests, and poor im-
pulse control as Klinefelter personality characteristics. Klinefelter's and
Turner's syndromes are among the most common sex chromosome aberrations.
They involve the addition or deletion of whole chromosomes with rather minor
somatic effects compared with those of Down's syndrome. Turner's syndrome
occurs with a frequency of about 2 per 10,000 among newborn. The frequency
of Klinefelter's syndrome is about 9 to 13 per 10,000 newborn.

The twins depicted in Figure 4-8 have Turner's syndrome and were 17
years old when this picture was taken. These girls were short, but they enjoyed
good mental and physical health (having graduated from high school as average
students), but neither had yet menstruated. No uterus could be palpated in
either girl, nor did thyroid therapy bring about menarche. According to Money
and Mitterthal (1970):

> As with many of the other symptoms that may be associated with Turner's syn-
> drome, space-form disability does not occur in 100 percent of cases (more likely in
> about 75 percent); and it occurs with varying degrees of severity. Its most likely
> explanation is that it is an effect of the genetic defect of the syndrome on the
> development and functioning of the cerebral cortex. On an intelligence test, this
> deficit shows up in the nonverbal and numerical items; verbal ability is not affected.
>
> There may be another direct effect of genetics on behavior in Turner's syn-
> drome, namely with respect to personality. There is no commonly agreed upon
> name for the feature of personality shared by many Turner girls, which might be
> identified as inertia of emotional arousal. It is constituted of compliancy, phleg-
> matism, stolidity, equability, acceptance, resignedness, slowness in asserting initia-
> tive, and tolerance of personal adversity.
>
> The indirect effects of genetics on personality in Turner's syndrome are medi-
> ated through body morphology and function, by way of the body image and the
> person's interaction with her social environment. Shortness of stature is the number
> one problem shared by all Turner girls. It makes itself felt at any early age. The
> number two problem, also shared by all patients, is pubertal failure, and the issue of
> deciding the timing of induced puberty. This decision requires a weighing of the
> demand for extra height against the demand for more maturity of appearance.
> Cosmetic deformities constitute, at their worst, a very severe problem: but they are
> at their worst in relatively few cases. The effects of cosmetic handicap, however
> bad, differ from shortness and pubertal infantilism in at least one dimension: short-
> ness and physical infantism elicit "infantilizing" social responses from other
> people of all ages. Thus, the major indirect effect of genetics on personality in

Turner's syndrome is to lower the threshold for retardation of social development. The closer she approaches teenage, the more the Turner girl encounters situations and pressures that tend to arrest or impede her social maturation.

In psychosexual differentiation, absence or impairment of the X chromosome, in all (or mosaically in some only) of the body's cells in Turner's syndrome, does not interfere with feminine gender identity. Nor is feminine gender identity impeded or impaired by the absence of the gonads and their hormones in fetal life. In adolescence, for the full maturation of psychosexual femininity, it is necessary first to give hormonal substitution therapy with estrogen, in order to bring about sexual maturation of the body. It is also necessary to give at least minimal psychologic guidance with respect to the postponement of adolescent estrogenization in favor of a possible extra increment in adult height.

The genetic and hormonal (fetal and pubertal) impairments of Turner's syndrome do not have any adverse direct influence on the patient's interest or ability to marry, nor to be motherly with children. On the contrary, Turner patients have a heterosexual inclination of and an interest in maternalism equal to that of matched normal controls.

Psychopathology is not a significant feature of Turner's syndrome, though it does occur, just as it does in a randomly selected population. Personality pathology in the parents, and/or their inability to cope with the implications of the diagnosis and prognosis of the syndrome, constitute in the long run more of a psychologic hazard to the girl herself than do the actual deficits and impairments of her body.

Campbell and coauthors (1972) observe: "The incidence of psychiatric abnormalities in Klinefelter's syndrome is far greater than in the general population." (Examples of this are seizure disorders, speech disorders, electroencephalographic abnormalities, schizophrenia, paranoid states, and deviant sexual behavior.) They tell the sad story of the youngest known Klinefelter patient, who exhibited psychopathology for which he was placed in a psychiatric hospital at the age of 3 years. His behavior involved temper tantrums, hyperactivity, withdrawal, drooling, brief attention span, grimacing, few, if any, vocalizations—mostly grunts, with aggressiveness directed against his disturbed (enough to be medicated) but intelligent parents or against his own body, for example, hair pulling and swallowing, skin picking to the extent of causing open lesions, and severe head banging. It was in the psychiatric hospital that the boy's XXY chromosomal complement was identified. When his parents were informed of the biological origin of their child's condition, they seemed relieved of some guilt.

More studies should surely be undertaken of the behavior of XXY tortoiseshell cats (mostly sterile) and of XO mice (apparently always fertile), as recommended by Morton [1972] (XX = female and XY = male in these mammals, as usual). One tortoiseshell XXY cat, Lucifer, was reported to have not the slightest sexual inclination. He was treated as a kitten by other tomcats; his presence did not disturb these toms even when in-season females were present (Bamber and Herdman, 1932). However, not all such cats are sterile (Jude and Searle, 1957; also see Thuline and Norby, 1961). One tortoiseshell tom fathered at least 65 offspring, so behavioral variability exists. Also note a report of testicular hypoplasia accompanied by the XXY sex chromosome complement in two rams—an ovine counterpart of Klinefelter's syndrome (Bruere, Marshall, and Ward, 1969). These two unrelated rams both exhibited testicular hypoplasia, were small, and "showed strong male libido to ewes in estrus and performed the physical actions of ejaculation." Aneuploid mice are available; Russell (1961), in an excellent review article on the genetics of mammalian sex chromosomes, reports on XO and XXY mice—on their existence, not on their behavior. Recent technical advances, for example, in autoradiography, allow the detection of relatively tiny deletions, duplications, and inversions. Now perhaps, new attempts should be made to localize genes that alter behavior without altering morphology and anatomy. As mentioned at the beginning of this section, a start has been made in human beings.

And what of the notorious XYY human male? In 1967, Price and Whatmore filed the following report about a maximum security hospital in Scotland:

All the patients admitted to this hospital have severely disordered personalities, some have brain damage which followed infections, others are epileptics, and others suffer from a psychosis. The largest group of patients have no known cause for their personality disorders. All the men with an XYY complement were classified in this category and eighteen other men have been randomly selected from this group for comparison with the nine XYY males. Seventeen of the eighteen control males were known to have an XY sex chromosome complement, the remaining being one of twenty-seven who had not been willing to be investigated when the chromosome survey was carried out.

There are three ways in which the XYY males differ importantly from the controls. First, although the patients in the two groups have penal records of comparable length, those of the XYY males include considerably fewer crimes of violence against persons. Thus, the nine XYY males had been convicted on a total of ninety-two occasions, but only eight of these convictions (8.7 percent) had been for crimes against persons, while eighty-one (88.0 percent) had been for crimes against property. In contrast, the eighteen control males had been convicted on 210 occasions, and forty-six of these (21.9 percent) had been for crimes against persons while 132 (62.9 percent) had been for crimes against property. Second, the disturbed behavior of the XYY patients showed itself at an earlier age. This is reflected in a mean age at first conviction of 13.1 yr., compared with a mean age of 18 yr. for the control patients, a difference which is significant at the 5 percent level. Third, in the families of these patients the incidence of crime among the siblings of the XYY patients is significantly less than among those of the control patients. Thus, only one conviction is recorded among thirty-one sibs of the XYY patients while no less than 139 convictions are recorded for twelve of sixty-three sibs of the control patients.

The distribution of intelligence quotient among the XYY males probably reflected the distribution among the patients of the hospital as a whole. Seven were considered to be mentally sub-normal, but it is worth noting that the pattern of behavior among the two whose intelligence quotients were not unusually low conformed with those of the other seven.

The picture of the XYY males that emerges from examination of those detained at the State Hospital is of highly irresponsible and immature individuals whose waywardness causes concern at a very early age. It is generally evident that the family background is not responsible for their behavior. They soon come into conflict with the law, their criminal activities being aimed mainly against property, although they are capable of violence against persons if frustrated or antagonized. Their failure to respond to corrective measures leads to a sentence of prolonged detention in safe custody at an earlier age than is usual for offences of this kind.

But is this a conclusion reached by too many too fast, as Levitan and Montagu (1971) carefully caution? Is the only relatively consistent characteristic shared by XYY males their greater than average height? Sutton (1975) points out that assorted surveys of male newborns suggest that XYY may occur within a range of 1 to 4 per 1,000 live births; such frequencies are in no way related to the incidences of troublesome or even just subnormally intelligent tall men. Note that this XYY frequency is occurring in the absence of transmission of the chromosomal abnormality from father to son (Melnyk et al., 1969). Then too, we are obliged to document the 1972 report by Gardner and Neu entitled "Evidence Linking an Extra Y Chromosome with Sociopathic Behavior," in which lawyers and their cohorts are advised to consider the evidence for this "linkage" in assessing legal responsibilities. (They ought simultaneously to consider the troublesome but very rare XXYY males.)

Hamerton (1976) concludes:

> To summarize the current state of our knowledge about the XYY male, we know that about 1/1,000 males in the general population have an XYY karyotype whereas in security settings the frequency is about 20/1,000. The original observation made by Jacobs *et al.* (1965) of an excess of XYY males in these population groups is thus amply confirmed. In addition, data are now available which indicate psychological differences between young noninstitutionalized adult XYY males when compared to XY controls. These differences indicated that XYY males were less able to control the normal male aggressive drive in frustrating or provocative situations and were more impulsive and immature than XY controls. There is also some evidence of an increased frequency of behavior problems and learning disabilities among children with this karyotype. At present, little can be said about the early childhood of XYY males; followup has not proceeded far enough on sufficient numbers of children to draw conclusions. Finally, it is now clear that perhaps only a small minority of XYY males spend part of their lives in security settings. There is little doubt, however, no matter which way the data is examined, that these males, or some of them, are at a greater risk than XY controls, due perhaps to adverse environmental influences interacting with the XYY genotype.

Clearly, as Figure 4-7 indicates, there are other sex chromosome abnormalities in addition to those already considered that have behavioral effects. XXXY and XXXXY individuals show Klinefelter's syndrome—IQ decline is positively associated with the number of additional X chromosomes beyond two. The same type of IQ decline occurs for XXX, XXXX, and XXXXX females. Thus the triple-X female, who occurs with a frequency of 6 in 10,000, tends to show subnormal mental capacity. The few females reported with

XXXX and XXXXX all showed severe mental deficiency. Triple-X females are fertile; one would expect them to have children in proportions ¼XX : ¼XY : ¼XXX : ¼XXY. However, it seems that the abnormal XXX and XXY karyotypes are rarely found in offspring of triple-X females, perhaps owing to directed meiotic segregation whereby two of the mother's X chromosomes segregate preferentially into functionless polar bodies and the third goes to the egg nucleus. A similar unexpected lack of abnormal karyotypes is found in the offspring of XYY males. Thus the effects of nondisjunction do not last long, in terms of generations, in the progenies of those fertile individuals having abnormal karyotypes.

The final category of chromosomal variants to be considered is the genetic mosaics that occur due to nondisjunction in somatic cells leading to cell lines with different chromosomal constitutions. Some are female mosaics (XO/XX), some male mosaics (XY/XXY), and some intersex mosaics (XO/XY). Table 4-3 gives some idea of the types known. The phenotypes are clearly variable depending on the proportions of the karyotypically differing tissues in individuals, and this depends on the time during development at which the abnormal cell division or divisions occurred, the localization of the differing tissues in the body, subsequent cell migrations, and finally, pure chance. For mosaics with cells of different sex, referred to as *gynandromorphs,* intersexuality may occur, depending upon the developmental factors noted above. In some cases this most unsatisfactory situation can be partly rectified by removal of a gonad or by hormone treatment supportive of the expression of but one sex.

Autosomal chromosome number changes appear to cause more severe effects on behavior than do X or Y chromosomal aberrations, as shown by the example of Down's syndrome. Furthermore, there is an apparent lack of individuals trisomic for the larger autosomes that are no doubt aborted as fetuses.

In a proportion of normal female cells (XX) but not in male cells (XY), there is a chromatin-positive DNA body situated at the nuclear membrane. This is referred to as a sex-chromatin body or a Barr body, named after its senior

Table 4-3   Human Sex Chromosome Mosaics

Female	Male	Gynandromorph (mixed sexes)
XO/XX	XY/XXY	XO/XY
XO/XXX	XY/XXXY	XO/XYY
XX/XXX	XXXY/XXXXY	XO/XXY
XXX/XXXX	XY/XXY/?XXYY	XX/XY
XO/XX/XXX	XXXY/XXXXY/XXXXXY	XX/XXY
XX/XXX/XXXX		XX/XXYY
		XO/XX/XY
		XO/XY/XXY
		XX/XXY/XXYYY

Source: Stern, 1973.

discoverer (Barr, 1959). In females, Barr bodies can be seen in many tissues, including the epidermis, oral mucosa, and the amniotic fluid surrounding female fetuses. It has been postulated by Lyon (1962) and others that the Barr body represents an inactivated X chromosome. Individuals with more than one X have sex-chromatin bodies and are called sex-chromatin positive, while individuals with only one X are sex-chromatin negative. In other words, irrespective of the number of X chromosomes, only one is fully active, the remainder being largely inactive, although as shown in Figure 4-7, individuals with three or more X chromosomes do tend to have a degree of mental deficiency. Generally, the rule is

Number of sex-chromatin (Barr) bodies = number of X chromosomes − 1

The other rule about sex determination in *Homo sapiens,* although not yet formally stated in this text, is that irrespective of the number of X chromosomes, the presence of a Y leads to a male phenotype (even if abnormal, as in the case of Klinefelter's syndrome).

Because few genes are known on the Y chromosome, it is not surprising that XYY individuals occur without gross morphological abnormalities. Individuals with Turner's syndrome are female without a Barr body, while those with Klinefelter's syndrome are male with a Barr body. In the case of XXXY Klinefelter's syndrome, two Barr bodies are expected and found. Because of the ease of staining a few cells in material scraped from the oral mucosa, in which Barr bodies can be readily studied, Barr bodies provide important population information on the frequency of abnormal males and females—at least those abnormal as regards sex chromosomes.

SUMMARY

Two major types of chromosome changes have effects on behavior:

1 *inversions,* where some genes in a chromosome are in reverse order compared with a standard situation.

2 *chromosome number* alterations, where extra chromosomes are present or chromosomes are deleted.

In certain *Drosophila* species, populations are *polymorphic* for inversions. Frequently the heterozygote for such inversions is fitter than the corresponding homozygote. This often applies to sexual selection as controlled by the genotype of the male.

Spontaneous chromosome number changes have mainly been studied in defective humans. When the material of one of the autosomes is present three times, rather than the normal two times, the resulting genomic imbalance results in retardation known as Down's syndrome. Considering sex-chromosomes, individuals with an extra X chromosome, XXY, are sterile males with Klinefelter's syndrome, and individuals deficient in an X, XO, are sterile females with Turner's syndrome. Other sex-chromosome abnormalities are XYY males who are suggested to have a tendency towards sociopathic behavior, XXX females who tend to have lower IQs than normal XX females, and genetic mosaics in whom cell lines have different constitutions.

Many Genes and Behavior

This chapter marks the entry into matters of greater genetic complexity than those discussed so far and involves the beginning of a consideration of traits under more complex genetic control than that of genes or chromosomes, which can be relatively easily followed from segregation data.

5-1 BIOMETRICAL GENETICS

In order to analyze the variations of such traits, the methods and techniques of quantitative genetics must be used. The aim is to separate the total variance of a trait into genotypic and environmental components. Some traits fall into an intermediate category, being partly controlled by genes whose segregation can be followed and partly the result of variation that makes tracing impossible. Essentially we are moving toward a consideration of traits for which the mode of inheritance is multifactorial or polygenic. In some cases, as will be seen, it is possible, using special breeding techniques, to localize genes controlling a quantitative trait to specific chromosomes and even to regions of chromosomes. One of the main methods comes from the *directional selection* experiment, in which individuals are selected at high or low extremes of a distribution in the hope of forming separate high or low lines in subsequent generations (Figure

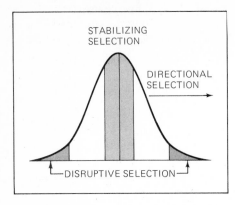

Figure 5-1 Directional, disruptive, and stabilizing selection. Sections of normal population distribution favored under these three selection regimens.

5-1). Provided a trait has some genetic basis, there should be a response, since the selection of extreme phenotypes implies the selection of a proportion of extreme genotypes. For theoretical considerations (inappropriate here) see Falconer (1960) and Lee and Parsons (1968).

Little work on localizing effects to chromosomes for quantitative traits has been done except in certain species of *Drosophila*. Some species of this genus have the advantages that (1) their chromosomes are well marked with precisely located genes and can therefore be used in genetic analyses, and (2) the rapid generation time of 2 to 3 weeks permits rather complex breeding programs to be carried out in reasonably feasible amounts of time. When we realize that mice produce only four or five generations per year, it becomes immediately clear why few detailed genetic, as opposed to biometrical, studies of behavioral traits have been carried out in mammals, including human beings.

As an indication of the diversity of behavioral traits that have been shown to be controlled polygenically, the following can be cited:

* *Drosophila* species: Locomotor activity, chemotaxis, duration of copulation, geotaxis (gravity-oriented locomotion), mating speed, optomotor response, phototaxis, preening, and level of sexual isolation within and between species. Evidence for actual localization of genetic activity to chromosomes exists for duration of copulation, geotaxis, and level of sexual isolation.
* Rodents (mainly mice, rats, and guinea pigs): Susceptibility to audiogenic seizures, running speed, activity, sexual drive, early or late onset of mating, emotional elimination (defecation and urination), fighting, alcohol preference, and various measures of learning such as maze brightness in running toward a food reward and the conditioned avoidance response wherein a mouse learns to avoid a shock by responding to a signal (light or buzzer). In only a few cases has any approach at the level of genetic loci been made; but some biochemical and physiologic variables associated with behavior have been studied.
* *Homo sapiens:* Specific major genes are known for some of the sensory perceptual traits, for example, taste deficiency and color blindness. In the normal range of traits such as intelligence, temperament, emotional behavior, specific abilities, and neuroticism, polygenic control has been established. Out-

side the normal range, major genes are known, as will be discussed in Chapter 11. With respect to behavior little or no localization to chromosomes has been carried out, one exception being color blindness. *Homo sapiens* is a species in which breeding experiments cannot be conducted.

This chapter discusses some examples of traits at least partly under polygenic control in *Drosophila,* lovebirds, rodents, and dogs. It concludes with a discussion of laterality, that is, whether there is a hereditary tendency toward right- or left-handedness. Unlike other quantitative traits, laterality is hardly if at all under genetic control.

5-2 GEOTAXIS IN *DROSOPHILA*

Gravity-oriented locomotion (geotaxis) in *Drosophila melanogaster* represents the most complete genetic analysis of a behavioral trait so far (Hirsch, 1962, 1963, 1967a). A vertical 10- or 15-unit plastic maze is used (Figure 5-2). Flies are introduced on the left-hand side and collected from one of the vials on the right after being attracted through the maze by the odor of food and by lighting in the form of a fluorescent tube on the right-hand side. Conditions of maximum objectivity are provided, since there is no human handling of the large number of flies involved, once they are introduced into the maze. Rapid responses to

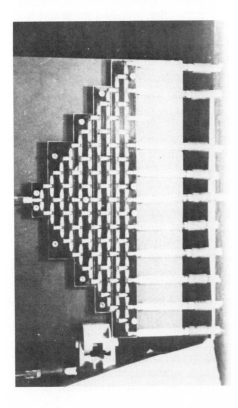

Figure 5-2 A 10-unit geotactic maze in a vertical position. Flies are introduced in the vial at the left and are collected from various vials at the right. Those moving to the top vials have negative geotactic scores; those moving to the bottom vials have positive geotactic scores.

selection for both positive and negative geotaxis are found (Figure 5-3), although the actual total response to selection is greater for negative geotaxis. Using methods developed by Mather (1942) and Mather and Harrison (1949) in bristle number selection experiments, Hirsch and Erlenmeyer-Kimling (1962) assayed the role of the three major chromosomes possessed by *D. melanogaster* in the response to selection over a number of generations.

A brief description of the method is appropriate. It is possible to combine in all ways chromosomes extracted from a selection line with those of the control line by using stocks carrying dominant genes with inversions inhibiting crossing over in the relevant chromosomes, one chromosome at a time. In this way the individual effects of chromosomes and their interactions can be studied with respect to a quantitative trait. Therefore, a multiple tester stock: $A/+, B/C, D/E$ is utilized. A is a dominant gene on the X chromosome, B and C are dominants on chromosome II, and D and E on chromosome III, all genes being associated with inversions. These segregate essentially as whole units, since the presence

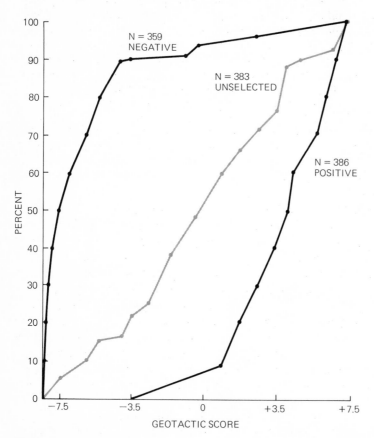

Figure 5-3 Geotactic scores of flies in a 15-unit maze. Cumulative percentages of flies achieving assorted scores for an unselected foundation population and for two selected lines. (*After Hirsch, 1963.*)

of inversions in the heterozygotes prevents or diminishes crossingover. When females of these stocks are crossed to the selection line S to be assayed, daughters of the type $A/S, B/S, D/S$ are obtained. These are then backcrossed to the selection line, and eight phenotypically distinguishable classes of progeny appear

- A B D
- A B
- A D
-       B D
- A
-       B
-             D
- Selected line itself (without A, B, or D)

These eight classes receive a representative S chromosome from their father and a variable number of S or tester (T) chromosomes (A,B,D) from their mother. Each of the three major chromosomes is therefore heterozygous or homozygous for an S chromosome in females. From these eight classes, the individual effects of the chromosomes and their interactions can be studied. One limitation is that the technique is fully efficient only in detecting recessive genes in the S chromosomes, because the comparison for each chromosome is between the heterozygote T/S and the homozygote S/S. This means that dominants are not detected and incomplete dominants are revealed to the extent of their level of dominance.

The above procedures have been applied to the lines selected for geotaxis (Figure 5-3). The mean effects of chromosomes X, II, and III in the unselected population are given in Table 5-1 and are compared with those selected for positive or negative geotaxis. In the unselected population, the X and II chromosomes have genetic activity for positive geotaxis, and chromosome III is negative by comparison with the standard tester chromosomes. Selection for positive geotaxis produced little effect for chromosome II, but a positive effect was increased for chromosome X; for chromosome III, a negative effect was

Table 5-1   Mean Chromosomal Effects (with Standard Errors) on Geotactic Scores after Selection Based on Maze Depicted in Figure 5-3.

Population	Chromosome		
	X	II	III
Selected for positive geotaxis	1.39 ± 0.13	1.81 ± 0.14	0.12 ± 0.12
Unselected	1.03 ± 0.21	1.74 ± 0.12	-0.29 ± 0.17
Selected for negative geotaxis	0.47 ± 0.17	0.33 ± 0.20	-1.08 ± 0.16

Means are based on 10 replicates in each case; each unit represents one notch on the maze.
Source: Hirsch, 1967a.

changed to slightly positive. The effect of these three chromosomes is greater for negative geotaxis considering the three chromosomes together, as expected, since the total response to selection obtained is greater for negative geotaxis. Selection for negative geotaxis reduced the positive effects of chromosomes X and II and increased the negative effect of chromosome III compared with the unselected stock. Therefore, the analysis so far confirms that there are genes distributed on all the three major *D. melanogaster* chromosomes that affect geotaxis. Although this analysis was not carried further, assuming extrapolation from the work of Thoday (1961) and his colleagues on bristle numbers, it is in theory possible to locate precise regions of genetic activity on chromosomes, of which there are likely to be several.

5-3 SEXUAL ISOLATION: *DROSOPHILA* SPECIES

Species of sexually reproducing organisms are genetically closed systems. They are closed systems because they do not exchange genes or do so rarely enough so that the species differences are not swamped. Races are, on the contrary, genetically open systems. They exchange genes by peripheral gene flow unless they are isolated by extrinsic causes such as spatial separation. The biological meaning of the closure of a genetic system is simple but important: it is evolutionary independence. Consider these four species—human beings, chimpanzees, gorillas, orangutans. No mutation and no gene combination arising in any one of them, no matter how favorable, can benefit any of the others. It cannot do so for the simple reason that no gene can be transferred from the gene pool of one species to that of another. On the contrary, races composing a species are not independent in their evolution; a favorable genetic change arising in one race is, at least potentially, capable of becoming a genetic characteristic of the species as a whole.

There are therefore problems in defining that pivotal concept race. According to Parsons (1972a), about all we can say is that:

> A race is a population in which the gene frequencies at some loci differ from one another. It is a *quantitative* and not a *qualitative* definition, as there are no biological isolating mechanisms between different populations. Thus gene pools of different races have differing gene frequencies. It must be stressed that because the definition is quantitative rather than qualitative, the amount of difference needed to accept that we have two different races is completely arbitrary.

Species are genetically closed systems because gene exchange between them is impeded or prevented by reproductive isolating mechanisms. The term *isolating mechanism* was proposed by Dobzhansky in 1937 as a common name for all barriers to gene exchange between sexually reproducing populations. According to Mayr (1963), isolating mechanisms are "perhaps the most important set of attributes a species has." It is a remarkable fact that the isolating mechanisms are physiologically and ecologically a most heterogeneous collection of phenomena. It is another remarkable fact that the isolating mechanisms that maintain the genetic separateness of species are quite different not only in different groups of organisms but even between different pairs of species in the same genus. They fall naturally into two primary divisions: (1) geographical or spatial isolation; and (2) reproductive isolation. In the case of geographical isolation, the populations involved are *allopatric,* which means that they are found in different territories, so gene exchange between them is minimal. Indeed, they may or may not be genetically similar. For a complete discussion of allopatric and *sympatric* (populations living in the same territory) speciation and of ecological factors in speciation see Mayr (1963).

The classification of reproductive isolating mechanisms outlined below is a composite one. It is based on the classifications published by Mayr (1942), Muller (1942), Patterson (1942), Allee et al. (1949), Stebbins (1950), and Dobzhansky (1951).

1 Barriers to gene flow preventing the meeting of potential mates
   • Habitat or ecological isolation: populations found in the same general territory but occupying different ecological niches
   • Seasonal or temporal isolation: sexual maturity or activity may occur at different times
2 Barriers to gene flow preventing the formation of hybrid zygotes
   • Mechanical isolation: as occurs if the genitalia of the two sexes do not correspond
   • Gametic isolation or prevention of fertilization: as occurs if eggs and spermatozoa do not meet or do not fuse normally
   • Sexual, psychological, or ethological isolation: greater mutual attraction of conspecific males and females than between males and females of different species
3 Barriers to gene flow that eliminate or handicap the hybrid zygotes that have been produced
   • Hybrid nonviability or weakness: lower viability of hybrid zygotes compared with either parental species
   • Hybrid sterility: hybrids unable to reproduce because of nonproduction of functional gametes
   • Selective hybrid elimination or hybrid breakdown: hybrid products eliminated in the F_2 or later generations because they are adaptively inferior

These eight items have a common function. They all have but one net effect—either singly or collectively—to prevent the exchange of genes between populations (Patterson, 1942).

There are behavioral components to a number of these mechanisms. Many of these are illustrated in Chapter 13 where the mechanisms involved in habitat selection in closely related species of *Drosophila* and the deer mouse *Peromyscus* are discussed. This section discusses sexual, psychological, or ethological isolation, incorporating the genetic detail *Drosophila* allows. (For more, see the whole of Chapter 8.) This particular isolating mechanism is an efficient barrier in that when effective, sexual isolation eliminates any wastage of gametes and, more importantly, eliminates the need for food and space for developing hybrids that may in some way be inferior to nonhybrids in fertility or viability. The sterile hybrids produced within the superspecies (species complex) *Drosophila paulistorum* are one example of this. The semispecies (or subspecies or races of incipient species) of *D. paulistorum* show a pronounced sexual isolation; matings *between* the semispecies succeed much less frequently than do those *within* a semispecies (Ehrman, 1961, 1965).

The genetic basis of the sexual isolation has been studied in the hybrids between the Centroamerican and Amazonian semispecies, among others. These are incipient species with overlapping geographic distributions (Figure 5-4); they are indistinguishable morphologically yet produce fertile female and sterile male hybrids when crossed. Crosses have been made in which the distribution of a certain pair of chromosomes was followed with the aid of mutant genes that served as genetic markers. The sexual preferences were studied in the F_1 hybrids between the semispecies and in a series of backcrosses to each of the parental semispecies. The evidence obtained shows that the sexual isolation is controlled by factors distributed on all of the three chromosomes the species possesses. The many genes controlling the sexual preferences produce additive effects, the sum of which makes the bar to crossing nearly complete between the semispecies. The plan followed in ascertaining the role of each of the two autosomes and of the X (the first) chromosome in the genetic basis of the sexual isolation barrier via a mating preference technique, is outlined in Figure 5-5 for one autosome (chromosome 3) as an example. The object was to transfer one marked chromosome into the nuclear and cytoplasmic background of an alien semispecies. Backcross progenies were obtained by crossing F_1 hybrid females carrying suitable marker genes in the third chromosome to males of one or the other of the parental semispecies. The backcrosses were repeated in each of three successive generations, always selecting as female parents the carriers of semispecies-foreign chromosomes. For each combination of parental semispecies, two series of recurrent backcrosses were made to each of these semispecies. The F_1 hybrid females between semispecies A and B obviously contain one A and one B chromosome in each pair; the F_1 hybrid males have the X of their mother, the Y of their father, and an A and a B autosome of each parent. In the backcrosses of the A/B hybrid females to A males, the B chromosomes, except the one with the gene marker, tend gradually to be replaced because only the one with the gene marker is selected for; in the backcrosses to B males, the A chromosomes tend to be eliminated. In the progeny of the third backcross, most of the flies carry chromosomes of one semispecies only, except the foreign chromosome with the genetic marker (and sometimes also a foreign Y chromosome), as itemized in Table 5-2.

Figure 5-4 The geographic distribution of the semispecies of *D. paulistorum*. Transitional may be the relict ancestor. (*From Dobzhansky and Powell, 1975.*) For one example note that on Trinidad as many as three semispecies are sympatric and reproductively isolated.

The control crosses involved the use of sisters of the same females used in the experimental series, but not containing the foreign chromosomes with the genetic markers. The progenies were, consequently, like the experimental ones, except that they did not contain the foreign marked chromosome.

In most instances, the foreign chromosome contained only a single mutant gene, which served as a marker. This was, nevertheless, deemed a satisfactory experimental technique for two reasons: (1) whenever more than one marker was present, crossingover between homologous pairs was found to be suppressed in the intersemispecific crosses; (2) the semispecies involved in these experiments differed in at least one inversion in each of their five chromosome "arms" (Dobzhansky and Pavlovsky, 1962), so that they did not pair intimately and in this way allow for crossingover between homologous pairs. This does not, of course, exclude the possibility that some undetected crossingover occurs in the hybrids, but such crossingover is not frequent.

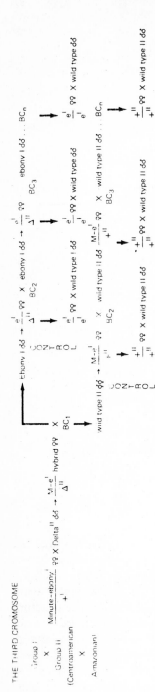

Figure 5-5 Crosses and backcrosses involving a marked third chromosome in the species complex *Drosophila paulistorum.*

Table 5-2 Fractional and Percentage Dilutions of Foreign Chromosomes in Repeated Backcrosses (BC$_1$ . . . BC$_n$) between Semispecies A and Semispecies B, where Semispecies B is the Recurrent Parent

	Marked* chromosomes		Unmarked chromosomes		Percent A in entire genome	Percent B in entire genome
	A	B	A	B		
F$_1$	$1/2$	$1/2$	$1/2$	$1/2$	50.0	50.0
BC$_1$	$1/2$	$1/2$	$1/4$	$3/4$	33.3	66.7
BC$_2$	$1/2$	$1/2$	$1/8$	$7/8$	25.0	75.0
BC$_3$	$1/2$	$1/2$	$1/16$	$15/16$	20.8	79.2
BC$_n$	$1/2$	$1/2$	$1/2^{n+1}$	$1 - (1/2^{n+1})$	$(0.5 + 1/2^n)/3$	$1 - [(0.5 + 1/2^n)/3]$
Controls						
BC$_2$	0	1	$1/8$	$7/8$	8.3	91.7
BC$_3$	0	1	$1/16$	$15/16$	4.2	95.8
BC$_n$	0	1	$1/2^{n+1}$	$1 - (1/2^{n+1})$	$(1/2^n)/3$	$1 - [(1/2^n)/3]$

* See Figure 5-4 for outline of these interpopulation crosses incorporating mutant marker genes.
Source: Ehrman, 1960a.

Although the role of each chromosome was analyzed in more than one interpopulational cross, only a single set of data will be presented as a sampler. To test the effects on the sexual preferences of the foreign third chromosomes, Amazonian females heterozygous for the dominant marker *Delta* (wing venation) in the third chromosome were crossed to Centroamerican males that carried on one of their third chromosomes the dominant *Minute* (bristles) and the recessive *ebony* (body color). In the F$_1$ generation, the *Delta/Minute-ebony* females were used as progenitors of the backcross progenies. Their sibs were used for tests of the mating preferences of the F$_1$ hybrids (*ebony* is approximately 50 crossover units from *Minute* and is used here as a check on the suppression of crossingover in the hybrids). The results are reported in Table 5-3. The F$_1$ hybrid females accepted Centroamerican males, while the F$_1$ hybrid males appeared to be neutral. The backcross progenies showed that the sexual preference of the hybrids was for the semispecies of the recurrent parent. It therefore appears that the third chromosome does not by itself determine sexual preferences in these crosses. Indeed in Amazonian × Centroamerican hybrids, sexual preference is decided by which semispecies contributes more than half the genome, no one chromosome being clearly more important than the others (Ehrman, 1961). The sexual isolation analyzed here, in which matings between females and males of the different *D. paulistorum* populations are much less likely to succeed than matings within a population, is apparently controlled by polygenes scattered in every one of the three pairs of chromosomes. As such, the similarity with geotaxis (discussed in Chapter 8) is obvious.

Table 5-3 Direct Observation of Matings Testing the Influence of the Third Chromosome in the Genetic Architecture of Sexual Isolation between Two Strains of *D. paulistorum*

	Number	Copulae with		χ^2	*p*
		CA	Am		
Tests of hybrid females					
F_1	19	17	2	10.3	<0.01
Backcrosses to Centroamerican parent					
BC_1	20	18	2	11.3	<0.01
BC_2	20	19	1	14.5	<0.01
BC_3	20	19	1	14.5	<0.01
Backcrosses to Amazonian parent					
BC_1	20	2	18	11.3	<0.01
BC_2	20	1	19	14.5	<0.01
BC_3	20	1	19	14.5	<0.01
Tests of hybrid males					
F_1	20	8	12	0.5	0.70–0.50
Backcrosses to Centroamerican parent					
BC_1	20	14	6	2.5	0.20–0.10
BC_2	20	19	1	14.5	<0.01
BC_3	20	19	1	14.5	<0.01
Backcrosses to Amazonian parent					
BC_1	20	2	18	11.3	<0.01
BC_2	20	2	18	11.3	<0.01
BC_3	20	2	18	11.3	<0.01

Source: Ehrman, 1961.

Extensions of these experiments illuminate a fascinating variation of hybrid behavior. F_1 hybrid females from a cross between Andean-Brazilian and Amazonian semispecies were observed to accept no males courting them (Ehrman, 1960b). Most crosses between these two semispecies fail because of the powerful sexual isolation barrier. However, after repeated and lengthy attempts, viable male and female hybrids were obtained. It should be emphasized that these are normal males and normal females as far as external and internal anatomy are concerned. Yet the genic endowments contributed by the parents of these hybrids are so discordant that the hybrids are virtually unable to perform successfully the mating rituals normal in this species.

A study of the behavior of living flies under a microscope in special observation chambers shows that the hybrid females will not accept any males that court them regardless of how vigorous or persistent the courtship is. They have been observed to reject consistently the males of both parental semispecies, as well as their own hybrid brother. They accomplish this by assuming the posture of rejection of the courtship that is characteristic of *D. paulistorum;* the female lowers her head and elevates the tip of her abdomen so that the vaginal orifice is inaccessible to an approaching male.

The hybrid males are of less interest in this respect because they are completely sterile. Even so, they are rarely successful in courting females, and they have been placed and observed with mature females of both parental semispecies, as well as with their own hybrid sisters. It is here suggested that the disharmonies in the sexual behavior of the hybrid females may serve as a very efficient isolating mechanism between the incipient species. These hybrid females, though potentially fertile in the sense that their ovaries do produce normal and mature eggs, would probably never mate. This would make the appearance of backcross progenies impossible solely on behavioral grounds.

If populations have diverged genetically, developing different coadapted complexes as a result of having adapted to different environments, then gene exchange between these populations is likely to produce ill-adapted genotypes. Natural selection acts to build and reinforce the barriers to gene exchange between populations whose hybridization results in reproductive wastage. The appearance of hybrids with inferior fitness is, in this way, minimized or avoided altogether (Fisher, 1930; Dobzhansky, 1940, 1970). Alternatively, Muller (1942) assumes that reproductive isolation arises as an accidental byproduct of genetic divergence. As populations become adapted to different environments, they become different in progressively more and more genes. Reproductive isolation arises because the action of many genes is pleiotropic. Some gene differences selected for different reasons or resulting from random genetic drift (Wright, 1955; Dobzhansky and Spassky, 1962) may thus have isolating side effects.

However, evidence that selection can strengthen reproductive isolation in wild populations comes from multiple-choice experiments using the Elens-Wattiaux mating chambers and scoring by direct observation (see Section 3-2). Joint isolation coefficients were computed for given pairs of semispecies that have been found to occur both sympatrically and allopatrically. In allopatric crosses the average isolation coefficient was +0.67 and in sympatric crosses +0.85 (Table 5-4). Thus pairs occurring sympatrically exhibit more sexual isolation than the same pairs occurring allopatrically, or semispecies coexisting geographically tend to be reproductively more isolated than those that do not, which is reasonable as the production of large numbers of hybrids would be very inefficient. Chapter 10 will show that such strengthening of reproductive isolation in sympatric situations by natural selection is almost certainly a general phenomenon, here illustrated by Ehrman (1965) and colleagues in *D. paulistorum.*

It is then not surprising that reproductive isolation can be strengthened by artificial laboratory selection (Koopman, 1950; Knight, Robertson, and Waddington, 1956). Koopman set up experimental populations in cages containing the two sibling species *Drosophila pseudoobscura* and *D. persimilis.* Each species was homozygous for a different recessive mutant gene with easily visible external effects, so that both species and their hybrids were readily dis-

Table 5-4 Numbers of Matings Observed and Isolation Coefficients Calculated for Sympatric and for Allopatric Crosses; the Total Number of Matings Observed Was 1,695

Races	Origin	Matings	Coefficient
1. Amazonian × Andean	Sym	108	0.86 ± 0.049
	Allo	100	0.66 ± 0.074
2. Amazonian × Guianan	Sym	104	0.94 ± 0.033
	Allo	109	0.76 ± 0.061
3. Amazonian × Orinocan	Sym	106	0.75 ± 0.065
	Allo	124	0.61 ± 0.070
4. Andean × Guianan	Sym	109	0.96 ± 0.026
	Allo	102	0.74 ± 0.066
5. Orinocan × Andean	Sym	100	0.94 ± 0.033
	Allo	111	0.46 ± 0.084
6. Orinocan × Guianan	Sym	104	0.85 ± 0.053
	Allo	100	0.72 ± 0.069
7. Centroamerican × Amazonian	Sym	102	0.68 ± 0.072
	Allo	103	0.71 ± 0.070
8. Centroamerican × Orinocan	Sym	110	0.85 ± 0.052
	Allo	103	0.73 ± 0.069

Average (sympatric) = 0.85
Average (allopatric) = 0.67

Source: Ehrman, 1965.

tinguishable. Adult hybrids were discarded each generation, and new populations were initiated with nonhybrid progenies. In this way Koopman was selecting the progenies of intraspecific, and excluding those of interspecific, matings. In a surprisingly small number of generations (five to six) he obtained strains of D. *pseudoobscura* and D. *persimilis* that showed nearly complete sexual isolation between the species. In what were essentially three replicate population cage experiments, Koopman recorded the following reductions in percentages of progenies from heterogamic matings in his experimental cages:

Replicate populations	Generation	Percent hybrids
I	1	22.5
	5	5.1
II	1	49.5
	5	1.4
III	1	36.5
	6	5.2

The results of Knight, Robertson, and Waddington (1956) are, perhaps, even more dramatic because they were recorded wholly within one species.

These authors obtained by selection a significant, though of course incomplete, sexual isolation of strains of *D. melanogaster* that originally showed no such isolation. After seven generations of selection by destroying the progeny of interstrain matings between the mutants *ebony* and *vestigial,* these authors were able to obtain an average reduction of 66 to 38 percent in the proportion of such hybrid offspring produced. Crossley (1975) also obtained divergence with these two mutants and, from a detailed analysis of mating behavior (see Bastock, 1956, and Section 3-2) concluded that female discrimination and changed female responses were the most important classes of sexual isolation.

Wallace (1954) obtained results similar to those of Knight, Robertson, and Waddington (1956) using a technique essentially like Koopman's, this time with the *straw* and *sepia* mutants of *D. melanogaster.* Seventy-three generations of selection against hybrid progeny produced a significant nonrandomness of mating. One type of female, *sepia,* gave a 9 : 1 ratio of homogamic to heterogamic matings with *sepia* males and with *straw* males, respectively. The *straw* females, however, mated as often heterogamically as homogamically. For another example of natural selection leading to reproductive isolation, in this case involving variant compound chromosomes, see Ehrman (1979) and the antecedent references therein.

Drosophila behavior is considered again at length in Chapter 8. Suffice it here to refer additionally to an analysis of incipient sexual isolation between two *D. willistoni* morphs, one preferentially pupating on moist food surfaces, the other on dry surfaces. De Souza, Da Cunha, and Dos Santos (1970, 1972) discovered that this behavioral polymorphism was fostered by a pair of autosomal genes and, furthermore, that under competitive conditions, males that pupate in dry spots are more successful in inseminating females that pupate in similar sites than they are in acquiring as mates females that had previously pupated directly in their humid food. This then is the spectrum: from natural or artificial directional selection forever strengthened until "complete" sexual isolation (many examples provided in this section); to sexual isolation constructed by disruptive (diversifying) selection (Gibson and Thoday, 1962, versus Scharloo, 1971), of the sort noted in Figure 5-1 and exemplified by the *D. willistoni* morphs; to the evolution of preferences for matings within rather than between strains of *D. melanogaster* raised so that they differ in numbers of stout hairs, bristles, on their integuments (Parsons, 1965a); and finally, to the appearance of rudiments of sexual isolation based upon no selection at all—as an accidental byproduct of adaptation to development and life at different temperatures in different environments (Ehrman, 1964, 1969).

5-4 ISOLATION IN LOVEBIRDS

Dilger (1962a,b) managed to obtain hybrids between two species of African parrots sometimes referred to as lovebirds, *Agapornis roseicollis* and *A. fischeri,* which breed and live well in captivity. These are the attractive birds employed in trained-bird acts to ride miniature trains, push small wagons, deliver mail, and such because they learn new behaviors quickly. They can also learn to open cage doors and evade capture. So can hybrids between them, but these same hybrids consistently have difficulty in preparing nests. *A. roseicollis* females carry strips of nesting material (paper, bark, or leaves) tucked between feathers on the lower back or the rump. Several such strips are carried at a time, and in a given trip to the nest pieces that slip out are retrieved. *A. fischeri* females, on the other hand, transfer pieces of bark, paper, or leaves, and more substantial materials such as twigs, one item at a time to the nest site in their bills. Hybrid females almost always attempt to tuck nesting material in the feathers, but are never successful—indeed when the hybrid first begins to build a nest it acts completely confused (Figure 5-6). Among the reasons for this are

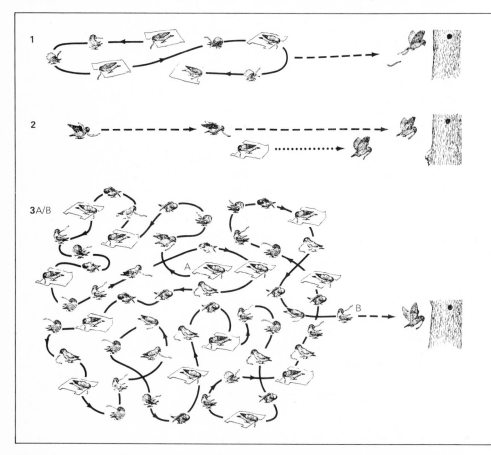

Figure 5-6 Nest building by hybrid lovebirds. Offspring of *A. roseicollis* × *A. fischeri* inherit patterns for two different ways of carrying nest-building materials. (1) The hybrid inherits patterns for carrying strips several at a time in its feathers from the peach-faced lovebird, *A. roseicollis*. (2) From Fischer's lovebird, *A. fischeri,* it inherits patterns for carrying strips one at a time, in its bill. (3) When the hybrid first begins to build a nest it acts completely confused.

that the hybrid bird seems unable to let go of the strip while tucking, those strips that are tucked fall out, the wrong spot for tucking is used, the strip is grasped awkwardly making proper tucking impossible, tucking movements gradually merge into preening ones, inappropriate objects are tucked, and sometimes the bird attempts to get its bill near its rump by running backward. In fact, the hybrids are successful only in carrying material in their bills, and it takes 3 years before the hybrids perfect even this behavior. Then they still are less efficient than *A. fischeri* (Figure 5-6). This long learning period is in contrast to the extreme rapidity these hybrids exhibit in learning the various other trick behaviors listed above.

No data are given for F_2 or backcross generations, but Dilger considers that the data so far suggest that this behavior is controlled polygenically. Clearly, a

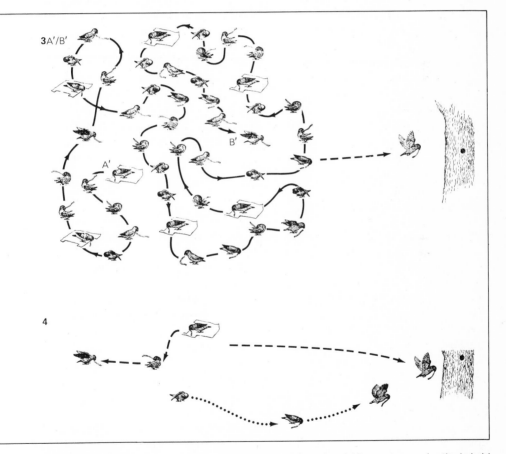

Lines from A to B and from A' to B' indicate the number of activities necessary for the hybrid to get two strips to the nest site, a feat achieved only when the strips are carried singly in the bill. (4) It takes three years for the hybrid bird to perfect bill-carrying behavior; even then, it makes unsuccessful efforts to tuck its nest materials in its feathers. (*From W. C. Dilger. The behavior of lovebirds. Copyright 1962 by Scientific American, Inc. All rights reserved.*)

detailed analysis needs to take into account the various components of behavior involved in nest building. The hybrids are intermediate for other behaviors too. "Switch sidling" is a common precopulatory display of males involving a male's initial lateral approach to his potential mate during which he moves first toward, then away from her, while reversing his own direction often for each approach. This composes about 32 percent of the precopulatory activities of *A. roseicollis* females mated with *A. roseicollis* males and 51 percent in the case of *A. fischeri* females and *A. fischeri* males, while the figure for hybrids mated with each other is intermediate (40 percent). However, when the hybrid males are mated to parental-type females, the situation is different: 33 percent for *A. roseicollis* females and F_1 hybrid males, and 50 percent for *A. fischeri* females and F_1 hybrid males. It seems that this is a situation where the female's response is very important in determining what sort of behavior is elicited from attentive males. Furthermore, the rest of the precopulatory displays of hybrid males shows the same general sort of pattern, but not as perfectly.

The nest-building and precopulatory behavior of lovebirds can therefore be interpreted at this point as being under polygenic control, although additional crosses are needed for more detailed genetic analyses. Learning is important for nest building, since hybrid behavior slowly changes over 3 years. For the precopulatory displays of males, it is not merely the male's behavior under study but also the reaction of males to differing females. The behavior of lovebirds therefore illustrates some of the unique complications that set behavior genetics aside from other subdivisions of genetics. This is surely an example worthy of more investigation.

5-5 EMOTIONALITY IN RODENTS

When confronted with a novel and unexpected situation, rats and mice may freeze, defecate, urinate, or simply explore their new environment. These behaviors, singly or in combination, are often used as measures of emotionality. Hall (1951) selected for high and low rates of urination and defecation in rats and produced two lines, which he called "emotional" and "nonemotional," according to their urination and defecation rates. Broadhurst (1960) carried out a selective breeding program over a number of generations in rats for high (which he referred to as reactive) and low (nonreactive) defecation scores. Defecation score was defined as the number of fecal boluses deposited in an arena in exactly 2 minutes. High and low lines were established that diverged rapidly (Figure 5-7a). The result is hardly surprising since various biometrical tests reveal heritabilities (see Section 2-2) between 0.5 and 1 for the trait. At the time of publication this work was recognized as being of considerable importance because it demonstrated more completely than earlier experiments the feasibility of applying biometrical methods to quantitative behavioral traits in experimental organisms. Although the analysis was not taken down to the chromosomal level, as has been done for some of the *Drosophila* examples, the

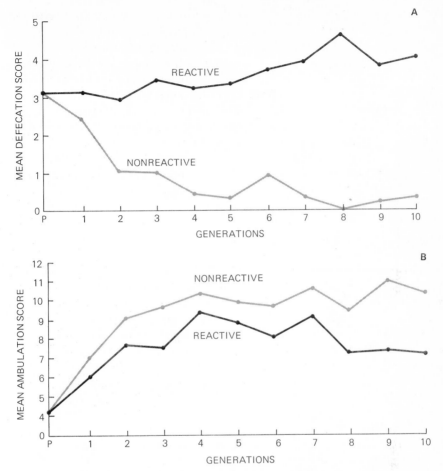

Figure 5-7 Emotionality in rats. (*a*) Mean defecation scores in rats selectively bred over 10 generations. (*b*) Mean ambulation scores. Correlated responses in the two lines selected for high and low defecation scores in (*a*). (*After Broadhurst, 1960.*)

basic assumption of biometrical genetics—a genetic architecture based upon a number of polygenes (multifactorial inheritance)—certainly is true here.

In a selection experiment it may be desirable to study any correlated responses to selection, as these may provide information of both behavioral and genetic importance. Broadhurst obtained simultaneous information on ambulation score, defined as the number of marked areas in the arena entered by a rat in exactly 2 minutes (Figure 5-7*b*). There is an increase in the score of both lines, but the increase in the nonreactive line is greater. Thus selection for defecation score has a marked effect on a trait not under direct selection. Two possibilities, those of pleiotropic effects and of linkage between polygenes affecting the two traits, are hypotheses worthy of consideration. In fact, one may conclude that a number of genes must affect both these behaviors: defecation and ambulation scores.

The same two behavioral traits have been studied extensively by others, particularly in mice by DeFries and Hegmann (1970), who applied sophisticated biometrical methods (Section 9-2). The conclusions can be regarded as similar to Broadhurst's work in that heritabilities were reasonably high, responses to selection were found, and a negative correlation between defecation and activity was found. A polygenic mode of inheritance explains the data best, just as is the case for the rat data above. It may in fact be concluded that many quantitative traits are under polygenic control, although some exceptions occur where behavioral traits can be associated with discrete loci. These mainly concern the pleiotropic effects of loci controlling coat color variations in mice and rats (see Chapter 9 for further details).

5-6 SOME BEHAVIORAL TRAITS IN DOGS

Scott and Fuller (1965) have published the results of a number of extended experiments on behavioral differences among breeds of dogs. Intrigued by great differences among the breeds as well as individual differences among specimens of the same breed, they set out to investigate the importance of heredity. In some instances they were able to come to some tentative conclusions about modes of inheritance. Their experimental design systematically varied the genetic constitution of the dogs while keeping all other factors as constant as possible. Five pure breeds were studied: wirehaired fox terrier, American cocker spaniel, African basenji, Shetland sheep dog, and beagle. Considerable breed differences were found for all behavioral traits studied. Perhaps the most interesting and extensive study was between the cocker spaniel and the basenji. The discussion in this section considers these two breeds and various derived crosses almost exclusively. It is based on the work of Scott and Fuller (1965).

Cocker spaniels have been selected in the past for nonaggressiveness and their ability to relate well to people. Basenjis on the other hand are highly aggressive, although not so much so as wirehaired terriers. In contrast with cocker spaniels, young basenjis reared under standard conditions are very fearful of human beings at 5 weeks of age, as shown by running away, yelping, snapping when cornered, and generally acting like wild wolf cubs. It is probable that in African jungle villages such wariness has considerable survival value. They were used in hunting by Pygmies and several other African tribes. The name *basenji* in the Lingala trade dialect of the Central Congo means "people of the bush"; the dogs were so called because they belonged to the bush people. They can be regarded as a general purpose hunting dog not fitting into any of the conventional divisions of the European breeds. Even so, under laboratory conditions, where they are observed to be timid at an early age, basenji puppies tame rapidly with handling and human contact. Another feature is that basenjis are relatively barkless dogs compared with other breeds. They bark only when

Table 5-5   Characteristics of Basenjis and Cocker Spaniels

Characteristic	Basenji	Cocker spaniel	Most likely mode of inheritance
Wildness and tameness			
Avoidance and vocalization in reaction to handling	High	Low	One dominant gene for wildness
Struggle against restraint	High	Low	One gene with no dominance
Playful aggressiveness at 13 to 15 weeks of age	High	Low	Two genes with no dominance
Barking at 11 weeks			
Threshold of stimulation	High	Low	Two dominant genes for low threshold
Tendency to bark a small number of times	High	Low	One gene with no dominance
Sexual behavior (time of estrus)	Annual	Semiannual	Basenji type as a recessive gene
Tendency to be quiet while weighed	Low	High	Two recessive genes for high tendency

Source: Scott and Fuller, 1965.

extremely excited and then soon terminate barking. At night in native villages they often produce a tremendous noise, referred to as yodeling or wailing.

This behavior was observed in Scott and Fuller's kennels. Figure 5-8 pictures cocker spaniel and basenji breeds and the hybrids between them. A summary of the characteristics of the two breeds appears in Table 5-5. The most likely modes of inheritance are based upon the following crosses:

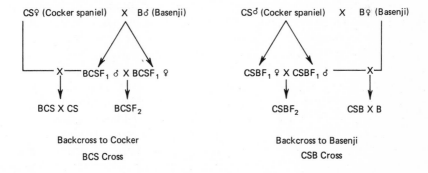

The two breeds were crossed reciprocally to give two F_1s from which were derived two F_2s. From the F_1 based on a cocker spaniel female parent, a backcross was made to the cocker spaniels, and for the F_1 based on a basenji female parent a backcross was made to the basenjis.

Figure 5-8 Hybrid dogs. *Upper left:* basenji male × cocker spaniel female. *Upper right:* their male and female offspring. *Lower left:* cocker spaniel male × basenji female. *Lower right:* their male and female offspring. Males are shown to the left of females. (From J. P. Scott and J. L. Fuller. *Dog Behavior: The Genetic Basis.* © 1965 by the University of Chicago. All rights reserved.)

The relative wildness of the basenjis is expressed in two behavioral characteristics. One is avoidance and vocalization in reaction to handling as young puppies; the other is the tendency to struggle against restraint, which is especially marked in leash training. For avoidance and vocalization in reaction to human handling, the handling test can be regarded as a mild to strong stimulation that may cause a young puppy to be fearful. A majority of cocker spaniels show no fearful behavior whereas all basenjis show some fearful behavior. Since the F_1 behavior is similar to basenji behavior, one or more dominant genes are likely to be implicated. The observed data, taking into account all crosses, fit a single dominant gene better than two dominant genes. Thus in basenjis wildness is controlled by a dominant gene. Its contrasting gene, tameness, is recessive in cocker spaniels. According to Scott and Fuller, the struggle against restraint during leash training can be explained by a one-gene difference showing no dominance. However, the situation is somewhat more complex, since there are large differences between the two possible F_1 populations—basenji female × cocker spaniel male compared with cocker spaniel female × basenji male—such that the hybrids tend to behave like their mothers, suggesting the possibility of maternal effects (see Section 6-6).

Playful aggressiveness was assessed by handling tests given at 13 to 15 weeks of age. The puppies characteristically rush toward the handler, leaping against him or up at his hands and nipping playfully. When patted, the puppy usually turns around and paws or wrestles with the hand presented, chewing on it gently. Scott and Fuller concluded that the mode of inheritance of playful aggressiveness could not be explained simply and that while two genes without dominance can explain the data, the possibility of a more complex type of inheritance cannot be excluded.

Barking capacities were assessed by a dominance test in which each pair of littermates was allowed to compete for a bone for 10 minutes. During this period, vocalizations were recorded including barking at 5, 11, and 15 weeks of age. Figure 5-9a shows that the maximum amount of barking occurs for all breeds tested at 11 weeks except for Shetland sheep dogs in which it is higher at 15 weeks. At 11 weeks cocker spaniels do most barking and basenjis least. The strain of basenji tested is not, however, completely barkless (Figure 5-9b). This offhandedly rather simple behavioral trait has two aspects as analyzed by Scott and Fuller: (1) the threshold of stimulation initiating barking, which is very high in the basenji and very low in the cocker spaniel; and (2) the tendency to bark only a few times (basenji) rather than to become excited and bark continuously (cocker spaniel). (The maximum number of barks recorded for a cocker spaniel in a 10-minute period was 907, or more than 90 per minute.)

For threshold of stimulation, the F_1 is similar to the cocker spaniel, indicating dominant inheritance for a low threshold. Assuming inheritance through a single dominant gene fits the data quite well, but somewhat better if two independent dominant genes are assumed as a theoretical expectation (Table 5-5). This does not rule out the possibility of a large number of genes, but a more complicated experiment incorporating a large number of marker genes

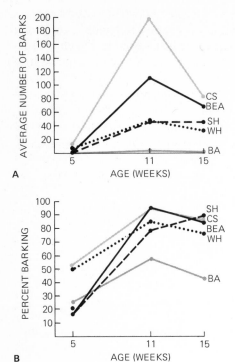

Figure 5-9 Barking differences among canine breeds. Occurrence of barking during dominance tests at different ages. (*a*) Average number of barks. (*b*) Percentage of animals barking. BA, basenji; BEA, beagle; CS, cocker spaniel; SH, Shetland sheep dog; WH, wolfhound. (*After Scott and Fuller, 1965.*)

would be needed for tests. For this trait, since there is no great difference between puppies born of basenji mothers and those born of cockers, the possibility of sex-linked inheritance or learning by example from the mother is not important. In the case of the tendency to bark in excess, the F_1 animals are intermediate between the two parental strains, and the F_2 dogs are very much like the F_1 animals. The data can be explained by a single gene without dominance. That two genetic mechanisms can be involved in barklessness is clear. A dog will not bark to excess if it does not bark at all, so the presence of one trait is conditional upon the presence of the other. An important principle emerges—that a greater understanding of the genetic architecture of quantitative traits may come from subdividing them into components prior to genetic analysis—and we recommend this.

Basenji females come into estrus annually close to the time of the autumnal equinox, while most domestic breeds have estrous cycles at any season of the year and about 6 months apart. It was concluded that the basenji type of estrous cycle is inherited as a single-factor recessive, but the possibility of a somewhat more complex situation cannot be definitely excluded.

For the inheritance of the tendency to be quiet while being weighed at 14 to 16 weeks of age, results compatible with two genes were obtained, the cocker spaniel tendency being recessive to the basenji.

Table 5-5 shows that the inheritance of these behavioral traits can be explained in many cases on the basis of one or two genes, although there are indications of greater complexity. In other words, the situation is intermediate between the simple mendelian inheritance patterns described in previous chapters and polygenic inheritance discussed earlier in this chapter. Scott and Fuller regard the result as rather surprising, arguing that a trait as complex as behavior might be expected to be affected by many genes, but evidence so far indicates this to be not necessarily so. The two dog breeds concerned, having been isolated from each other for so long, no doubt were exposed to highly different selective pressures in quite different habitats, leading to the behaviors observed. If this led to genotypes homozygous for the traits discussed (and with little or no segregation occurring within breeds for the traits), then assuming only one or two major gene loci involved for each trait, the results obtained are reasonable. It should be noted, too, that other behavioral traits in dogs—generally of a more complex nature than those discussed—are difficult to interpret on a simple mendelian basis because of the complex interactions with the environment that are frequent for behavioral traits.

Even in an environment standardized as comprehensively as possible, a large portion of variance may be ascribed to nongenetic sources. If the environment is largely uncontrolled, only the effects of major genes will be revealed via experimentally manipulated selection.

We applaud Scott and Fuller and their many students and collaborators. They are to be congratulated for their long-term efforts expended in watching behaviors of interest to them develop and differentiate in a complex animal, a distinctly fruitful method in approaches to the genetics of behavior akin to developmental genetics. More recently Scott, Stewart, and DeGhett (1973) and Corson et al. (1979) have considered depression and hyperkinesis and appropriate therapies in dog breeds (see Section 11-9), especially in Telomian × beagle hybrids. The parents of these two, for but a single example, respond quite differently behaviorally to the drug amphetamine (a central nervous system stimulant that produces mood elevation; see Section 9-7) and their hybrids' responses combine parental ones.

In Chapters 6 and 7 and later in the book, the traits exclusively controlled by many genes are more extensively discussed. The material in this chapter provides a descriptive bridge between traits clearly controlled by major genes and those controlled polygenically; the examples selected for this chapter cover this spectrum.

5-7 LATERALITY IN *DROSOPHILA*, MICE, AND HUMAN BEINGS

We shift now to an instance where the very existence of the genetics is unclear. Asymmetrical lateral biases have been found for both morphological and behavioral traits. An obvious morphological example (with behavioral consequences) is the asymmetry of the pair of dimorphic claws of the lobster *Homarus americanus*. Behavioral asymmetries (pawedness, clawedness, handedness) are known in various birds and mammals including human beings. Human handedness may be associated with asymmetrical brain development (reviews in Dimond and Blizard, 1977; Harnad et al., 1977). Questions on the possibility of genetic components of laterality have been mainly asked in *Drosophila,* mice, and human beings.

Drosophila

Section 5-1 showed that, provided a trait has some genetic basis, there should be a response to selection simply because the selection of extreme phenotypes implies the selection of extreme genotypes. Ehrman et al. (1979) chose two behaviors as selection targets: wing folding and maze direction choice. *Drosophila* characteristically fold one wing over the other when at rest; an individual will usually be quite consistent in folding the left wing over the right or vice versa. If there is a genetic component to this behavior, then selecting left-over-right males and females to form a selection line and similarly selecting right-over-left males and females should give diverging responses to selection in the two lines. In other words left-over-right males and females mated together should produce more left-over-right offspring than if mating occurs at random. The same expectation applies for matings among right-over-left individuals. Maze direction choice was followed in a Y-maze, so that flies could make left or right turns. If there is a genetic component in maze direction choice, then responses to selection would also be expected.

The results in Figure 5-10 are clear. Selection for increased expression of wing folding and maze direction choice did not produce a positive response over nine generations, even though some individuals show predominantly left or right behaviors. This suggests that the genetic component for laterality for behavioral traits in *Drosophila,* if it exists, must be slight.

We do not feel that a genetic component can be entirely eliminated, since in selection experiments for asymmetry of sternopleural bristle number, Beardmore (1965) obtained some slight divergence between lines selected for an excess of bristles on the right- and left-hand sides respectively. By comparison with the experiments discussed earlier in this chapter describing rapid responses to selection, such was not the situation for this quantitative morphological asymmetry measure where only slight divergence occurred after many generations of selection.

Mice

The paw preference of hungry mice can be readily determined by requiring them to reach into a narrow tube to withdraw pieces of food. In this test situation, most mice show definite right or left paw preferences, such that in a wide variety of inbred strains and hybrids, the probability of right or left preferences is about 0.50. For example, inbred strains such as DBA/2J and C57BL/6J that differ in many behavioral traits (Chapter 9) are almost identical for paw preferences (Figure 5-11). Results such as these suggest that a significant genetic component in paw preference can be ruled out (Collins, 1968).

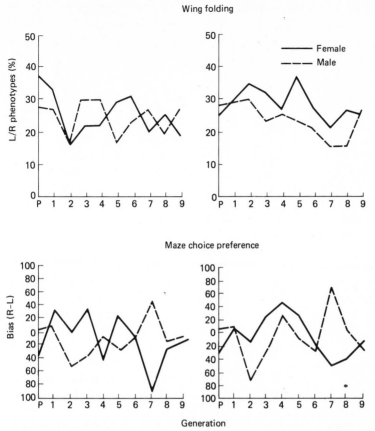

Figure 5-10   Responses to selection for increased left or right preference in *D. melanogaster*. Solid line = female responses; dashed line = male responses. The upper two graphs clearly indicate that no increase in wing-folding behavior was obtained over nine generations of selection. The lower two graphs, displaying maze direction choice behavior indicate an increase in lateral preference for left turn preference females and right turn preference males, both peaking in the seventh generation. The apparent increase is entirely spurious, however, for the preference is completely eliminated by the ninth generation. (*From Ehrman, Thompson, Perelle, and Hisey, 1978.*)

    Collins conducted his experiments in an unbiased world (U-world), wherein the feeding tube in the test apparatus was located equidistant from the right and left sides of the cubicle. The apparatus was modified to form "right-biased" and "left-biased" worlds by making the right and left feeding tubes more or less easily accessible according to the world being considered. Most mice then exhibited paw preferences consistent in direction with the world bias, although about 10 percent exhibited an opposite laterality. Further testing of mice in worlds now biased the reverse compared to that of the first tests, indicated that the pawedness of mice was not likely to have been learned in the

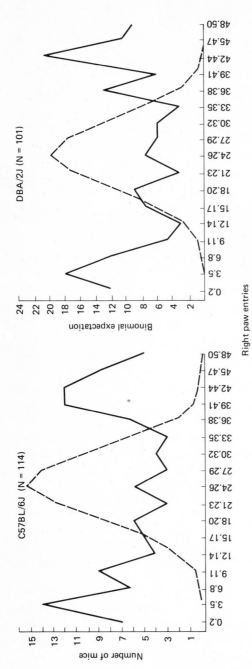

Figure 5-11 Frequency distribution (solid lines) of lateral preference scores for males of two inbred strains of mice. Preference was measured by the number of times the right paw was used to withdraw food from a narrow tube. The dashed lines represent the expected binomial frequencies if individual preferences did not exist. (*From Collins, 1968.*)

process of testing. Rather, mice arrived for testing with an already established sense for differentiated pawedness (Collins, 1977a).

In unbiased worlds, female mice are more strongly lateralized than are males, although the proportions of right- and left-pawedness between the sexes were similar. However, when naive mice were tested in biased worlds, male mice became more strongly lateralized than did female mice. Does this suggest a possible sex-chromosomal effect influencing the strength or weakness of lateralization? Even if this is true, there remains no evidence for a significant genetic component for asymmetry as measured by the alternatives of right versus left choices.

Human Beings

Human beings are predominantly right-handed as defined in terms of skills and preferences. The bias to the right appears to occur in all present-day cultures as well as in ancient extinct ones.

Section 2-4 indicated that a possible test for genetic control is a comparison of monozygotic and dizygotic twins (see also Chapter 7), in that monozygotic twin pairs are expected to be more alike than are dizygotic twin pairs. The complicating effect of the differing environments of monozygotic and dizygotic twins was also noted.

Collins summarized data for handedness by tabulating the numbers of R-R, R-L, and L-L monozygotic and dizygotic twin pairs (Table 5-6). The expected numbers are those based on chance where p = number of right-handed individuals and L = number of left-handed individuals. In both cases, the distributions are very close to those expected from the binomial expectations of p^2 R-R : $2pq$ R-L : q^2 L-L. These results are exactly what would be expected by taking unrelated individuals from one population. A genetic component for handedness would be expressed as a deviation from binomial expectations in twins where there is a deficiency of R-L pairs. In addition, the deviation of monozygotic twins from this binomial should be greater than that calculated for dizygotic twins if a genetic component existed, but this is also not so. The twin data are therefore incompatible with expectations based on mendelian inheritance. Twins are no more similar in hand preferences than one would expect on the basis of chance. Collins (1977a) commented during a discussion of twin data: ''I urge that these data be faced squarely. It would be unwise to continue to

Table 5-6 Summed Data for Distributions of Right- and Left-Handedness in Human Monozygous and Dizygous Twin Pairs

	Number of pairs			Binomial expectation		
	RR	RL	LL	RR	RL	LL
Monozygous	782	244	37	771.70	264.60	26.70
Dizygous	812	224	18	811.66	224.69	17.66

Source: Collins, 1977a.

develop genetic models of handedness applicable to singletons only.'' He is of course referring to the array of laterality models put forward in the literature that include (1) single gene models, (2) two pairs of genes, (3) a nongenetic model, and (4) a modification of the nongenetic model wherein right-handedness is argued to be the basic human character with sinistrality (left-handedness) arising from various environmental causes (Annett, 1978, also see Levy, 1977). In the nongenetic model, laterality could arise from cultural inheritance or more likely from some unknown extrachromosomal predisposition to asymmetry. Since laterality is manifested so early, both at behavioral and anatomical levels, the latter possibility appears likely.

Finally, if handedness itself does not have a genetic component, we should ask, as was done in mice, whether there is a genetic component determining degrees of lateralization. As in mice a sex difference occurs, whereby adult males show a greater average degree of lateralization than do adult females in terms of verbal, visuospatial, and overall lateralization (Collins, 1977b). Does this mean that human males possess inherently greater cerebral specialization, or do males possess a weaker, more diffuse cerebral organization that leads to a greater lability to adapt to persistent environmental biases?

In conclusion, the problem of whether there is a genetic component for laterality is far from solved, even in organisms such as *Drosophila* and mice where sophisticated behavior genetic experiments can be performed. For *Homo sapiens,* approaches as comprehensive as some of the studies on mental illness and intelligence (Chapters 11 and 12) appear necessary. We believe that if phylogenetic behavior geneticists can demonstrate a genetics of laterality, such genetics should apply to human beings. But this remains to be accomplished. Yet laterality is exceptional in that the demonstration of a genetic component for all the other traits discussed in this chapter is simple and without controversy. Even in the case of IQ, as will be seen especially in Chapter 12, the controversy is on the relative influence of genotype and environment (and interactions between them), not on whether or not a genetic component exists!

SUMMARY

Most behavioral traits are under more complex genetic control than can be followed easily from segregation data. The methods and techniques of quantitative genetics, where the aim is to separate the total variability of a trait into genetic and environmental components, are important. Essentially we are moving towards a consideration of traits where the genetic control is polygenic. It is possible with specialized breeding techniques, to locate the genes controlling quantitative traits to regions of chromosomes and ultimately genetic loci, as shown by studies on geotaxis in *D. melanogaster,* and sexual isolation levels in *D. paulistorum.* This latter observation demonstrates that levels of reproductive isolation are under genetic control, as is also shown by studies on two species of lovebirds and their hybrids.

An enormous diversity of behavioral traits have been shown to be under polygenic control. Data have been collected on *Drosophila,* rodents, dogs, and human beings. Many of these findings were established in the 1950s and 1960s, when prevailing thought discounted heritable components for behavioral traits.

However, unlike other quantitative traits, laterality (handedness) has not been shown to be under genetic control. By comparison, the demonstration of a genetic component for many other quantitative traits appears simple and without controversy.

6

Quantitative Analysis: Experimental Animals

6-1 QUANTITATIVE GENETICS

Quantitative genetics is the study of variation of those traits where the genes responsible for the observed variation cannot normally be recognized individually. Some examples were presented in Chapter 5, and some basic principles were given in Section 2-2. The aim of quantitative genetics is to subdivide the phenotypic value measured into component parts—the genotypic and environmental components. From this point of view, behavioral traits differ in no way from the more conventionally studied quantitative morphological traits in animals. Complete accounts appear in various texts on quantitative genetics, for example, in Falconer (1960) and Mather and Jinks (1977). These authors employ different mathematical symbolism; in this discussion we use mainly that credited to Falconer.

The level of variability for many behavioral traits within and between different environments is greater than for many morphological traits. For this reason, more attention must be paid to controlling the environment in which behavioral traits are studied than is necessary in the analysis of morphological traits. In many cases the effect of the environment itself is of direct interest. In addition, learning and reasoning must be taken into account—one of the features that differentiates behavior genetics from other subdivisions of genetics.

The possibility of previous experience makes it essential to standardize experimental conditions extremely carefully; otherwise it may be difficult to obtain accurate genetic interpretations. In *Homo sapiens,* previous experience cannot be controlled as in experimental animals because of the impossibility of defining environments or of carrying out breeding experiments. Therefore, assessing the meaning of behavioral data in people is often very difficult. This is a major reason for the hotly debated issue of racial differences in intelligence (see Chapter 12). In many ways then, *Homo sapiens* can be regarded as a special case. We consider experimental animals in this chapter and, armed with this knowledge, discuss human beings in Chapter 7. Sections 6-8 to 6-10 in this chapter are, however, directly applicable to people.

6-2　INTERACTION OF GENOTYPE AND ENVIRONMENT (GE INTERACTIONS)

The simplest models of quantitative genetics assume that the joint effects of genotype and environment are additive. Under this assumption, if one genotype has a greater value for a trait than another genotype in a given environment, it has a greater value in all environments. This is an assumption of most of the simple theoretical quantitative inheritance models; it is not necessarily valid in practice. Haldane (1946) discusses the relations that may exist between two genotypes, A and B, measured for a quantitative trait in two environments, X and Y. The four corresponding values of the trait are numbered from 1 to 4 in order of magnitude. There are $4! = 4 \times 3 \times 2 = 24$ ways of arranging four items in a sequence. But if, arbitrarily, AX (genotype A, environment X) is designated as the largest measurement, there are only six distinguishable logical arrangements, as illustrated in Table 6-1. From this we see

- Arrangements 1a and 1b. $A > B$ in both environments. In 1a both values of $A >$ the highest of B, that is, A is always $> B$. In 1b, $X > Y$, although within them A and B have the same relative ranking.
- Arrangement 2. $A > B$ in X, but $A < B$ in Y although $X > Y$. Haldane points out a possible example of this, where domesticated species (A) and wild (B) are in synthetic (X) and natural (Y) habitats. Both species fare better in synthetic habitats that offer protection than in the natural habitat, although wild types do relatively better in natural habitats than do domestic species.
- Arrangement 3. $A > B$ in X and Y, but $AX > AY$ and $BX < BY$. The environments X and Y have opposite effects on the two types of individuals. Haldane gives the example of normal (A) and mentally retarded (B) individuals in normal (X) and special (Y) schools.
- Arrangements 4a and 4b. The environments again have opposite effects on the two types of individuals, as in 3, but represent specialization such that A and B are each optimally adapted to their respective environments, X and Y. This represents the situation of evolutionary adaptation of individuals to their respective environments referred to as *habitat selection*. The temperature preferences of races of *Peromyscus maniculatus* can be cited as a behavioral example. Races *P. maniculatus bairdii* (prairie mice) and *P. maniculatus gracilis* (deer

Table 6-1   Relations of Measurements of a Quantitative Trait in Two Genotypes (*A* and *B*) in Two Environments (*X* and *Y*)*

Arrangement	Genotype	Environment		
		X	*Y*	
1a	*A*	1	2 ⎫	
	B	3	4 ⎬	*A* > *B* in *X* and *Y*
1b	*A*	1	3 ⎭	*X* > *Y* for *A* and *B*
	B	2	4	
2	*A*	1	4 ⎫	*A* > *B* in *X*, *B* > *A* in *Y*;
	B	2	3 ⎭	*X* > *Y*
3	*A*	1	2 ⎫	*A* > *B* in *X* and *Y*, but
	B	4	3 ⎭	*BX* < *BY* and *AX* > *AY*
4a	*A*	1	3 ⎫	
	B	4	2 ⎬	*A* > *B* in *X*; *B* > *A* in *Y*
4b	*A*	1	4 ⎭	
	B	3	2	

* Measurements are numbered 1 to 4 in order of magnitude. *AX* is always assumed to be the largest in each of the four types.

Source: Haldane, 1946.

mice) each display a preference for the artificial habitat most closely resembling its natural one (Harris, 1952). Furthermore, descendants of these mice raised in the laboratory also select similar environments, suggesting a genotypic role in the selection. Ogilvie and Stinson (1966) found the thermotactic optima of *P. maniculatus bairdii* and *P. maniculatus gracilis* to be 25.8°C and 29.1°C, respectively, showing that *P. maniculatus gracilis* has been selected for its warm, wooded environment and *P. maniculatus bairdii* for its cooler, field environment.

To finish with laboratory examples where environments can be defined, we first consider some data of Henderson (1970) on the effect of early experience. The mean number of minutes required to reach food was assessed in six inbred strains of mice reared in both standard and enriched environments (Figure 6-1). Some of Haldane's interactions emerge here if we consider the strains in pairs. Look at the data taking the strains in pairs and try to see the types of interactions they represent.

Another type of genotype-environment (GE) interaction, little investigated behaviorally, is the frequent occurrence of a high level of heterokaryotype (and heterozygote) advantage in extreme environments (often extreme temperatures) compared with optimal environments. A good example comes from *Drosophila pseudoobscura*. Large fitness differentials are found between heterokaryotypes and homokaryotypes at the extreme temperature of 25°C leading to heterokaryotype advantage and a stable polymorphism, but at 16.5°C the fitness differentials disappear and at 22°C an intermediate situation occurs (Wright and Dobzhansky, 1946; Van Valen, Levine, and Beardmore, 1962). Parsons and Kaul (1966) found substantial GE interactions between 20°C and 25°C for certain karyotypes of *D. pseudoobscura* for mating speed, such that there was more change between temperatures for homokaryotypes than for

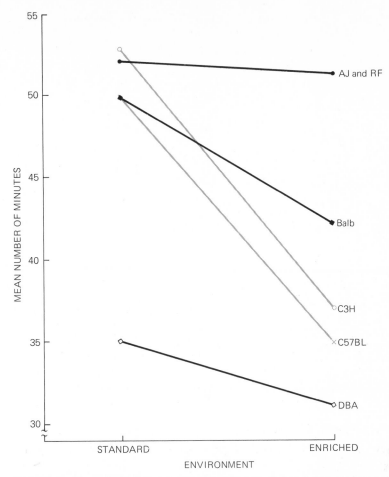

Figure 6-1 Time required to reach food for six inbred strains of mice reared in standard and enriched environments. Two strains, AJ and RF, achieved identical means. (*From Erlenmeyer-Kimling, 1972.*)

heterokaryotypes, leading to heterokaryotype advantage at the extreme temperature. We can say, because of this GE interaction, that the heterokaryotypes display more *behavioral homeostasis* than do the homokaryotypes. Analogous results have been found for many fitness factors in many species (Parsons, 1971). For example, traits such as larval survival and viability show more homeostasis in heterozygotes than in homozygotes in *D. melanogaster* and *D. pseudoobscura* (for references, see Parsons, 1971, 1973).

Our final example of GE interaction comes from McKenzie (1978), who studied the number of matings occurring within 30 minutes at 25°C for the sibling species *D. melanogaster* and *D. simulans* raised at temperatures ranging between 12°C and 30°C. McKenzie used ten strains of each species derived from single inseminated females collected from wild populations; in Figure 6-2

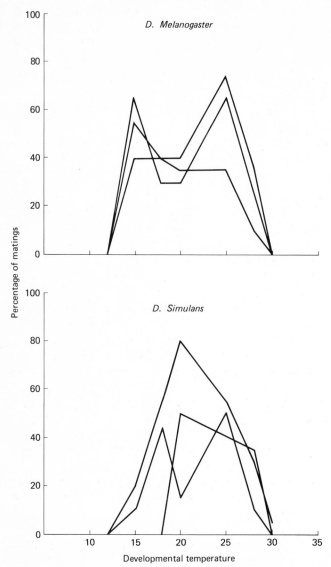

Figure 6-2 The effect of development temperature on percentage of matings in three representative isofemale strains of the sibling species *D. melanogaster* and *D. simulans*. (*From McKenzie, 1978.*)

three representative strains are plotted for each species. *D. melanogaster* maintains a more consistent mating propensity over a greater range of temperatures than does *D. simulans,* a point that will be more generally discussed in Section 13-2. Inspection of Figure 6-2 shows that the ordering of the strains is not the same at every temperature. This suggests GE interactions in both species. The total variability in data of this type can be analyzed with a statistical procedure called *analysis of variance* (see Appendix 6-1). This procedure permits the total variability in a set of data to be attributed to specific causes and assessment of their significances. In this example, we can analyze the total variability into the effects of temperatures, strains (i.e., genotypes), and interactions between temperatures and strains. It is obvious that there is a highly significant effect due to temperature since matings are minimal for flies bred at the extreme temperatures and maximal when they are bred at intermediate temperatures. In addition, there are significant strain-specific effects and interactions between strains and temperatures. All three sources of variability are significant at $p < 0.001$. This example of GE interaction is of great interest since it relates directly back to variability in nature for an environmental variable of great importance to *Drosophila*.

The basic models of quantitative inheritance assume no GE interactions. Quantitative genetic theory becomes extremely complex if this assumption cannot be made. It should be clear by now that a major deficiency in quantitative genetic theory, as far as the behavior geneticist is concerned, lies in the frequent assumption of no GE interaction.

As outlined in Section 2-2, the phenotypic value P of an individual consists of two parts, a genotypic value G determined by the individual's genetic constitution and an environmental deviation E, which may be positive or negative, so that

$$P = G + E$$

A fundamental feature of this formulation is that G and E are not associated. The other parameter describing populations is variance; the phenotypic variance, assuming no GE interaction, is

$$V_P = V_G + V_E$$

where V_G and V_E represent the genotypic and environmental variances respectively. The presence of GE interactions will increase total phenotypic variance, a result that is not surprising since these interactions increase the level of variability observed.

It is of great importance to distinguish GE interactions as discussed from another situation often confused with GE interactions. This is where there is a *correlation between genotype and environment*. If there is a positive correlation between genotype and environment, the genotypic variance is overestimated; if the correlation is negative, it is underestimated. For example, in human beings it has been suggested that in a favorable environment genetic effects might be expressed more fully than otherwise. If this is so, then it would be reflected as a positive correlation between genotype and environment. It will be shown, especially in Chapters 7 and 12 where human data are considered, that positive correlations between genotype and environment frequently occur. An agricultural situation giving such correlations is where the better animals are given more food. In natural situations it would occur if animals sought out the microhabitat for which they are best suited. A detailed consideration of such habitat selection is presented in Chapter 13. It is important to appreciate the subtle differences between GE interactions and GE correlations. In the main, such complications will be ignored in this chapter, where we look at some of the simpler models of quantitative inheritance. This is because in experimental animals data can be collected using experimental designs planned to minimize these complications. As will be seen, however, in the next chapter when we turn to *Homo sapiens,* it is impossible to ignore these issues—simply because in people the situation consists of analyzing data as presented directly from human populations. The same problem occurs for the study of animals directly in their natural habitats (Chapter 13).

6-3   VARIATION WITHIN AND BETWEEN INBRED STRAINS

Inbred strains maintained by brother-sister (sib mating) and other mating systems have been developed in laboratory species such as *D. melanogaster,* mice, and rats. Sib mating, being a form of inbreeding, leads to a progressive increase in homozygosity in each generation. Figure 6-3 gives the percentage of

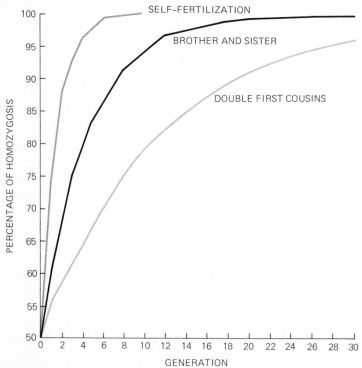

Figure 6-3   Homozygosis in successive generations under three different systems of inbreeding. (*From Fuller and Thompson, 1960.*)

homozygosis expected under three different systems of inbreeding: self-fertilization, sib mating, and double first-cousin mating. For a system such as sib mating, the expected rate of increase in homozygosity is high, and in fact the proportion of heterozygotes per generation is expected to fall by 19.1 percent of those in the previous generation, compared with 8 percent for double first-cousin mating. For self-fertilization the equivalent figure is 50 percent, that is, the proportion of heterozygotes falls by one-half every generation.

The rate of inbreeding, or the degree to which an individual is inbred, can be measured by the *inbreeding coefficient*, F (Falconer, 1960). Its detailed method of calculation need not concern us here; it expresses the probability that the two alleles at a locus in an individual are derived from a common ancestral allele. This means that the more remote the common ancestor, the smaller the value of F. F is calculated for any specific individual by tracing his lines of descent to the common ancestor of his parents. If we designate the number of steps from the inbred individual through each of his parents to the common ancestor as n_1 and n_2, respectively, his inbreeding coefficient can be expressed most simply as

$$F = (\tfrac{1}{2})^{n_1 + n_2}$$

Thus, examining the pedigrees in Figure 6-4 for the progeny of first cousins $n_1 = n_2 = 2$ giving $F = \frac{1}{16}$, and for the progeny of second cousins $n_1 = n_2 = 3$ giving $F = \frac{1}{64}$. For aunt-nephews or uncle-nieces the pedigree is asymmetrical with $n_1 = 1$, $n_2 = 2$ giving $F = \frac{1}{8}$, and for first cousins once removed $n_1 = 2$, $n_2 = 3$ giving $F = \frac{1}{32}$. In a large random-breeding population $F = 0$.

Whether inbred strains ever become completely homozygous (*isogenic*) is debatable because the approach to homozygosity may be retarded if the heterozygotes are fitter than the corresponding homozygotes. However, the level of homozygosity is normally very high after many generations of inbreeding. Assuming that complete homozygosity is attained, all individuals within an inbred strain are genetically identical. This means that all variation within an inbred strain is environmental. Between strains, however, there are variations due to the different genetic constitutions of the strains as well as to the environment. Even if several inbred strains are set up from the same population, the genetic constitution of the strains is expected to differ, since by chance different loci are likely to be made homozygous in the different strains.

Table 6-2 gives some data for a behavioral trait in six inbred strains of *D. melanogaster*. The trait is the number of times out of 10 observations at 6-second intervals that each fly runs along (i.e., does not stand still) an observation tube. There are in total six sets of 10 observations per strain. To what extent is the variability in these data made up of variation within strains and variation between strains? Since the strains are inbred and are assumed homozygous, the variation within strains must be regarded as environmental. The variation between strains contains a genotypic component as well as the environmental

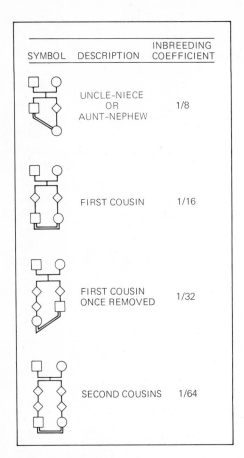

SYMBOL	DESCRIPTION	INBREEDING COEFFICIENT
	UNCLE–NIECE OR AUNT–NEPHEW	1/8
	FIRST COUSIN	1/16
	FIRST COUSIN ONCE REMOVED	1/32
	SECOND COUSINS	1/64

Figure 6-4 Inbreeding coefficients under different breeding systems.

component. To assess the variation within and between strains, an analysis of variance (see Section 6-2) must be carried out. An analysis of variance enables the total variability in a set of data to be attributed to specific causes. The detailed treatment of the data in Table 6-2 is given in Appendix 6-1. From this

Table 6-2   Number of Times Out of 10 Observations at 6-second Intervals that Individuals of Six Inbred Strains of *D. melanogaster* Ran along Observation Tube

	Oregon	Samarkand	Florida	6C/L	Edinburgh	Wellington
	4	3	7	8	7	5
	6	1	5	10	4	7
	8	1	6	6	7	9
	6	3	6	10	7	6
	7	3	6	9	9	8
	5	5	6	8	6	9
Total	36	16	36	51	40	44

Six flies were examined per strain. These 36 flies represent one testing session.
Source: Hay, 1972.

we see that the genotypic variance $V_G = 3.53$ and the environmental variance $V_E = 2.05$.

It is reasonable to compute the proportion of the phenotypic variance that is genotypic in this manner

$$\frac{V_G}{V_G + V_E} = \frac{V_G}{V_P} = 0.63 = h_B^2$$

This is referred to as the *heritability in the broad sense, h_B^2,* or the *degree of genetic determination* of the trait. The value of h_B^2 must range between 0 and 1. If it were 0, $V_G = 0$ and the trait would be determined entirely environmentally. If it were 1, the trait would be determined entirely genetically. The figure of 0.63 above indicates a relatively high value for h_B^2 and is in the range likely for quantitative traits, whether morphological or behavioral.

It must be stressed that the heritability in the broad sense so estimated is a characteristic of the actual inbred strains under the environmental conditions prevailing. If the experiment were done under different conditions, or with different strains, or both, different values might be obtained. This should be clear from the discussion on GE interactions in the previous section. To make inferences about the components of variance and the heritability of a trait in a given population of an outbred species, inbred strains at random from the population must be set up. Then, in theory, the estimates obtained will relate to the parameters of the parent population rather than to the strains in the sample. Unfortunately this condition may not often be directly fulfilled, but to all intents and purposes a set of inbred strains should permit more or less realistic estimates to be made. However, in all cases, but especially for behavioral traits, the environment must be defined as accurately as possible and should be specified together with any estimates made.

6-4 COMPONENTS OF GENOTYPIC VARIANCE

Let us now look at the genotype itself. If we consider two alleles at a locus A_1 and A_2, there are three possible genotypes—A_1A_1, A_1A_2, and A_2A_2—two of which are homozygotes and one a heterozygote. If the average measurement (*genotypic value*) of the heterozygote A_1A_2 is the mean of the two homozygotes or $(A_2A_2 + A_1A_1)/2$, we can say there is no dominance. Alternatively we could write the mean of the values of the homozygotes as zero, to represent the situation when there is no dominance (Figure 6-5). This figure shows the three genotypes on a linear scale with the origin (value 0) at the midpoint between the two homozygotes. The homozygotes are given values $-a$ and $+a$, and the heterozygote a positive or negative value d on either side of the origin, depending on the size and magnitude of the heterozygote effect. It should be stressed that the averages are taken within defined environments.

We wish now to look at the expected contribution of this locus to the genotypic variance of the F_2 derived from a cross between two homozygous

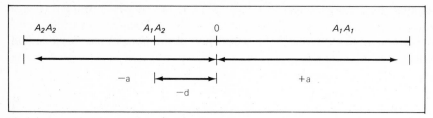

Figure 6-5 Genotypic variance. Values for three genotypes A_1A_1, A_1A_2, and A_2A_2 arranged on a linear scale. Origin is at the midpoint between the two homozygotes. The heterozygote is on either side of the origin depending on the sign and magnitude of the heterozygote effect (d).

parental strains P_1 and P_2 carrying different alleles at this locus, that is, the parents are A_1A_1 and A_2A_2. Since the F_1 must be A_1A_2, the frequencies of the three genotypes in the F_2 are $\frac{1}{4} A_1A_1$, $\frac{1}{2} A_1A_2$, and $\frac{1}{4} A_2A_2$. The mean measurement of the F_2 is

$$\Sigma p_i x_i = \frac{1}{4}a - \frac{1}{2}d - \frac{1}{4}a = -\frac{1}{2}d$$

where p_i is the frequency of each class and x_i is its phenotypic value. The contribution of the locus to the variance of the F_2 is then

$$\Sigma p_i x_i^2 - (\Sigma p_i x_i)^2 = \frac{1}{4}a^2 + \frac{1}{2}d^2 + \frac{1}{4}a^2 - (-\frac{1}{2}d)^2 = \frac{1}{2}a^2 + \frac{1}{4}d^2$$

If, as in a cross between two inbred strains, there are many such loci contributing independently to the genotypic variance of the F_2, this expression can be written

$$\frac{1}{2}\Sigma a^2 + \frac{1}{4}\Sigma d^2 = V_A + V_D$$

where summation is over the various loci. The term $V_A = \frac{1}{2}\Sigma a^2$, being a function only of the difference between homozygotes, is the *additive genetic variance*. For $d = 0$, when there is no dominance, the effect of the genes A_1 and A_2 is said to be additive. The quantity d is often referred to as the *dominance deviation,* and $V_D = \frac{1}{4}\Sigma d^2$ is the *dominance variance.* Therefore, where there are deviations of the heterozygote from the mean of the two homozygotes, a dominance variance component occurs. The total phenotypic variance of the $F_2(V_{F_2})$ can therefore be written

$$V_{F_2} = V_A + V_D + V_E$$

This same model permits the assessment of the contribution of such a set of loci to the variance components of other crosses, such as the various backcrosses to the parents and the F_3 generation. If data on enough crosses are available, the quantities V_A, V_D, and V_E can be estimated. This provides a basic description of the genotypic and environmental components of variation with respect to a quantitative trait in a given population. The reader is referred to a standard text such as Mather and Jinks (1977) for these further elaborations.

Before proceeding further, mention must be made of one of the intractable problems of quantitative genetics: this concerns the *scale* on which a trait is measured. If a normal distribution cannot be assumed, it may be possible to find a suitable algebraic transformation that converts the data to an approximate normal distribution. The problem of scaling has not been satisfactorily resolved, although Mather (1949) has set up some scaling criteria based on relations between certain generations. [The issue is too complex for detailed considerations in a text of this nature, so the reader is referred to various specialized texts (e.g., Falconer, 1960; Mather and Jinks, 1977).] In theory, therefore, before any calculations on the relative importance of genotype and environment are carried out, the adequacy of the scale must be tested; if it is found to be inadequate, a search should be made for a more suitable scale. The process of searching for a suitable scale is rather arbitrary. In some cases a suitable scale cannot be found, so that genetic interpretations become difficult.

An example of a scaling criterion is that the variances should be independent of the mean in nonsegregating generations. For behavioral traits, which are often very sensitive to the environment, this criterion may be more difficult to satisfy than for less environment-sensitive characteristics such as morphological traits, but more experimentation is needed. Occasionally a suitable transformation is obvious. For example, if the variance is proportional to the mean in the P_1, P_2, (P = parental) and F_1 generations, taking logarithms provides a suitable transformation.

Broadhurst and Jinks (1961) summarize various behavioral experiments applying biometrical methods. As an example, we can take the data of Dawson (1932), who tested the inheritance of "wildness" in mice. He defined this in terms of the speed animals showed in running down a straight runway. A movable partition was used by the experimenter to prevent test subjects from moving backward. Unfortunately, this introduces a subjective element, but even so the experiment can be used to illustrate genetic points. Two strains were used, wild and tame, which although not highly inbred were very different, so it is reasonable to assess how they differ if we assume that they are likely to be homozygous for genes for wildness and tameness. The means and standard errors used in the analysis are given in Table 6-3. A transformation of the data was found to be necessary, and the best fit was given by the logarithmic transformation. Variances estimated after taking logarithms are

$$V_A = 0.026 \pm 0.012$$
$$V_D = 0.002 \pm 0.008$$
$$V_E = 0.020 \pm 0.005$$

These variance components show a common feature of many but not all traits: the additive genetic variance (V_A) is substantially greater than the dominance variance (V_D).

The ratio of the additive genetic variance to the phenotypic variance (V_A/V_P) can now be computed and is referred to as the *heritability in the narrow sense* (h_N^2) in contrast to the heritability in the broad sense h_B^2 defined as V_G/V_P in

Table 6-3 Running Speed (in Seconds) of Various Generations of Wild and Tame Mice

Generation	P_1 (wild)	P_2 (tame)	F_1	F_2	BC_1	BC_2
Males	6.7 ± 0.3	24.5 ± 1.0	7.6 ± 0.3	13.0 ± 0.6	6.6 ± 0.3	20.8 ± 1.6
Females	5.3 ± 0.3	25.3 ± 1.2	6.9 ± 0.3	11.8 ± 0.5	6.2 ± 0.5	18.7 ± 1.5
Both	5.9 ± 0.2	24.9 ± 0.8	7.2 ± 0.2	12.4 ± 0.4	6.4 ± 0.4	19.7 ± 1.4

Source: Broadhurst and Jinks, 1961.

the previous section. Thus

$$\frac{V_A}{V_P} = h_N^2$$

The parameter h_N^2 is, therefore, a measure of the proportion of variation due to additive genes. It is a more useful concept than the heritability in the broad sense because it provides a measure of the gametes and the genes they carry from one generation to the next. From the predictive point of view, therefore, the ratio V_A/V_P is more useful than h_B^2. In animal and plant breeding programs, for example, the heritability in the narrow sense h_N^2 is a measure of the amount of genetic variability available as a basis for selective breeding. Comparisons between inbred strains (Section 6-3) do not provide an estimate of h_N^2 because it is not possible to obtain estimates of V_A and V_D by this method; as seen, this procedure only provides h_B^2.

By taking the two homozygous inbred strains P_1 and P_2, and various crosses to give the F_1, F_2, BC_1, and BC_2 (backcrosses of F_1 to P_1 and P_2, respectively) generations, as well as all possible reciprocal crosses, a total of 14 crosses can be set up, which permits the separation of the genotypic variation into additive effects, dominance effects, epistasis, and reciprocal effects. *Epistasis* refers to interactions between genes at different loci, and a *reciprocal effect* occurs when a cross between two phenotypes differs according to which parental phenotype was male and which was female. Thus a reciprocal effect is obtained if P_1 female × P_2 male differs from P_2 female × P_1 male. Such effects may be due to sex-linkage or to maternal effects. This more extensive analysis enables sex-linkage effects to be distinguished from maternal effects. Epistatic effects are normally small in quantitative data and are frequently ignored completely (see Mather and Jinks, 1977, for details). Maternal effects in particular will be considered in Section 6-6.

6-5 DIALLEL CROSSES

An extremely useful technique in behavior-genetics work in laboratory species is the *diallel cross*. This is the set of all possible matings between several strains or genotypes. For 4 strains there are 16 possible combinations:

Strain of male parent	Strain of female parent			
	A	B	C	D
A	AA	AB	AC	AD
B	BA	BB	BC	BD
C	CA	CB	CC	CD
D	DA	DB	DC	DD

These are made up of six crosses, AB, AC, AD, BC, BD, and CD; their six reciprocals, BA, CA, DA, CB, DB, and DC, where the sex of the parents is transposed; and four kinds of offspring derived from the four parental strains, AA, BB, CC, and DD, which are arranged along the leading diagonal. In general, if there are n strains, the diallel table has n^2 entries made up of n parental strains, $n(n-1)/2$ crosses, and $n(n-1)/2$ reciprocals. Quite often not all these crosses are included; for example, the reciprocal F_1 and/or the parental strains may be omitted.

A variety of theoretical methods of analysis of diallel crosses is available depending somewhat on the information desired (see, for example, Griffing, 1956; Kempthorne, 1969; Mather and Jinks, 1977). The first detailed analysis of a behavioral trait was carried out on a 6×6 diallel cross between inbred strains of rats (Broadhurst, 1960) for defecation and ambulation scores. The analysis employed the methods of Mather and Jinks (1977) and Hayman (1958) and their colleagues. A heritability in the narrow sense h_N^2 of 0.62 was obtained for the defecation score, and 0.59 for the ambulation score. Since the maximum value of h_N^2 is unity, when the phenotypic variance is equal to the additive genetic variance, these values of h_N^2 show that the additive genetic variance component is relatively high for these traits.

Another good example comes from Fulker (1966), who studied the mating speed of male *D. melanogaster* by taking single males from each of six inbred strains and testing them with six virgin females, one female from each of the inbred strains. The number of females fertilized in 12 hours was scored as assessed by the production of progeny. Since each male was given the same array of six females, the females can be regarded as a standard testing set so that we are dealing with the behavior of the male alone. (This contrasts, for example, with earlier data of Parsons, 1964, which studied single-pair matings within inbred strains and hybrids, a situation where genetic interpretations are made more difficult by behavioral interactions between the two sexes.) Five males were tested for each of the six inbred strains and all possible hybrids between the inbred strains, thus making up a 6×6 diallel cross (Table 6-4). The entries in the table represent the number of females fertilized out of six possible. Two entries appear for each genotype because the diallel cross was replicated 2 weeks after the first cross. The diallel cross was analyzed by the rigorous methods of Hayman (1958) and Mather and Jinks (1977). Conveniently, it turned out that the only parameters of importance were V_A, V_D, and V_E; furthermore, no reciprocal effects were found. The estimated variance compo-

Table 6-4   Replicated Diallel Cross of Mating Speed (Number of Females Fertilized Out of Six Possible) for Male *D. melanogaster*

Lines of fathers of males tested	Lines of mothers of males tested					
	6C/L	Edinburgh	Oregon	Wellington	Samarkand	Florida
6C/L	*1.4**	3.6	2.2	3.2	2.6	3.0
	1.2	2.6	2.6	3.8	3.4	3.2
Edinburgh	4.0	*3.0*	3.7	3.4	3.2	3.2
	3.2	*3.8*	4.6	4.0	2.8	4.2
Oregon	2.3	3.4	*1.8*	3.4	2.4	2.8
	1.6	4.6	*0.8*	4.0	1.6	3.8
Wellington	3.2	4.4	3.8	*3.0*	2.4	3.6
	3.4	3.0	3.2	*2.2*	3.6	4.2
Samarkand	2.4	3.6	2.0	2.4	*1.2*	2.4
	3.2	4.0	2.2	4.6	*1.2*	3.8
Florida	3.3	4.0	3.2	4.6	2.0	*2.8*
	3.8	4.2	2.8	3.4	3.6	*1.8*

* Italic indicates inbred strain speed.
Source: Fulker, 1966.

nents are: $V_A = 0.345$, $V_D = 0.328$, $V_E = 0.260$, so that $V_P = 0.933$, giving $h_B^2 = 0.72$ and $h_N^2 = 0.37$.

The proportion of additive genetic variance is relatively low compared with many traits; the proportion of dominance variance is relatively high. Dominance in these data is in the direction of rapid mating speed. In fact, this is so marked that there is generally overdominance, or heterosis, in this direction. This is shown in Table 6-5. In all cases the means for the hybrids exceed the corresponding means for the inbred strains, showing the general occurrence of heterosis for fast mating speed. This result suggests that in natural populations,

Table 6-5   Mean Mating Speed Scores (Numbers of Females Fertilized Out of Six Possible) for Male *D. melanogaster*

	Hybrids of females of given strain with males of remaining five strains	Hybrids of males of given strain with females of remaining five strains	Inbred strains
6C/L	3.04	3.02	1.3
Edinburgh	3.74	3.63	3.4
Oregon	3.03	2.99	1.3
Wellington	3.68	3.48	2.6
Samarkand	2.76	3.06	1.2
Florida	3.42	3.49	2.3
Overall mean	3.93	3.93	2.4

Source: Fulker, 1966. Based on data in Table 6-4.

there is likely to be strong natural selection in the direction of fast mating (Parsons, 1974a). The importance of rapid mating speed as a component of fitness is discussed later, especially in Section 13-2.

Another method related to the diallel cross, the simplified triple test cross (TTC), has been developed for the analysis of quantitative traits (see Fulker, 1972). This method is more economical in that usually fewer crosses are needed. In its simplest form, the design involves crossing n inbred strains to the two extremes of the n strains, producing a $2 \times n$ table. From the analysis of variance of this table, tests of significance for additive and dominance variation are obtainable. If scores for the n strains themselves are also available, a test of significance for epistasis is possible. When the sizes of the simplified TTC, the full diallel cross, and the half diallel cross (which omits reciprocal crosses) are compared for $n = 3$ (Table 6-6), the simplified TTC requires as many crosses as the full diallel, and in fact the numbers are the same, but thereafter the TTC is more economical than the full diallel, the relative economy increasing as n increases. The simplified TTC and half diallel require an equivalent number of crosses at $n = 5$; thereafter the TTC is more economical.

The simplified TTC scheme is particularly useful when the aim is to survey genetic mechanisms widely; many inbred strains can be used, facilitating inferences about the base population. A limitation is that the n strains are being assayed only in relation to the genes carried by the two tester strains; but, provided the tested strains are extreme, there should be no loss of information about important loci. An obvious further advantage for behavior-genetics work comes from the possibility of replicating over a variety of environments, so that ambitious genotype-environment studies become feasible.

The disadvantage is the need to choose two strains that are extreme as far as phenotypic effects are concerned. This means that if a given genotype is extreme for only one of a number of phenotypic traits, the study is restricted to that one trait. In a diallel cross there is no such restriction; many traits can be studied simultaneously, irrespective of which strains are extreme for them. On the other hand, it frequently turns out that certain strains are extreme for a number of traits, thus giving a complex behavioral phenotype. This issue is discussed further with reference to mice in Section 9-3.

Fulker considers empirically that for $n < 8$ the half diallel represents the best all-round design for the investigation of gene action controlling a number of behaviors. However, in comparison with the complete diallel cross, the

Table 6-6 Comparison of Minimum Numbers of Crosses Required to Test n Strains for Three Test Designs

		Number of strains n					
	Number of crosses	3	4	6	8	12	20
Full diallel	n^2	9	16	36	64	144	400
Half diallel	$n(n + 1)/2$	6	10	21	36	78	210
TTC	$2n + n$ parents $= 3n$	9	12	18	24	36	60

simplified TTC and the half diallel do not permit the estimation of reciprocal differences. Even though reciprocal differences are not common in the behavior-genetics literature, it seems desirable to test for them routinely, as will be discussed further in the next section.

As an example of the simplified TTC, we can return to Fulker's (1966) male mating speed data as presented in Tables 6-4 and 6-5. The analysis of variance using the TTC scheme shows significant additive genetic and dominance components of variance as found for the complete diallel arrangement. The test for epistasis is not significant. The variance component estimates are: $V_A = 0.415$, $V_D = 0.3$, $V_E = 0.32$, giving $V_P = 1.035$, so that $h_B^2 = 0.69$ and $h_A^2 = 0.40$. These figures are very close to those estimated previously from the 6×6 diallel cross.

Table 6-7 presents some data gathered on rats by Broadhurst (1960) that Fulker, Wilcock, and Broadhurst (1972) analyzed by means of both the TTC and the diallel cross for comparative purposes. Of the four traits analyzed, defecation and ambulation are discussed in Section 5-5 and the diallel is a complete 6×6 cross. The other two examples (avoidance and intertrial crossing) are taken from an 8×8 diallel cross. The subjects were given 30 one-minute trials of escape-avoidance training in a shuttle box divided into two equal compartments, one compartment of which was shocked after 8 seconds during which a buzzer was sounded. Crossing from one side of the box to the other terminated the buzzer or both buzzer and shock. If animals failed to cross, shock terminated automatically after 10 seconds. Intertrial intervals averaged 1 minute and were varied unsystematically between 40 and 80 seconds. The number of avoidances out of 30 was taken as the measure of avoidance. In the same experiment, a crossing from one compartment to the other was recorded as an intertrial crossing.

Table 6-7 Components of Variation for Behavioral Traits in Rats Analyzed by Means of TTC and Diallel Cross

Component of variation	Ambulation (Broadhurst, 1960)		Defecation (Broadhurst, 1960)		Avoidance (Fulker et al., 1972)		Intertrial crossing (Fulker et al., 1972)	
	TTC	Diallel	TTC	Diallel	TTC	Diallel	TTC	Diallel
V_A	15.2	19.7	0.083	0.131	24.12	19.44	0.28	0.22
V_D	2.5	1.5	−0.006*	−0.034*	−0.028*	−1.97*	0.03	−0.03*
V_E	5.8	4.9	0.166	0.160	5.42	3.94	0.24	0.17
Directional dominance	None	None	None	None	None	None	None	For low expression
Epistasis	None	None	None	None	None	None	None	None
h_B^2	0.75	0.81	0.32	0.38	0.82	0.82	0.56	0.52
h_N^2	0.65	0.75	0.34	0.51	0.82	0.91	0.50	0.62

* Not significant.
Source: Fulker, 1972.

Considering the variance components, the agreement between the simplified TTC and the diallel cross is striking for ambulation, but somewhat less so for defecation, suggesting that the simplified TTC may be somewhat limited in value when the heritability is low. Both methods are in reasonable agreement for avoidance and intertrial crossing, although the diallel method indicates significant directional dominance for low intertrial crossing. For all traits the additive genetic variance V_A accounts for by far the highest proportion of the genotypic variance, the dominance variance V_D being much smaller and in some cases negative, although not differing significantly from zero in these cases. No significant epistasis is detected. It is not, then, surprising that correspondences between designs for h_A^2 and h_B^2 are quite close in most cases.

The diallel cross also provides a useful method for ascertaining the sex important in determining mating speed or duration of copulation. This can be illustrated with a 3×3 diallel table constructed from data on duration of copulation in three *D. pseudoobscura* strains—ST/ST, ST/CH, and CH/CH (Table 6-8). Here the durations of copulation for males of each strain with females of each strain are studied. The experimental procedure was to shake unetherized flies together into vials as single pairs and record the duration of copulation immediately following the commencement of mating (Kaul and Parsons, 1965). Inspection of the marginal frequencies shows that males of the CH/CH karyotype have the shortest duration of copulation, followed by ST/CH and ST/ST with the longest. The differences between strains are much smaller for females. Analysis of variance of the data in Table 6-8 shows significant effects for both sexes, but that for males is far greater than that for females. Therefore, the diallel method enables us to say that duration of copulation is primarily male-determined. MacBean and Parsons (1967) arrived at similar conclusions for *D. melanogaster*. Mating speed data in *D. melanogaster* based on a 5×5 diallel cross are presented in Parsons (1965b); these show males to be important for rapid mating, but with females assuming progressively more importance over time. The possibly general significance of this observation is emphasized by Blizard and Fulker (1978) who came to a similar conclusion when studying all possible combinations of a series of rat strains.

Table 6-8   Mean Duration of Copulation (in Minutes) in
D. pseudoobscura

Female	Male			
	ST/ST	ST/CH	CH/CH	All strains
ST/ST	5.08	4.22	3.17	4.16
ST/CH	5.49	4.47	3.82	4.59
CH/CH	5.95	4.38	3.55	4.63
All strains	5.51	4.36	3.51	4.46

Each based on 78 observations.
Source: Kaul and Parsons, 1965.

6-6 MATERNAL EFFECTS

A complete diallel cross permits the assessment of reciprocal effects that have not often been found in behavioral data. However, especially early in the life of offspring of animals such as rodents, the influence of the mother may well be apparent. Some comments on the designs needed to detect maternal effects are therefore necessary. Indeed, a major biometrical study that does not incorporate means of testing for maternal effects is perhaps limited. We are more concerned here with mammals than with insects, in which there is no appreciable parental care of offspring.

There are two possible times at which a maternal effect can be exerted. The first is the prenatal period when the animal is in the mother's uterus and is physiologically dependent upon her; the second is the postnatal period, the time before weaning when the animal is in intimate contact with the mother and is still to some extent dependent on her. At this time too, learning may take place, both from the mother and from littermates.

The prenatal effect can be assessed from reciprocal crosses and comparisons of the offspring. A difference between reciprocal crosses may be due to prenatal intrauterine environment. The complete diallel cross is suited to assess this possibility, since it involves breeding reciprocal crosses throughout. Even so, there is the possibility that sex linkage may lead to discrepancies between reciprocal crosses. Fulker (1970) reanalyzed a study based on two parental strains and their reciprocal F_1 generations in mice. The study compared open-field exploration in mice subjected to prenatal stresses and in unstressed mice. Stressful experiences such as mechanical vibration, swimming, and loud noise were applied to half the pregnant mice (DeFries, 1964). Maternal effects were found that acted in opposition to the additive genetic effects. This suggested to Fulker the possibility of a buffering mechanism moderating the expression of the offspring phenotype. It also shows the likely complexity of maternal effects. Fulker goes on to discuss models for assessing such maternal effects.

Another possibility, not much discussed with relation to behavioral traits, is extranuclear or cytoplasmic inheritance. The amounts of cytoplasm contributed by sperm and ovum are very different, the contribution of the ovum being much greater, so that the cytoplasmic contribution of the male parent is negligible. If a maternal effect is established phenotypically, therefore, it is necessary to investigate the possibility that the effect results from cytoplasmic involvement rather than from an intrauterine factor. To elucidate this, the technique of transplanting fertilized ova between strains is useful; the relevant techniques for mice are described by McLaren and Michie (1956, 1959). DeFries et al. (1967) recorded a small maternal effect for open-field behavior by means of ovarian transplantation, but a much larger effect was found for body weight.

Postnatal effects in rodents can be detected by transferring part or whole litters to a foster mother of an appropriate genotype, who then rears them until weaning. Three main types of postnatal environment can be visualized, assuming two strains, A and B: (1) rearing by natural mothers; (2) rearing by foster mothers of the same strain as the natural mother; and (3) rearing of offspring of strain A by mothers of strain B and vice versa. Comparisons of traits from these various groups of offspring reveal the presence or absence of postnatal maternal effects (see Broadhurst, 1967a, for further discussion). Here then we have animal analogies of human-adoption studies, which as will be seen in later chapters play a crucial role in the disentangling of genetic and environmental effects of complex traits such as intelligence.

The study of postnatal maternal effects by fostering could well be incorporated into routine behavior-genetics analyses especially given the importance of human-adoption studies. Indeed, it may be possible to examine such matters by means of a diallel crossing system if designed appropriately. Any procedure would entail an extensive breeding program to ensure that the appropriate litters were available for fostering at birth or within a few days of each other, otherwise age differences would lead to complications; in some cases, however, the age differences themselves would be of interest. Clearly, by careful breeding and experimental designs, maternal effects can be investigated.

6-7 ISOFEMALE STRAINS

Few investigators have measured the total range of polygenic variation in natural populations and its evolutionary significance. In such studies, *isofemale strains,* which are derived from single inseminated females in the wild, are beginning to play an increasingly important role, especially when combined with the methods discussed so far in this chapter and the analysis of the effects of selection as discussed for geotaxis in *Drosophila* in Section 5-2. Initially, variation among isofemale strains derived from natural populations, known essentially for all measurable traits—behavioral, morphological and physiological (laterality is presumably an exception, Section 5-7)—tended to be put into the nuisance category. However, since the variation between strains persists consistently over a number of generations, this directly implies polymorphic differences in natural populations originating from the founder females from which isofemale strains were set up. As a specific example, isofemale strains set up from females collected in Victoria, Australia, were individually found to be different for three quantitative traits—scutellar bristle number, mating speed, and duration of copulation (Parsons, Hosgood, and Lee, 1967). Parsons (1977a) showed that studies of isofemale strains permit inferences about natural populations to be made directly and quickly. In particular, this is important for species that can be cultured in the laboratory but for which there is little genetic information. For example, comparisons have been made between closely related sibling species of *Drosophila* for larval reactions to alcohol (Section 8-5) and for phototaxis and dispersal toward light (Section 13-2).

Using simple diallel crosses among isofemale strains, it is possible to obtain information concerning the additive and dominance components of quantitative traits in one generation. For example, in *D. melanogaster,* isofemale strain heterogeneity has been found for duration of copulation and mating speed

(Hosgood and Parsons, 1967a). Diallel crosses among strains showed that duration of copulation was controlled by male-controlled additive differences, whereas for mating speed both additive and nonadditive effects were relevant, the male being more important than the female. The nonadditive effects were mainly in the direction of rapid mating—agreeing with Fulker's data discussed earlier in Section 6-5. Once again, therefore, we have evidence for directional selection for rapid mating. In this case the result is from strains direct from nature meaning that the results are directly applicable to natural populations. Although these are simple behaviors, they illustrate an approach applicable to more complex problems in natural populations.

In Figure 6-2 the number of matings in 30 minutes for three representative isofemale strains of each of the sibling species *D. melanogaster* and *D. simulans* are given for developmental temperatures ranging from 12 to 30°C. It was pointed out that there is significant heterogeneity among the total of ten strains tested for each species, indicating genetic differences as expected. In addition, there was a significant interaction between strains and temperatures. Quite clearly, isofemale strains can be readily studied over a range of environments in one generation without the complication of complex crossing procedures. This permits obtaining information on the effects of many environments on a series of strains that differ genetically. Since they are derived directly from natural populations, an idea of the overall responses of natural populations to important environmental variables such as temperature are readily measurable. This approach would appear to be particularly valuable for behavioral traits, being very labile environmentally as compared with morphological traits.

Isofemale strains can be used to assay differences among populations within species. The approach has been used successfully in *D. melanogaster* for sensitivity to alcohol within the wine cellar at the Chateau Tahbilk winery in Victoria, Australia, immediately outside the cellar, and distant from the cellar, demonstrating genetic heterogeneity due to natural selection over a remarkably short distance (McKenzie and Parsons, 1974). Hence isofemale strain studies permit conclusions on microdifferentiation of populations as the result of ecological heterogeneity. In Section 8-5 this approach will be explored for larval behavior in reaction to various possible metabolites, especially alcohol, in the sibling species *D. melanogaster* and *D. simulans*. In conclusion, the isofemale strain approach is of great value if rapid and quick inferences about natural populations are required for any species that can be cultured in the laboratory. This applies for any measurable trait, including behavioral ones, over a multiplicity of environments (Parsons, 1977a).

6-8 COMPONENTS OF THE GENOTYPIC VARIANCE IN RANDOM MATING POPULATIONS

In Section 6-4 the genotypic variance was partitioned into additive genetic and dominance components in the F_2 for a cross between two inbred strains. Three genotypes—A_1A_1, A_1A_2, and A_2A_2, which have frequencies or proportions of

$\frac{1}{4} : \frac{1}{2} : \frac{1}{4}$ in the F_2 generation—were considered. In a random-mating popula-tion, however, the proportions are $p^2 : 2pq : q^2$, where p is the gene frequency of A_1 and q of A_2 such that $p + q = 1$. This follows from the Hardy-Weinberg law discussed in Section 2-3.

As in Section 6-4, we give the homozygotes A_1A_1 and A_2A_2 genotypic values of $-a$ and $+a$ and the heterozygote A_1A_2 a value d. Once again the genotypic variance (V_G) can be expressed as the sum of the additive genetic variance (V_A) and the dominance variance (V_D), as derived in Appendix 6-2. Not unexpectedly, the expressions for V_A and V_D are now more complex and depend on the gene frequencies, but they become identical to those previously derived when $p = q = \frac{1}{2}$.

6-9 RELATIONS BETWEEN RELATIVES: CORRELATION APPROACH

Much research is concerned with the relations between relatives. For example, consider a trait in a sample of brothers and sisters. In Section 2-2 the formula for the variance $V(x)$ of a trait x_i was given. If the trait is to be measured on brothers and sisters and we let the values of the trait for the brothers be x_i and for the sisters y_i, a formula analogous to that in Section 2-2 applies to the sisters

$$V(y) = \frac{1}{n-1} \sum (y_i - \bar{y})^2$$

This tells us nothing about the possibility of associations between brothers and sisters for the trait. To obtain this information, we need the sum of prod-ucts between the two variables x_i and y_i. From this a quantity analogous to the variance, which is referred to as the *covariance,* can be calculated

$$W(x,y) = \frac{1}{n-1} \sum (x_i - \bar{x})(y_i - \bar{y})$$

From this we obtain the *correlation coefficient* between the two variables; which as shown in elementary statistics texts can be written

$$r = \frac{W(x,y)}{\sqrt{V(x)V(y)}}$$

Table 6-9 presents some data for heights of 11 pairs of brothers and sisters, all university students in Melbourne, Australia. The means $\bar{x}$ and $\bar{y}$ show brothers to be taller than their sisters on average as is usual. The correlation coefficient between their heights was $+0.57$. This shows that in families where the boy is tall relative to other members of his sex, so is his sister relative to other females. In theory the range of r is from -1 when the correlation is completely negative to $+1$ when the correlation is completely positive. Graphically, a positive correla-

Table 6-9 Heights (Centimeters) of 11 Pairs of Brothers and Sisters and Computation of Correlation Coefficient (r)

	Family										
	1	2	3	4	5	6	7	8	9	10	11
Brother (x)	180	173	168	170	178	180	178	186	183	165	168
Sister (y)	175	162	165	160	165	157	165	163	168	160	157

Thus

$$\bar{x} = 175.36, \bar{y} = 162.45$$

$$\Sigma x_i^2 - \frac{(\Sigma x_i)^2}{n} = 478.54$$

$$\Sigma y_i^2 - \frac{(\Sigma y_i)^2}{n} = 428.73$$

$$\Sigma x_i y_i - \frac{\Sigma x_i \Sigma y_i}{n} = 259.18$$

Hence

$$r = \frac{259.18}{\sqrt{478.54 \times 428.73}} = +0.57$$

For mode of computation, see Appendixes 6-1 and 6-3.

tion gives a positive slope between x and y, and a negative correlation a negative slope (Figure 6-6). In the absence of any correlation between x and y, the plotting of the data shows no obvious slope in which case $r = 0$. Therefore the value of $+0.57$ indicates a positive correlation between sibs. This value differs significantly from zero at the 5 percent level and indicates a highly heritable trait as will be seen from the theoretical expected correlation between sibs soon to be derived.

We are now ready to consider correlations between relatives in more detail—a valuable analytical tool in organisms that lack a ready supply of inbred strains. The method also applies to people, but correlations between relatives are most frequently calculated in experimental organisms. Considering the correlation between one parent and offspring, summing over all loci, it can be shown (Falconer, 1960) that the covariance between one parent and offspring is

$$W_{OP} = \tfrac{1}{2} V_A$$

This is intuitively reasonable, since half the genes of any offspring are on average the same as those of a given parent and half are different. Thus, of the additive genetic variance (V_A) in the parent, half goes to the offspring. Note that there is no dominance component in this expression. This is logical since in transmission from parent to offspring, the gametes carrying genes, not the genotypes, are passed on from generation to generation (Section 6-4). A situation where a dominance variance component (V_D) is expected is for covariances between full sibs. Full sibs, unlike other relatives, have two parents in common

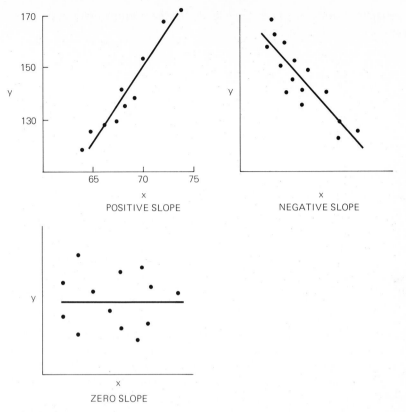

Figure 6-6 Positive, negative, and zero correlations illustrated by slopes plotted for various data.

and thus have some genotypes in common. The covariance between full sibs can be shown to be

$$W_{SS} = \tfrac{1}{2}V_A + \tfrac{1}{4}V_D$$

The $\tfrac{1}{4} V_D$ component occurs because of the genotypes in common between full sibs. Numerically, therefore, W_{SS} is expected to be a little greater than W_{OP} but not much since V_A is usually much greater than V_D as we saw especially when discussing diallel crosses.

From the above two covariances, correlations between relatives can be obtained by dividing the covariances by the total phenotypic variance (V_P). This comes directly from the formula given above for the correlation coefficient. In that formula we can generally regard $V(x) = V(y) = V_P$. Hence the covariance is merely divided by the total variance. For the correlation between one parent and offspring, we therefore have

$$r_{OP} = \frac{\tfrac{1}{2}V_A}{V_A + V_D + V_E} = \frac{\tfrac{1}{2}V_A}{V_P}$$

which is equal to $\frac{1}{2}h_N^2$, the heritability in the narrow sense. The correlation between sibs is

$$r_{SS} = \frac{\frac{1}{2}V_A + \frac{1}{4}V_D}{V_A + V_D + V_E}$$

which slightly overestimates $\frac{1}{2}h_N^2$. In Section 12-1, Table 12-1, these correlations are used in a wide-ranging assessment of the degree of genetic control of intelligence test scores.

Finally it should be noted that full-sib analysis alone is difficult to use, especially from the behavioral point of view, because of the possibility of bias due to the common environment of sibs reared together. Thus, for completeness

$$W_{SS} = \frac{1}{2}V_A + \frac{1}{4}V_D + V_{EC}$$

where V_{EC} is the variance component due to the common environment of sibs reared together. A full-sib analysis alone can do little more than set an upper limit to h_N^2. For behavioral traits where V_{EC} could be high due to early experience, this mode of analysis should be viewed with caution and interpretations made accordingly. These problems are further amplified in the next chapter where *Homo sapiens* is specifically considered.

In theory, correlations are obtainable between any sets of relatives, and as the relation between the relatives becomes more remote, the coefficient of V_A in the covariance becomes smaller:

- Half sibs, aunt-nephew, uncle-niece $\quad 1/4$
- First cousin $\qquad\qquad\qquad\qquad\qquad 1/8$
- First cousin once removed $\qquad\qquad\quad 1/16$
- Second cousin $\qquad\qquad\qquad\qquad\quad 1/32$

The coefficient of V_A is referred to as the *coefficient of relationship* and reflects the share of genes due to common ancestry. It is related to the inbreeding coefficient F discussed in Section 6-3 and is twice the value of F for the relatives considered so far.

Before leaving the subject of correlations between relatives, for completeness we should consider the correlation between midparent and offspring. The midparent ($\bar{P}$) is defined as $\frac{1}{2}(P_1 + P_2)$, where P_1 and P_2 are the values of the two parents. The correlation between midparent and offspring can be shown to be

$$r_{O\bar{P}} = \sqrt{2} \times r_{OP}$$

This method is used less than the one parent–offspring relations because there is the possibility of maternal effects. This applies of course to mother-offspring relations as well, so the comparison of mother-offspring data with father-offspring data is often of interest. Furthermore, the midparent approach assumes that the variances are equal in both sexes. For sex-behavior traits, many of which are sex-limited, the approach is clearly invalid. Also the presence of assortative mating, which seems common for behavioral traits especially in people, may lead to bias.

6-10 RELATIONS BETWEEN RELATIVES: REGRESSION APPROACH

Relations between relatives can be looked at in another way. The early work of Galton and Pearson in England showed that the sons of tall men tend to be tall—but not so tall as their fathers and yet not so short as the average of the population; in fact the height of sons tends to be about halfway between that of their fathers and the population average. Similarly, the sons of short men tend to be short, but not so short as their fathers; on average they have heights about halfway between those of their fathers and the population average. This trend back toward the population mean is exactly what is expected on the basis of additive genes. To look at this situation, some additional statistics must be mentioned. The correlation coefficient so far discussed does not imply any causal relation between the variables x and y even if such a relation exists. However, in some cases we can look upon y as a variable dependent on x. Frequently both approaches can be used, for example, in looking at parent-offspring data. The correlation coefficient allows us to test only whether two variables are related. However, a procedure called *linear regression,* discussed in most elementary statistics texts, allows us also (1) to predict the value the dependent variable y should have for any value of the independent variable x; and (2) to test how much of the variation in y actually depends on x.

Essentially we aim to find the values of a and b in the *regression line,* which has the formula

$$y_i = a + b(x_i - \bar{x})$$

This line is so constructed that the squared distance between it and all points on a graph is minimized. In Figure 6-7 Connolly's (1966) data on locomotor activity in *D. melanogaster* are plotted. The activity was assessed by measurements in an open-field apparatus. The procedure was to take 25 pairs of parents selected from a wild-type (Pacific) strain and mate them in single pairs. From among the offspring of each of these matings, two were selected and measured. Figure 6-7 gives the regression of offspring on midparent. The equation of the line is

$$y = 15.56 + 0.51x$$

showing a positive association between the midparent's and offspring's activities. The estimate of $b = 0.51 \pm 0.10$, which measures the slope of the line,

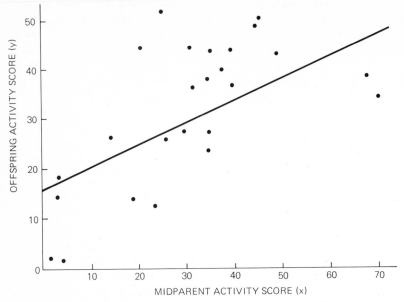

Figure 6-7 Activity scores in *D. melanogaster*. Regression of offspring scores on midparent scores. (*From Connolly, 1966.*)

is significantly greater than zero ($P < 0.01$). Therefore the activities of the offspring depend in some way on the parental values.

The quantity b is called the *coefficient of regression* of y on x and can be shown to be written as

$$b = \frac{W(x,y)}{V(x)}$$

This may be compared with the correlation coefficient r between the two variables in the previous section. The regression coefficient of y on x as the independent variable has $V(x)$ as the denominator; if on the other hand x is regressed on y, then $V(y)$ is the denominator. For the correlation coefficient between x and y, where neither variable is considered dependent on the other, it is not unreasonable that the denominator should be $\sqrt{V(x)V(y)}$.

Section 6-9 showed that the covariance between one parent and offspring is $\frac{1}{2} V_A$, so that from the above equations the regression of one offspring on parent is

$$b_{OP} = \frac{\frac{1}{2} V_A}{V_P} = \frac{1}{2} h_N^2 = r_{OP}$$

As in the previous section $V(x)$ is written as equal to V_P in this equation.

Finally, it is possible to show that the regression of offspring on midparent $\bar{P}$ is

$$b_{O\bar{P}} = h_N^2$$

In other words, the regression coefficient is equal to the heritability in the narrow sense. Thus it can be concluded that h_N^2 for locomotor activity is 0.51 ± 0.10. Additional examples of the regression approach are given in Section 12-1, where intelligence in human beings is discussed.

6-11 DIRECTIONAL SELECTION EXPERIMENTS FOR QUANTITATIVE TRAITS

The selection experiment consists of selecting and manipulating various chosen genotypes with respect to a trait from a population. We are here concerned with *directional* selection (see Figure 5-1), where extreme individuals of a population are selected in the hope of forming separate high or low lines in subsequent generations. Examples discussed in Chapter 5 included geotaxis in *D. melanogaster* and emotionality as measured by defecation scores in rats.

If a quantitative trait has some genetic basis, there should be a response to directional selection, since the selection of extreme phenotypes means that extreme genotypes are selected. Initially the *response to selection (R)* can be assessed by

$$R = b_{O\bar{P}}S$$

where $b_{O\bar{P}}$ is the regression of offspring on midparent as discussed in the last section and S is the *selection differential*. The selection differential is defined as the mean phenotypic value of the individuals selected as parents, expressed as a deviation from the mean phenotypic value of all individuals in the parental generation before selection was begun (see Figure 6-8). It can be shown that the magnitude of S depends on both the proportion of the population included among the selected group and the standard deviation of the trait.

Since in the previous section we showed that $b_{O\bar{P}} = h_N^2$, then it follows that

$$R = h_N^2 S$$

This is not a surprising result, since the response to selection must be based on a component representing the selection differential combined with the heritability of the trait under selection. From this equation, if $h_N^2 = 0$, no response is possible since the trait is entirely environmentally determined. Conversely, the larger h_N^2 is, the greater is the expected response as indicated in Figure 6-8. In theory, the prediction of response is valid for one generation only, because a major effect of selection is to change gene frequencies and hence the genetic properties of offspring. Even so, in many experiments the

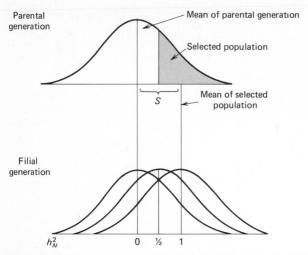

Figure 6-8 Diagram to show the meaning of the selection differential S where all individuals in the shaded region of the distribution are selected in the parental generation. The response to selection depends on the heritability h_N^2, as shown in the filial generation curves.

predicted response is maintained for five or more generations. Much effort and statistical sophistication have been devoted to the estimation of the predicted response, which depends on an accurate estimate of the heritability h_N^2. If the heritability had not been estimated before selection, the above equation provides an estimate of it as R/S. Given a response to selection over a number of generations, the response obtained gives an estimate of R/S that is referred to as the *realized heritability*.

It is clear from the above equation that there are two main methods for improving the response to selection. The first is by increasing the heritability, which may be possible by reducing the environmental variation by selecting a trait that can be measured objectively and easily, and generally trying to minimize random effects. Repeated measurements on an individual can be useful under some circumstances. For a detailed consideration of the accuracy gained from repeated measurements, Falconer (1960) should be consulted. As long as there is some correlation between measurements as would be expected for a trait with a genetic component, the greatest gain comes from simple replication with little gain likely beyond three to five repeated measurements. Since the repeatability of many behavioral traits is likely to be low, multiple measurements to increase reliability may be worthwhile in animals such as rodents where obtaining adequate numbers presents difficulties. However, complications due to the possibility of learning between trials must be borne in mind, since some individuals may learn more rapidly than others, as will be discussed in Chapter 9. Because of this, repeated measurements are probably most reliable for traits without a learned component. Most behavioral traits in a species such as *D. melanogaster* have no detectable learning component. However, since it is easy to breed large numbers of this species, repeated measurements are not usually carried out.

The second method of improving the response is by reducing the proportion selected so that those that are selected are extremes. There are some limitations to this. One important consideration is that the population size sets a lower limit to the number of individuals to be used as parents, since extremely high numbers need to be measured to reduce the proportion selected to a very low level. Furthermore, if population sizes are very small, this automatically leads to inbreeding and hence homozygosity, which reduces the variability on which selection can act. Another factor limiting the response is that many direct fitness traits such as fertility and viability are often adversely affected during selection. This can be explained by the appearance of extreme genotypes not previously exposed to the action of natural selection that often tend to be of rather low fitness.

There is, in addition, a third and potentially powerful method of improving responses to selection that is not usually considered. It involves a consideration of the base population before directional selection is begun. In Section 6-7, it was pointed out that isofemale strains set up from natural populations provide a method of making quick inferences about natural populations. Since the variation among strains persists for many generations, this suggests that a method of obtaining rapid responses to directional selection is to base selection upon the extreme strains out of a set of isofemale strains. The value of this approach has been shown for scutellar bristle number in *D. melanogaster* where remarkably rapid responses to selection for high bristle numbers were obtained by basing selection on a hybrid population of the most extreme 4 of 16 isofemale strains (Hosgood and Parsons, 1967b, Parsons, 1975), as compared with a much slower response when the population was not sampled in this way. See Figure 6-9 for highly variable responses to selection for scutellar bristle number over the first seven generations of selection according to the isofemale strain combinations selected prior to selection. MacBean and Parsons (1967) also showed the value of the approach in directional selection for duration of copulation in *D. melanogaster,* a trait of rather low heritability.

During the directional selection process, extreme phenotypes are continuously favored. This leads to an increasing proportion of extreme genotypes likely to be homozygous. Ultimately, therefore, the rate of response to selection is expected to diminish. Frequently a plateau is attained, and for a variable number of generations there is no response; occasionally, after some generations at the plateau, there is a rapid response to selection. The rapid response is likely the result of recombination between linked genes controlling a trait. Some of the recombinants would be favored by selection and would rapidly increase in frequency (Thoday, 1961).

The research value of the selection experiment lies in the evidence it may provide on the genetic basis of a trait, including behavioral traits. This was shown for geotaxis in Section 5-2. Furthermore, it may make possible studies on behavior itself, especially if the behavioral trait under analysis can be divided into components, some of which may be differentially affected by selection. Early studies showing that responses to selection can be obtained for behavioral traits in rodents are reviewed by Broadhurst (1960). Traits examined

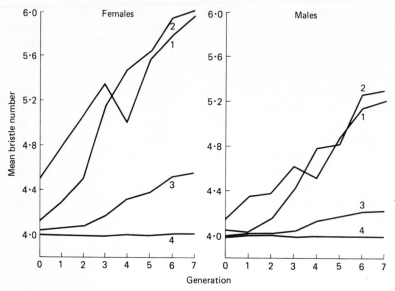

Figure 6-9 Mean scutellar bristle number in *D. melanogaster* in four directional selection lines set up from 16 isofemale strains as follows: *Line (1):* from the strain with the highest mean scutellar bristle number (4.18). *Line (2):* from a hybrid population derived from the four strains with the highest means (4.18, 4.05, 4.10, 4.06). *Line (3):* a hybrid population of all 16 strains (mean 4.03), corresponding to normal directional selection without prior sampling of the population for variability among isofemale strains. *Line (4):* the strain having the lowest mean (4.00); this strain rarely if ever has flies deviating from four bristles, so the lack of response to selection is predictable. Indeed, in wild populations most flies have this number of bristles.

include audiogenic seizures in rats and mice; running speed in mice; and sex drive, maze-learning ability, cage activity, early or late onset of mating, and emotional elimination in rats (see Section 5-5). The occurrence of responses shows that there are heritable components for these traits, but further conclusions are difficult without additional genetic analyses on selectively bred strains. In *Drosophila,* traits such as activity, duration of copulation, geotaxis, mating speed, and phototaxis show responses to selection. In some cases genetic analyses have been carried out (see Section 5-2).

As pointed out in Section 5-5, it is desirable to study correlated responses to selection. These should provide information on behavior itself as well as on its genetic control. Broadhurst obtained simultaneous information on ambulation score in his selection experiment for defecation score (Figure 5-6*a*). Eysenck and Broadhurst (1964) have taken this procedure further and list the results of more than 50 tests, some behavioral and some physiological, many of which show correlated responses. Many of the correlated responses agree with what would be predicted from the dichotomy of emotionality occurring in the reactive and nonreactive lines as shown in Figure 5-6*b*. Thus the selection experiment supplies not only information of genetic importance but also some hints about the biochemical and physiological bases of behavior.

Correlated responses to selection may be positive or negative. The frequent negative correlation between defecation and ambulation scores in rodents has already been noted. It is therefore important to know how a change of one trait by selection is associated with simultaneous changes in other traits. In genetic studies it is necessary to distinguish two causes of correlation between traits—genetic and environmental. The directly observable correlation between two phenotypic values for traits x and y is the *phenotypic correlation* (r_P). Similarly, we can take r_A as the *genetic correlation* (the correlation between the additive genetic values of x and y) and r_E as the *environmental correlation* between the traits. Details of the methods for calculating these correlations can be found in Falconer (1960). However, to compute genetic correlations, the method of correlations between relatives or data from a directional selection experiment can be used. For the latter approach, correlations from five generations of selection for high or low activity in mice with defecation score and body weight can be cited (DeFries and Hegmann, 1970). For activity and defecation, $r_A = -0.86 \pm 0.14$, which is not surprising in view of data already cited indicating a negative genetic correlation. For activity and weight, $r_A = 0.34 \pm 0.22$, which is positive but not significant.

SUMMARY

The major aim of quantitative analysis is to separate the continuously varying phenotype that we measure into genotypic and environmental component parts. Compared with morphological traits, much more attention must be paid to the environment when measuring behavioral traits. In many cases the effect of the environment itself is of direct interest. While the simplest models of quantitative inheritance assume that the joint effects of genotype and environment are additive, a more realistic situation assumes interactions or correlations between genotype and environment. In natural situations, habitat selection would represent a correlation between genotype and environment. The relative effects of genotype and environment can be readily analyzed if such interactions and correlations can be assumed to be of no consequence; however, the environment must be specified as accurately as possible in all cases.

In experimental animals, the genetic manipulations for estimating genetic components of traits include an analysis of variation within and between inbred strains, and an analysis of the diallel cross and the related triple test cross, which are both based upon inbred strains. However, few researchers have considered the total *range* of polygenic variation in natural populations and its evolutionary significance. In *Drosophila* the use of isofemale strains (each derived from a single female inseminated in the wild) is becoming increasingly important. In addition, analyses of statistical associations between relatives, and the responses after several generations of selecting extremes for a trait, provide estimates of the amount of genetic variation in natural populations.

The results of a quantitative analysis can provide information on the genetic architecture of traits. For example, the proportion of additive genetic variance is relatively low and the proportion of dominance variance is relatively high for mating speed in *Drosophila*. This suggests that there is likely to be strong selection in natural populations for fast mating speed. Other traits, for example defecation and ambulation in rats, show quite high additive genetic variances reflecting selection in natural populations for intermediate optima rather than extremes.

APPENDIX 6-1 Analysis of Variance Within and Between Inbred Strains

The analysis of variance is a procedure whereby the total variability in a set of data can be attributed to specific causes. A measure of variability or the variance comes from

$$V(x) = \frac{1}{n-1} \sum (x_i - \bar{x})^2$$

as defined in Section 2-2. For greater ease in computation this can be shown to be equivalent to

$$\frac{1}{n-1} \left[\sum x_i^2 - \frac{(\sum x_i)^2}{n} \right]$$

For the data under discussion in Table 6-2 we therefore have

$$\sum x_i^2 - \frac{(\sum x_i)^2}{n} = 4^2 + 6^2 + 8^2 + 6^2 + 7^2 + 5^2 + 3^2 + 1^2 + 1^2 + \ldots - \frac{223^2}{36} = 177.6488$$

which is referred to as the *corrected sum of squares* of the total data (Table 6-10). (Dividing by $n - 1 = 35$ gives the overall phenotypic variance $= 5.0757$.)

Since there are six totals for the strains, we can examine the variability between strains by computing

$$\frac{1}{6}(36^2 + 16^2 + 36^2 + 51^2 + 40^2 + 44^2) - \frac{223^2}{36} = 116.1389$$

The sum of the squared quantities in this corrected sum of squares must be divided by 6; otherwise it is too large, since in obtaining it the summed scores for each inbred strain

Table 6-10   Analysis of Variance of Data in Table 6-2

Source of variation	Degrees of freedom	Corrected sum of squares	Mean square (variance)	Expected mean square
Between strains	5	116.1389	23.2277	$M_1 = V_E + 6V_G$
Within strains	30	61.5099	2.0503	$M_2 = V_E$
Total	35	177.6488		

are squared, making them on average six times as large as single observations (simply because the summed scores are made up of six observations).

The variation within strains comes from subtracting the total corrected sum of squares from that for strains, thus giving a value of 61.5099. The analysis of variance is merely a table setting out this subdivision of causes of variation (Table 6-10). There are five degrees of freedom for strains (see Section 2-4 for definition) since there is a total of six strains. Similarly, the total number of degrees of freedom is 35, based on 36 observations. The number of within-strains degrees of freedom (30) is obtained by subtraction.

The variances are obtained by dividing the corrected sum of squares by the number of degrees of freedom. These are frequently referred to as *mean squares* in analyses of variance. Clearly, the variance between strains is greater than that within strains, and the ratio of the variances is 23.2277/2.0503 = 11.33. This is the basis of the *variance ratio* or *F test*. Values of F = [larger variance/smaller variance] have been tabulated based on n_1 degrees of freedom for the larger variance and n_2 for the smaller variance. In our example, $n_1 = 5$ and $n_2 = 30$. Consultation of standard statistical tables reveals that the above value of F is significant ($P < 0.001$), confirming the high variability between strains. Hence the strains differ behaviorally.

In this example we can interpret the values of these variances further, since the base material consists of inbred strains. The variance within strains is expected to be purely environmental; thus $V_E = M_2 = 2.0503$ (Table 6-10). However, the variance between strains is expected to contain a genotypic component as well and can be shown to equal

$$V_E + rV_G = M_1$$

where r = the number of replicates within each strain, namely 6. From this we obtain

$$V_G = \frac{1}{r}(M_1 - M_2) = \frac{1}{6}(23.2277 - 2.0503) = 3.5296$$

APPENDIX 6-2   Components of Genotypic Variance in Random-Mating Populations

The genotypes A_1A_1, A_1A_2, and A_2A_2 have frequencies $p^2 : 2pq : q^2$ following the Hardy-Weinberg law. Giving the homozygotes A_2A_2 and A_1A_1, genotypic values of $-a$ and $+a$, and the heterozygote A_1A_2 a value d, which may be positive or negative as in Figure 6-5, the mean of the population (m) is

$$m = ap^2 - 2pqd - aq^2 = a(p - q) - 2dpq$$

since $p^2 - q^2 = (p - q)(p + q) = p - q$
The variance due to segregation at this locus is given by

$$\begin{aligned}
p^2 \times a^2 + 2pq \times d^2 + q^2 \times a^2 - m^2 &= a^2(p^2 + q^2) + 2pqd^2 - [a(p - q) - 2pqd]^2 \\
&= 2pq[a^2 + 2ad(p - q) + d^2(1 - 2pq)] \\
&= 2pq[a + d(p - q)]^2 + 4p^2q^2d^2
\end{aligned}$$

As in the cross between inbred strains, if there are many such loci each acting independently, the total contribution to the genotypic variance can be written

$$V_G = \Sigma 2pq \, [a + d \, (p - q)]^2 + \Sigma 4p^2q^2d^2 = V_A + V_D$$

where $V_A = \Sigma 2pq \, [a + d \, (p - q)]^2$, and $V_D = \Sigma(2pqd)^2$, the summation being over all the polymorphic loci in question. V_A and V_D are, as before, the additive genetic and dominance variances. If $d = 0$ at every locus, then $V_D = 0$ as expected. Thus when there is no dominance, $V_A = 2pqa^2$, where a is half the difference between homozygotes.

It is not surprising that both V_A and V_D depend on the gene frequencies. Thus V_D has a maximum value when $p = q = \frac{1}{2}$, which can easily be checked arithmetically. However, the terms of V_A are at a maximum when $p = q = \frac{1}{2}$ only if as well $d = 0$. When $p = q = \frac{1}{2}$, V_A and V_D are the same as those derived for the variance of the F_2 between two inbred strains (in Section 6-4). This is to be expected, since an F_2 between two inbred strains is equivalent to a Hardy-Weinberg population with the gene frequencies of p and q equal to $\frac{1}{2}$. This is because at all segregating loci in an F_2, $\frac{1}{4}A_1A_1$; $\frac{1}{2}A_1A_2$: $\frac{1}{4}A_2A_2$ is expected.

APPENDIX 6-3   Computation of Correlation Coefficient

In Section 6-9 the covariance between two sets of measurements x and y is given as

$$W(x,y) = \frac{1}{n - 1} \sum (x_i - \bar{x})(y_i - \bar{y})$$

This can be shown to be equivalent to

$$W(x,y) = \frac{1}{n - 1} \left[\Sigma x_i y_i - \frac{\Sigma x_i \Sigma y_i}{n} \right]$$

a form that is easier for computation. Note the analogous form for the variance given in Appendix 6-1. Using the variance formula in Appendix 6-1, the correlation coefficient

$$r = \frac{W(x,y)}{\sqrt{V(x)V(y)}}$$

can be expressed as

$$r = \frac{\Sigma(x_i - \bar{x})(y_i - \bar{y})}{\sqrt{\Sigma(x_i - \bar{x})^2 \Sigma(y_i - \bar{y})^2}}$$

For computation, the forms in Appendixes 6-1 and 6-3 are normally used. See example in Table 6-9.

GENERAL READINGS

Falconer, D. S. 1960. *Introduction to Quantitative Genetics*. Edinburgh: Oliver & Boyd. A well-presented account of principles, mainly using the notation of this chapter.

Hirsch, J. (ed.). 1967. *Behavior-Genetic Analysis*. New York: McGraw-Hill. Most of the topics in this chapter are discussed.

Mather, K., and J. L. Jinks. 1977. *Introduction to Biometrical Genetics*. London: Chapman & Hall. A text useful for those with statistical training.

Parsons, P. A. 1967a. *The Genetic Analysis of Behaviour*. London: Methuen. Some aspects of quantitative inheritance are discussed, using behavioral traits as examples.

Quantitative Analysis:
Homo sapiens

7-1 TWIN ANALYSIS: GENERAL CONSIDERATIONS

In this chapter, the concepts discussed in Chapter 6 are applied to human beings. Francis Galton was one of the first to emphasize the significance of twins for studies on human inheritance. Since then, twins have been studied extensively with a view to defining the relative importance of genetic and environmental effects on a wide variety of traits: morphological, behavioral, and pathological. Therefore, the study of twins seems a reasonable starting point for the genetic analysis of quantitative traits in human beings. The basic comparison is between monozygotic (MZ) or identical twins, who arise from one fertilization and therefore are identical genetically, and dizygotic (DZ) or nonidentical (fraternal) twins, who arise from two fertilizations and therefore are genetically comparable to sibs. MZ twins are always of like sex, but DZ twins can be of like or unlike sex. Because MZ twins are the only human beings who have identical genotypes, numerous twin studies have been carried out. In many experimental animals, as we have seen, inbred strains are commonly developed, each strain composed of individuals with identical or nearly identical genotypes. There are problems in dealing with MZ twins, since there are likely to be special factors, especially relating to personality development and other behavioral traits of learning and reasoning types, that may influence MZ more than DZ twins. This problem can be further examined by comparing differences between the members of a pair of MZ twins reared apart in different homes with the differences between MZ twins reared together in the same home. Such comparisons provide an assessment of the effect of the environment on twins reared in the same home. Although twin data have been extensively used in human genetic research, it is important to realize that twin studies can provide only limited information on the degree of genetic determination of a trait and can provide no information about modes of inheritance.

MZ twins are derived from a single fertilization, but even so, four different types of pregnancies are possible as far as uterine configurations of the various fetal membranes are concerned, and two of these occur in DZ twins also:

- MZ or DZ with separate amnions, chorions, and placentas
- MZ or DZ with separate amnions and chorions, and fused placentas
- MZ with separate amnions and a single chorion and placenta
- MZ sharing a single amnion, chorion, and placenta

Overall, the total frequency of twin births is between 1.0 and 1.5 percent, with some racial variation, Japan having an unusually low rate of 0.65 percent and U.S. blacks having a relatively high rate, with Africans higher still (Morton, Chung, and Mi, 1967). Almost all the racial variation is due to variations in DZ twinning rates. Maximum twinning rates occur in the 35- to 40-year maternal age group, owing mainly to variations in the DZ rate; there is little variation in the MZ rate. There may be a small genetic component for twinning rates, particularly for DZ twins, but Cavalli-Sforza and Bodmer (1971) regard much of the data as inconclusive.

For genetic studies it is critical that the determination of whether a twin is MZ or DZ be absolutely objective. Often the outward similarities of MZ twins are obvious compared with DZ twins, who are genetically no more alike than are other sibs. Even so, such a diagnosis could involve some subjectivity, and the only true criterion is genotypic identity. However, there are so many known polymorphisms (e.g., for blood groups, enzymes, serum proteins, red-green color blindness, ability to taste phenylthiocarbamide) that when twins are classified with respect to an adequate number of polymorphisms, the probability that a set of DZ twins will be identical for all is so low that it can safely be discounted (see Mittler, 1971, for further details). Therefore, if identity is obtained for a large number of traits, the twins are likely to be MZ. Furthermore, various dermatoglyphic traits can be used as an aid in diagnosis. Since there are known to be a large number of polymorphic loci for histocompatibility antigens (HLA) responsible for graft rejection, these loci (HL-A,-B,-C,-D) may be profitably used for accurate diagnoses of zygosity (Wagner, Judd, Sanders, and Richardson, 1980).

The principle of zygosity diagnosis using various polymorphic markers is as follows. The probability of DZ twins being identical is calculated separately for each locus. The nature of the calculation depends on the information available on the parental types. If the genotypes of the parents and twins are known exactly (taking into account other relatives where available), exact probabilities can be computed. If, however, the parental genotypes are not known, probabilities can be worked out based on the gene frequencies in the population to which the twins belong. Detailed examples of the procedure appear in several texts (see especially Mittler, 1971; and Stern, 1973).

Even though critical studies differentiate with almost complete certainty between MZ and DZ twins, the simple use of visual criteria is almost as efficient as is the use of blood groups and other polymorphisms. In large-scale twin studies, such as one carried out in Denmark, a simple questionnaire about the degree of similarity between twin pairs has been found to be 90 to 95 percent accurate in diagnosing zygosity. The typical questions relate to eye color, hair color and texture, height, weight, body build, the tendency to be mistaken by parents and by close and casual friends, and the twins' own opinion (Harvald and Hauge, 1965). The probability that they would be identical for all these factors is negligible if they are not MZ.

7-2   TWINS IN GENETIC STUDIES: THRESHOLD TRAITS

We begin with a consideration of *threshold traits*—traits for which organisms can be phenotypically classified into those having a given trait and those not having it. Table 7-1 gives the approximate population incidence of some common congenital malformations, excluding the chromosomal abnormalities discussed in Chapter 4. The former add up to about 1.2 percent of total births and thus constitute a very important source of illness in present-day Western industrialized societies, where relatively few die in infancy. Anencephaly and spina bifida, abnormalities of the central nervous system, are disorders with behavioral consequences. Harelip with or without cleft palate, and clubfoot, especially if not corrected surgically, must necessarily have behavioral consequences. For harelip the frequency among sibs is 35 times the population incidence, and for anencephaly and spina bifida about 8 times. From these figures and information on other relatives, it can be argued that genetic factors

Table 7-1   Population Incidence of Some Common Congenital Malformations, Based on British Surveys

Malformation	Incidence per 1,000 births
Anencephaly	3
Spina bifida	3
Heart malformations	1
Harelip with or without cleft palate	1
Clubfoot	1
Pyloric stenosis	3
Congenital dislocation of the hip	1

Source: Carter, 1965.

play a part in the cause of these conditions. Even so, in no case does the incidence among sibs rise above 5 percent, which is low compared with the expected incidence of 25 percent for a simple recessive defect among the sibs of a person having the defect.

There is no clear-cut evidence for specific environmental effects as causative agents, but there are some associations between the incidences of the defects and certain socioeconomic and demographic parameters. In Scotland, Edwards (1958) showed that the frequency of anencephaly differs from 0.9 per 1,000 among professional people to 3.6 per 1,000 among skilled workers. There are significant variations among localities; the variation among seasons in which birth occurs is 30 to 50 percent. Many congenital malformations vary in incidence for different birth orders and between sexes. Factors such as prenatal exposure to radiation, chemical agents, infection, and injury at birth, may all be relevant, especially if we can base our argument upon a number of animal experiments (see Penrose, 1961, for review). For example, fairly heavy therapeutic irradiation during pregnancy has been shown to lead to severe microcephaly in the child, and of 205 children exposed to the atomic blast at Hiroshima during the first half of intrauterine life, 7 were microcephalic and mentally retarded. All these factors make a simple basis for the inheritance of these disorders unlikely.

We now consider how threshold traits can be handled in twin data. For threshold traits, a pair of twins is *concordant* if both members have the trait or both do not, that is, if they are alike. The concordance frequency is the proportion of concordant twin pairs among all those that include at least one having the trait. Therefore, a significantly higher concordance frequency in MZ than in DZ twins is considered evidence for a significant genetic component in the determination of the trait. The significance of data can be tested by a 2×2

Table 7-2 Twin Concordance for Various Mental Disorders

| Disorder | Concordant pairs | | Discordant pairs | Total pairs | χ_i^2 | H |
	Number	%				
Mental deficiency						
MZ	12	66.67	6	18	35.39*	0.67
DZ	0	0	49	49		
Epilepsy						
MZ	10	37.04	17	27	9.76†	0.30
DZ	10	10.00	90	100		
Manic depressive psychosis						
MZ	10	66.67	5	15	20.84*	0.65
DZ	2	5.00	38	40		

* $P < 0.001$.
† $P < 0.01$.
Source: Harvald and Hauge, 1965.

contingency χ_1^2 (see Section 2-4), as shown for some data of Harvald and Hauge (1965) collected from the extensive Danish twin study for such human behavioral disorders as mental deficiency, epilepsy, and manic-depressive psychosis (Table 7-2). Writing the percentage concordances of MZ and DZ twins as CMZ and CDZ respectively, in all cases CMZ is greater than CDZ. The χ_1^2 values are all highly significant, indicating the likelihood of a heritable component. Conveniently, in these data, like- and unlike-sexed DZ twins could be combined because they showed no significant differences in concordance; however, many analyses must consider data separately for like and unlike sex.

An index for estimating the degree of genetic determination (Holzinger, 1929) has been widely used. It is the equation

$$H = \frac{CMZ - CDZ}{100 - CDZ}$$

that has been referred to in the literature as the heritability. However, to avoid confusion we refer to it as the *H statistic,* since it is a completely arbitrary quantity and difficult to relate to the estimates of heritability or degree of genetic determination based on quantitative traits discussed in Chapter 6. Cavalli-Sforza and Bodmer (1971) have however, provided a method for obtaining estimates of the degree of genetic determination from H statistics as upper and lower limits based on two extreme assumptions. One is where the dominance variance is absent and the other is where the additive genetic variance is absent. Such limits are given for data (Table 7-3) from Harvald and Hauge (1965) based on the Danish Twin Registry. At these limits, the degrees of genetic determination differ by about 10 percent at most. The only nonsignifi-

Table 7-3   Twin Concordance and Upper and Lower Limits for Degrees of Genetic Determination

Disease	Percent concordance		Limits of genetic determination	
	MZ	DZ	Upper $(V_D = 0)$	Lower $(V_A = 0)$
Cancer at same site	6.8	2.6	0.33	0.23
Cancer at any site	15.9	12.9	0.15	0.1
Arterial hypertension	25.0	6.6	0.62	0.53
Mental deficiency	67.0	0.0	1.0	1.0
Manic-depressive psychosis	67.0	5.0	1.05	1.04
Death from acute infection	7.9	8.8	−0.06	−0.06
Tuberculosis	37.2	15.3	0.65	0.53
Rheumatic fever	20.2	6.1	0.55	0.47
Rheumatoid arthritis	34.0	7.1	0.74	0.63
Bronchial asthma	47.0	24.0	0.71	0.58

All comparisons except those for cancer and acute infection are highly significant.

Source: Data of Harvald and Hauge (1965), analyzed by Cavalli-Sforza and Bodmer (1971). From *The Genetics of Human Populations* by L. L. Cavalli-Sforza and W. F. Bodmer. W. H. Freeman Company. Copyright 1971.

cant comparisons between MZ and DZ twins are for cancer at any site and death from acute infection, which are very nonspecific categories. There is such a high level of heterogeneity in the category "cancer at any site" that little difference in concordance rate is expected. When cancer is defined as being at the same site, a higher degree of genetic determination is found, which is expected since this is a more homogeneous category.

The high degrees of genetic determination for mental deficiency and manic-depressive psychosis are in contrast to the other diseases listed in Table 7-3, even though these figures are thought to be overestimates (Cavalli-Sforza and Bodmer, 1971, advance possible mathematical reasons for this). These extremely high values contrast with a series of rather specific diseases—arterial hypertension, tuberculosis, rheumatic fever, rheumatoid arthritis, and bronchial asthma—that cluster in the relatively high 0.5 to 0.7 region. The latter estimates are compatible with other comparable studies (Cavalli-Sforza and Bodmer, 1971) in suggesting important genetic components for these diseases. Environmental factors are known that may affect their incidence; for example, the incidence of arterial hypertension and of bronchial asthma is undoubtedly affected by stress. The high degrees of genetic determination for the behavioral traits, mental deficiency and manic-depressive psychosis, are considered further in Chapters 11 and 12.

7-3 TWINS IN GENETIC STUDIES: CONTINUOUSLY VARYING TRAITS

Let us look first at the mean intrapair differences between MZ and DZ twins for an anthropometric trait, standing height. From studies by Newman, Freeman, and Holzinger (1937) we obtain:

	Number	Intrapair difference, cm
MZ	50	1.7
DZ	52	4.4
Sibs	52	4.5

MZ twins are much more alike than DZ twins and sibs. Similar results have been obtained for many other continuously varying anthropometric traits, such as weight, and for behavioral traits, such as assorted measures of intelligence.

Intelligence quotient (IQ) is determined by a standard testing procedure, for example, the Stanford-Binet test. The IQ consists of the quotient of the mental age of an individual as defined by the test multiplied by 100, divided by his chronological age. A score of about 100 is the mean of the population, so that higher and lower scores represent higher or lower intelligence—assuming of course that the IQ test is a true measure of the essentially indefinable trait, intelligence. Figure 7-1 depicts curves of intrapair differences in the Stanford-Binet IQ

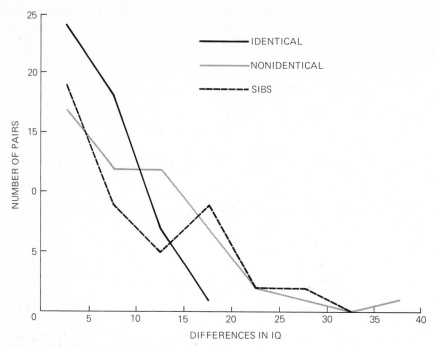

Figure 7-1 Genetic basis of intelligence. Curves are based on Stanford-Binet IQs of 50 pairs of identical (MZ) twins, 47 pairs of nonidentical (DZ) twins, and 52 pairs of sibs. (*After Newman, Freeman, and Holzinger, 1937.*)

for MZ twins, DZ twins, and pairs of sibs. The smaller difference between MZ twins compared with DZ twins and sibs is clear, as is the similarity of the DZ twins and sibs. The probability that the Stanford-Binet IQ is at least partly under genetic control must be rated high at this stage.

Twin data can also be assessed from correlations between members of a pair. We use a different type of correlation coefficient from that in Section 6-9, the *intraclass correlation coefficient*, which treats pairs symmetrically. It is defined as

$$r = \frac{2\Sigma(x_i - \bar{x})(x_i' - \bar{x})}{\Sigma(x_i - \bar{x})^2 + \Sigma(x_i' - \bar{x})^2}$$

The measurements x_i and x_i' are the pairs of measurements taken in a completely arbitrary order. This procedure is used because if a twin has a measure x and another y, it is not possible as in the calculation of the correlation coefficient in Section 6-9 to decide which twin is x and which is y. Therefore the only valid approach is to take paired measurements x and x' in an arbitrary order when it can be shown that the above formula is appropriate.

The H statistic formula in Section 7-2 has been widely used where it is expressed in terms of concordances of MZ and DZ twins. An exactly analogous H statistic can be derived from intraclass correlation coefficients. Thus if r_{MZ}

and r_{DZ} are the intraclass correlation coefficients of MZ and DZ twins respectively, then an H statistic is

$$H = \frac{r_{MZ} - r_{DZ}}{1 - r_{DZ}}$$

Obviously, if r_{MZ} is very much greater than r_{DZ} and is close to unity, H is close to unity; conversely, as in the case of an infectious disease, when r_{MZ} and r_{DZ} are expected to be about equal, H is close to zero.

The existence of MZ twins who have been reared apart permits an extension of this analysis so that the effect of two different environments on the same genotype can be investigated by comparing MZ twins reared apart (MZA) with those reared together (MZT). If r_{MZA} and r_{MZT} are the appropriate intraclass correlation coefficients, it is possible to estimate the effect of different environments on the same genotype. This estimate is calculated in a relation analogous to the above equation

$$E = \frac{r_{MZT} - r_{MZA}}{1 - r_{MZA}}$$

where E represents the environmental effect.

In addition H and E statistics can be shown to be expressible in terms of variances of the differences between members of twin pairs, thus

$$H = \frac{V_{DZ} - V_{MZ}}{V_{DZ}} \quad \text{and} \quad E = \frac{V_{MZA} - V_{MZT}}{V_{MZA}}$$

providing an alternative method of calculation.

We conclude with the cautionary note made when the H statistic was introduced in Section 7-2. A significant value for the H statistic is likely to indicate that there is genetic variability in a population for a trait, but it is not possible to estimate more definable quantities such as the degree of genetic determination and the heritability. Furthermore, little can be said about the genetic basis of the trait under study; for example, nothing can be said about the relative dominance of the genes controlling the trait.

7-4 HEREDITY AND ENVIRONMENT IN HUMAN BEINGS

Previous sections have presented indications of a high degree of genetic determination of some behavioral abnormalities in people. There is, especially for twins, the problem of a lack of control over the environment. This is a drawback in all studies on quantitative traits in human beings. In twin studies this problem is aggravated because of the difficulties inherent in comparing the environmental variation within twin pairs with that of unrelated individuals chosen at random and that of sibs who are not twins. The environment of relatives, especially sibs, is, of course, likely to be similar.

A method of assessing the effect of the similar environments of MZ twins, described in Section 7-3, has come from the small proportion of MZ twins who are separated at birth or soon after and then brought up apart (Table 7-4). These unusual MZ pairs provide a unique natural experimental situation for comparing the performance of two identical genotypes in different families, that is, in different environments. From two surveys, values of H and E as defined above are obtained, providing estimates of hereditary and environmental determination (Table 7-4). Generally H for height $>$ H for weight $>$ H for the various behavioral measures (IQ and personality). In particular, the values for personality assessments are low, probably because of the low precision and arbitrary nature of the tests. The E values are rather erratic, some being negative and some positive, in contrast with the H values, which are all positive. These values suggest the greater importance of genotype than environment for all traits, including IQ and personality, but does not mean that the environment is irrelevant. Variations in E may be partly explained by differences between the samples. Another behavioral example showing a greater genetic component than an environmental one is smoking habit (see Section 2-4), since MZ twins brought up together and apart are in close agreement, but collectively differ

Table 7-4 Estimates of Hereditary (H) and Environment (E) Determination of Traits of MZ Twins Reared Apart and Together, and of DZ Twins

Trait	H		E	
	Shields (1962)	Newman, Freeman, Holzinger (1937)	Shields (1962)	Newman, Freeman, Holzinger (1937)
Height				
Females	+0.89		+0.67	
Males			+0.89	−0.54
Both sexes		+0.81		−0.64
Weight				
Females	+0.57			
Both sexes		+0.78	−0.62	+0.27
			+0.68	
IQ				
Dominoes and vocabulary	+0.53			
Binet		+0.68		+0.64
Other			−0.04	
Personality				
Extroversion	+0.50		−0.33	
Neuroticism	+0.30		−0.36	
Woodworth Mathews Neuroticism Questionnaire		+0.30		−0.06

Number of pairs studied: Shields, 44; Newman, Freeman, and Holzinger, 19.
Source: Modified from Cavalli-Sforza and Bodmer (1971).

from DZ twins. Twin studies may, therefore, give valuable information on the genetic and environmental components of quantitative traits, especially in the rare cases where MZ twins separated at birth are included.

Convincing evidence for genetic similarity of twins comes from the work of Wilson (1972, 1975, 1977) based on the Louisville twin study that has recruited newborn twins for participation in a longitudinal study on growth and development. Wilson concluded from tests carried out at 3, 6, 9, 12, 18, and 24 months of age that only an unusual environmental situation would upset an infant's mental development, which is primarily determined by the twin's genetic blueprint. Such situations might include pairs where development of one or both twins is suppressed by serious prematurity or an impoverished enrivonment or where prenatal birth traumas occur (Wilson, 1972). Wilson (1975) subsequently assessed patterns of cognitive development at 4, 5, and 6 years of age. He concluded that within a broad range of home environments the genetic blueprint made a substantial contribution to cognitive patterning and development. A graphic representation constitutes Figure 7-2, where two pairs of monozygotic twins are shown for verbal and performance IQs. In the example the IQ values are similar, but also note the high level of concordance for the subtests making up the verbal and performance IQs. Even so, at this stage genotype × environment correlations begin to appear since measures of family socioeconomic status and parental IQ gave a correlation of 0.28 to 0.32 with the

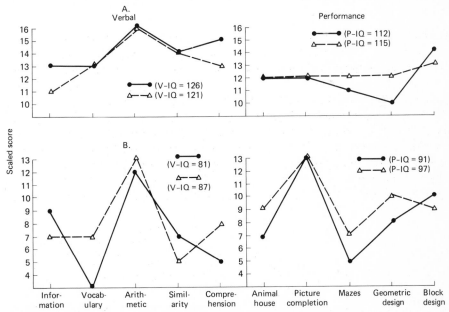

Figure 7-2 Profiles of subtest scores for two pairs of monozygotic twins for components of verbal IQ (V-IQ) and performance IQ (P-IQ). The twins in Figure 7-2A show a relatively flat profile of subtest scores, while those in Figure 7-2B show a marked scatter or dispersion among the subtests of both scales. The verbal and performance IQs are given in each case. (*Modified from Wilson, 1975.*)

twins' IQ at the age of 6. Not unexpectedly, similar conclusions are given in Wilson (1977) for the twins at 7 and 8 years of age. The continued reporting of this unique developmental study will be awaited with interest, for it will certainly assist in our interpretations of adult twin data. In 1977 Wilson's conclusion was: "Individual differences in intelligence will never be abolished, no matter how intensive the instruction or enthusiastic the instructor. The variation coded in the genotype is too deeply rooted to be swept aside by special training. But the maximum realization of each child's intelligence is a plausible goal, and educational efforts guided by this goal should be a matter of first priority."

Such an approach to education will be expected to lead to the genotype-environment correlation observed above. This is a common feature of studies on the relative roles of heredity and intelligence, however the latter is defined. Note how informative longitudinal studies are, as compared to surveys completed at one time.

We now extend the discussion beyond consideration of twins as such and consider family groups more widely. Considering the genotypic variance, in human beings we can include an additional component V_{am}, the variance due to positive assortative mating (Section 2-3), which has the consequence that homozygous genotypes tend to be somewhat more common than under truly random mating. As a consequence, V_A is inflated by increasing the frequency of individuals who have extreme expressions for a trait (normally homozygotes). The genotypic variance (ignoring epistasis, as in Section 6-4) can then be written

$$V_G = V_A + V_{am} + V_D$$

Using appropriate breeding techniques in experimental animals, V_{am} can be made equal to zero. One effect of V_{am} in human beings is to increase the heritability because V_{am} inflates the observed additive genetic variance.

We turn now to estimating the environmental variance. It is possible in experimental animals, but not for human beings, to control the environment precisely. One way of looking at the environmental variance in human beings is by subdividing it following Cavalli-Sforza and Bodmer (1971)

$$V_E = V_{ind} + V_{fam} + V_{soc} + V_{rac} + V_{GE}$$

where the variance components are defined as follows:

- V_{ind} is the variance among individuals within families. It is included in all families, but may vary from family to family. For example, the environmental variance for MZ twins may be less than for DZ twins; MZ twins, because of their genotypic identity, may choose similar environments. The environmental variance between DZ twins may differ from that between nontwin sibs, the latter including a birth-order component. Also, there may be variations for families of different sizes.

- V_{fam} is the variance among families within socioeconomic strata. It inflates the covariance between parent and offspring. Some idea of its importance can be obtained from the correlation between foster parents and adopted children, but the selective placement often carried out by adoption agents biases results.

- V_{soc} is the variance among socioeconomic strata. Cultural differences among families or social groups may be maintained by sociocultural inheritance that leads to correlations between relatives that are very difficult to distinguish from those that are due to genetic determination. Such factors are important in comparisons among racial groups. The geographic isolation in different environments that has permitted the development of genetic differences among races has also created a parallel, but probably largely independent, development of cultural differences.

- V_{rac} is the variance in environmental conditions accompanying racial differences, included in which are the sociocultural differences above. In some communities V_{rac} may be high, as are the differences between black and white Americans (Chapter 12).

- V_{GE} is the variance due to the genotype-environment interaction that occurs when given genotypes show different phenotypes in different environments (Section 6-2). It is difficult to give examples in people, but in addition to those cited in Section 6-2, it is instructive to look at an experiment on maze-learning ability in rats (Cooper and Zubek, 1958). Two lines were successfully selected by directional selection to be "bright" or "dull" at finding their way through a maze. For rats in the typical laboratory environment the difference in the mean number of errors between the maze-dull and maze-bright strains was about 50 (Figure 7-3), but in a restricted environment both strains were equivalent, since the brights were reduced to the level of the dulls. This reflects a genotype-environment interaction such that maze-brights are much more affected in the restricted environment than the maze-dulls. Conversely, in a stimulating environment the relative improvement of the maze-dulls compared with the maze-brights was much greater: creating a better environment improved the maze-dulls relatively more than the maze-brights. Therefore, there is a complex genotype-environment interaction in rats that can be assessed because both genotype and environment are definable more precisely than is normally possible for human beings. Indeed, it is not normally possible to define the genotype or environment of a human population. This means that the isola-

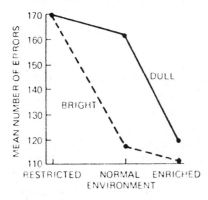

Figure 7-3 Genotype-environment interaction. Mean number of errors in a closed-field maze for bright and dull rats reared in enriched, normal, and restricted environments. (*After Cooper and Zubek, 1958.*)

tion of the interaction between the two components is intractable; hence, separating V_{GE} from V_G and V_E in *Homo sapiens* is a problem of extreme complexity. As is shown in Chapter 12, this conclusion is of fundamental significance in issues such as the interpretation (but not the existence) of racial differences in IQ scores and other behavioral traits.

Cattell (1965) has suggested an approach to the heredity-environment issue in *Homo sapiens* that uses several diverse family types and so has potentially more generality than the methods so far discussed. It is the multiple abstract variance analysis method (MAVA). The method entails a major difficulty of time and cost, since Cattell estimates that some 2,500 pairs of children are needed for a comprehensive analysis. This method is prohibitive from the practical point of view, so it is not surprising that few analyses have been carried out. The main categories of family types are (1) identical twins reared together; (2) identical twins reared apart; (3) sibs reared together; (4) sibs reared apart; (5) half-sibs reared together; (6) half-sibs reared apart; (7) unrelated children reared in the same family; and (8) unrelated children reared apart. From these categories, information on the correlation between heredity and environment can be obtained. For example, a correlation of +0.25 was obtained between hereditary and environmental effects on intelligence; this value is close to the values of +0.22 to +0.30 repeatedly obtained for correlations between intelligence and social status, as Wilson's data show above.

7-5 CAN RANDOM MATING BE ASSUMED IN HUMAN BEINGS?

In the derivation of correlations between relatives discussed in Chapter 6, it was assumed that mating occurs at random. As soon as there are deviations from random mating, such as inbreeding or assortative mating, the formulas given in Section 6-8 are no longer strictly accurate, as indicated in Section 7-4. In human beings, as long ago as 1903, Pearson and Lee found positive correlations between partners at marriage for physical traits such as stature and arm span. The correlation coefficients between mates were usually around +0.2. For example, Spuhler (1968) surveyed 105 physical traits in 40 human population samples. Correlation coefficients in the range of +0.1 to +0.2 are the most common for measurements of body size in Europeans and Americans of European descent, although coefficients smaller than +0.1 and in the range +0.2 to +0.3 are quite common. Correlation coefficients greater than +0.5 are rare. Data from Spuhler's paper are presented in Table 7-5. Studies of assortative mating in non-European populations are few. In two populations studied, the Rama-Navajo Indians and the Japanese, homogamy for body size was not found (Spuhler, 1968).

For behavioral traits, the tendency toward strong positive assortative mating occurs in some cases (Spuhler, 1962). One estimate of intelligence, the Raven Progressive Matrices Test gave a correlation coefficient of +0.399 ($P < 0.01$ for deviation from zero). A test on verbal meaning based on the selec-

Table 7-5 Correlation Coefficients Found in Various Studies for Various Physical Traits in a Human Population

| Trait | Correlation coefficients | | | | | | Total number of studies |
	<0	0–0.1	0.1–0.2	0.2–0.3	0.3–0.4	>0.4	
Stature	1	6	8	7	4	1	27
Sitting height	1		3	3			7
Weight		1	2	3	1		7
Chest circumference		2	5				7
Head circumference	2	3	1	2			8
Cephalic index	2	12	5	3			22
Facial index	4	7	3		1		15
Nasal index	3	2	1	2			8
Hair color			2	2	1		5
Eye color	1	1	1	1		1	5

Source: Spuhler, 1968.

tion of one of four words that best completes the meaning of each of 40 sentences gave $r = 0.305$ and 0.732 ($P < 0.01$ in both cases) for the total number of right answers and the proportion of right answers, respectively. Spuhler also reported data showing substantial positive assortative mating for such psychological traits as association, neurotic tendency, and dominance, as did Beckman (1962) for musical ability. Generally, the tendency for assortative mating seems stronger for behavioral traits than for physical traits.

Furthermore, a positive correlation is frequently found between socioeconomic status and stature. Similarly, there is a strong correlation for socioeconomic status between husband and wife. Therefore it can be argued that the positive correlations between mates with regard to stature may be partly due to a correlation with socioeconomic status. Also, since stature is correlated with a number of other physical traits, such as weight and chest circumference, similar correlations are expected for these traits and are found. However, in the Rama-Navajos no assortative mating for physical traits was found (Spuhler 1968). It could be argued that the discrepancy compared to Caucasians is due to a different social structure.

Stature has been increasing in most Western societies during this century. From data on Italian conscripts, Conterio and Cavalli-Sforza (1959) estimated the mean increase of stature as 0.1 cm per year or 3.0 cm per generation during this century. General improvement in living conditions, particularly with respect to nutrition and disease control, is undoubtedly of great importance, as indicated by a significant positive correlation of stature and socioeconomic group in the Italian data. The relative importance of genetic factors is more difficult to assess, but heterosis (hybrid vigor) as a result of the integration of formerly isolated communities has been postulated. For age at marriage there is a correlation of 0.8 between mates. Hence mates could show a correlation for stature merely because of a tendency to be born at the same time, if stature itself is increasing over the time period being considered. The literature contains other examples where positive correlations are removed when secular trends are taken into account. For example, Beckman (1962) found that the correlation between wives and husbands with respect to number of sibs disappeared when comparisons were restricted to single time periods.

For continuously varying traits, excluding nongenetic factors as discussed above, one of the main effects of positive assortative mating is to increase the additive genetic variance (V_A) compared with the random-mating situation (see also Section 7-4). If V_A is the additive genetic variance under random mating, $\hat{V}_A$ is that under positive assortative mating, and r is the correlation coefficient between mates, then, for a large number of genes Crow and Felsenstein (1968) have shown that

$$\hat{V}_A \approx \frac{V_A}{1 - r}$$

so that for $r > 0$ then $\hat{V}_A > V_A$. For example, for $r = 0.2$, $\hat{V}_A = 1.25V_A$. This shows that assortative mating may be of genetic importance as regards quantitative traits in human beings, including behavioral traits, and must be included in models devised to examine them.

So far we have largely considered the possibility of nonrandom mating within supposedly homogeneous groups for which the above genetic statement is valid. Unfortunately it is almost impossible to prove a group to be homogeneous since heterogeneity may appear under finer analyses. The north-south gradient of blood groups found in the British Isles (Mourant, 1954) is reproduced in Victoria, Australia, when people are classified according to their ethnic origin (Hatt and Parsons, 1965). Random mating should rapidly alter this situation; the persistence of the gradient suggests that there may be strong positive assortative mating according to place of origin. Using surnames as indicators of ethnic origin, the distributions of marriages in Victoria over a 3-month period in 1963 were obtained. The four classes used were based upon individuals with English (E), Scottish (S), Irish (I), and other surnames (X). Omitting the very large class X, the marriage distributions are given in Table 7-6. There is a considerable excess of marriages between people having Irish surnames and an almost equal excess for Scottish surnames, associated with a substantial deficiency in the Scottish × Irish class.

It is curious that Australians with Scottish and Irish surnames are the most isolated from each other in view of the ethnic similarity between these groups, both being predominantly Celtic. Apparently the spatial isolation between the groups in their homelands is being maintained in Australia partly because of cultural differences, since people of Irish descent are predominantly Roman Catholic while those of Scottish descent are not. In addition, there is possible imprinting in the choice of one's spouse based upon traits of one's relatives, especially one's parents (Parsons, 1967a). Even so, amalgamation will no doubt occur relatively rapidly. However, when skin color variations occur across groups, then amalgamation is usually slow. In the United States there is

Table 7-6 Frequencies of the Six Possible Marriage Classes from English (E), Scottish (S), and Irish (I) Surnames with Expectations Based on Random Mating

Marriage class	Observed frequencies	Probability* assuming random mating	Expected frequencies	$\dfrac{\text{Expected}}{\text{Observed}}$
E × E	141	p^2	138.785	0.984
E × S	148	$2pq$	151.354	1.023
E × I	100	$2pr$	101.077	1.011
S × S	50	q^2	41.265	0.825
S × I	41	$2qr$	55.116	1.344
I × I	26	r^2	18.404	0.708

* Letting p, q, and r be the frequencies of E, S, and I surnames, these probabilities come from the Hardy-Weinberg proportions $(p + q + r)^2$ where $p + q + r = 1$, a trinomial expansion.
Source: Hatt and Parsons, 1965.

little tendency toward white-nonwhite marriages. Only 2.3 percent of the marriages between whites and nonwhites that would be expected assuming random mating occurred in 1960, although a slow increase in mixed marriages is occurring with time. Not unexpectedly there are conspicuous differences among states, with Hawaii approaching midway toward random mating in the period 1959–1964 (Cavalli-Sforza and Bodmer, 1971). In general, however, skin color differences among races are likely to be preserved for many generations due to nonrandom mating; this, as will be seen in Chapter 12, makes cross racial comparisons difficult and of limited validity.

7-6 THRESHOLD TRAITS

Section 7-2 discussed threshold traits in twins. We now extend this to relationships between relatives generally. (This section can be omitted on a first reading.)

Wright (1934) introduced a model for dealing with threshold traits in a study of the number of toes of guinea pigs. The assumption made is that threshold traits are inherited polygenically in the same way as the quantitative traits discussed in Chapter 6. We can look upon quantitative traits such as those discussed in Chapter 6 as threshold traits for which we can all be assessed. Regarding intelligence, for example, individuals can be classified as normal, borderline, or defective, according to predetermined cutoff scores on a test. Therefore, in the analysis of threshold traits it is reasonable to assume an underlying continuous variable, which is inherited in the same way as the continuously varying traits discussed in Chapter 6. In Figure 7-4 all individuals with a value of x greater than some threshold value (T) are assumed to have the trait. The added complication is that we are obliged to specify the relation between the distribution of x and the underlying continuous variable, and P, the proportion of people having the trait.

Threshold traits range all the way from those that show hardly any evidence of familial concentration to those that can be explained in terms of a

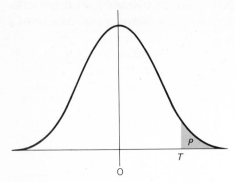

Figure 7-4 Basic model for threshold trait. All individuals with a value of *x* exceeding *T* are affected. The proportion of affected individuals (*P*) is the area under the curve beyond *T*.

single gene with reduced penetrance. A method of dealing with human threshold traits was developed by Falconer (1965). He used some of the concepts developed by animal and plant breeders to predict the outcome of directional selection experiments (Section 6-11). In such experiments, a proportion of a population is selected to provide subsequent generations. In the analysis of threshold traits, the analogy is the proportion of relatives of affected individuals who are themselves affected. If we take the *liability* to a certain disease in a population at large as represented by individuals above the threshold value (*T*), we then compare this liability with that among the relatives of those affected. In other words, the liability to the defect rather than the defect itself is being considered. In Figure 7-5, distribution 1, a vertical line indicates the threshold (*T*) in the

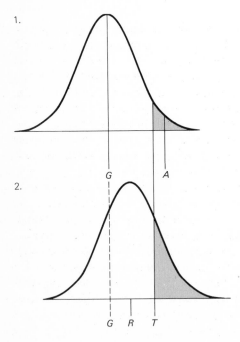

Figure 7-5 Inheritance of liability to diseases. Distribution 1 represents the general population. Distribution 2 represents the relatives of affected individuals compared with a fixed threshold (*T*). *G* is the mean liability of the general population. *A* is the mean liability of affected individuals in the general population. *R* is the mean liability of the relatives.

population at large; distribution 2 gives the liability distribution of the relatives of affected individuals. The mean in distribution 2 is shifted toward the threshold (T) indicating that the liability to the disease has a genetic component. We are therefore interested in the liability among the relatives derived from all those individuals with liability above T in Figure 7-5, distribution 1. The directional selection experiment analogy is to find the gain in liability due to this selection process of breeding only from such affected individuals. The difference between the means of the two distributions ($R - G$) gives the realized gain in liability based on selecting the relatives. In Figure 7-5, the difference between the mean of the general population and that of the selected individuals (A), or $A - G$, is equivalent to the selection differential of the directional selection experiment (Section 6-11).

The ratio of these two differences

$$\frac{R - G}{A - G}$$

is the regression of relatives on affected subjects with respect to liability. This regression coefficient comes from the slope of the line obtained by plotting a number of $A - G$ values against $R - G$ values. We are therefore dealing with a simple parent-offspring regression situation for which we have already seen (Section 6-10) that

$$b_{OP} = \frac{\tfrac{1}{2}V_A}{V_P} = \tfrac{1}{2}h_N^2$$

which is also equal to the above ratio. Therefore, the regression of relatives on affected subjects gives an estimate of the heritability of the liability to a disease.

Falconer (1965) prepared a graphic representation (Figure 7-6) of the incidence in the general population and in relatives who are sibs, parents, or children (first-degree relatives). The incidence among them is represented along the vertical axis. The population incidence is plotted along the horizontal axis. Both the horizontal and vertical incidence scales are logarithmic. To estimate heritability, first read along the horizontal axis to determine the point representing the population incidence; then read up to the point representing the familial incidence. The point where the two incidences intersect is the heritability.

Generally we can write, for relatives of various degrees of relationship

$$b = rh_N^2$$

where r is the coefficient of relationship. This equation is strictly true only when the dominance variance (V_D) = 0, that is, when dominance plays no role. For human populations this applies to many but not all those relationship categories readily available for analysis. Full sibs (Section 6-9) are an exception, but in any case V_D is usually much lower than the additive genetic variance. Some values

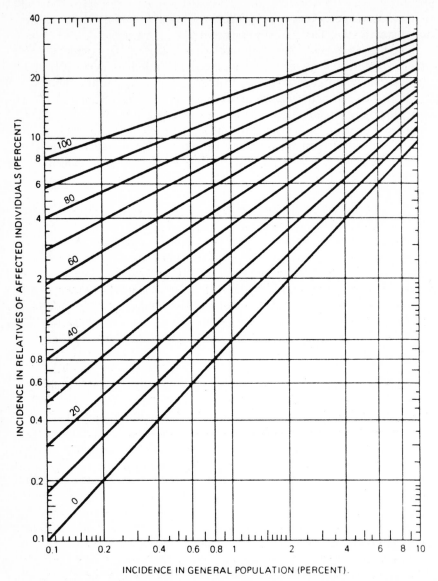

Figure 7-6 Heritability of liability for a threshold trait from two observed incidences when the relatives are sibs, parents, or children. Numbers on the lines are the heritabilities (h^2) in percent. (*After Falconer, 1965.*)

for r for more remote relatives are given in Section 6-9. The above equation, together with other considerations presented in this section, show that if mean liabilities are available in a population, then twin data will provide estimates of heritability or the degree of genetic determination unlike the H statistic.

A detailed analysis using this method has been carried out on the disease diabetes mellitus (Falconer, 1967). Diabetics show a normal IQ distribution, but

Table 7-7 Prevalence of Diagnosed Diabetes Mellitus, Based on U.S. National Health Survey, 1960

Age (years)	Males, %	Females, %
0–24	0.11	0.07
25–44	0.49	0.38
45–54	1.12	1.37
55–64	2.52	3.15
65–74	3.44	5.03
75 and over	3.15	3.88

Source: Rosenthal, 1970.

diabetic children seem to be relatively superior in verbal expression and retarded in performance (Rosenthal, 1970). Some studies purport to show an emotional basis for the onset of the disease, but others do not. Some have claimed a distinctive personality in diabetics, and the question of whether the disease produces abnormal psychological behavior has been raised. As Rosenthal comments, it is difficult to reach conclusions, especially as culture may be one determinant in its manifestation (Section 3-3). A major problem in dealing with the disease is that its incidence is age-dependent, as shown in Table 7-7. The incidence is about 0.1 percent for persons up to 24 years of age and 3 to 5 percent for persons over 60 years of age. The incidence is to some extent a matter of the criteria used to define the disease, which may vary from an almost complete lack of insulin activity to a mild but consistently elevated blood sugar level of little clinical significance. The incidence increases if diagnostic criteria are extended to include more than just the clinical manifestations of the disorder. Thus the incidence of clinical diabetes is about 3.5 percent in the United States, but this figure increases to over 6 percent if a glucose tolerance test is employed as the diagnostic criterion.

Neel et al. (1965) have written: "Diabetes mellitus is in many respects a geneticist's nightmare. As a disease, it presents almost every impediment to a proper genetic study which can be recognized." Some authors have suggested that single-gene inheritance, particularly that due to a single recessive gene with incomplete penetrance, may be the genetic basis—a view not held by Neel et al. (1965), who argue for multifactorial inheritance. However, Edwards (1960) points out that there is little difference between a model of a single gene with incomplete penetrance and multifactorial inheritance, unless the variability in penetrance is entirely environmental. Even though variations in diagnostic precision occur that would increase the environmental variance, it seems reasonable to consider diabetes a threshold trait.

It is clearly desirable to have both subjects and relatives close to the same age in carrying out analyses of diabetes incidence; a sib analysis therefore is convenient. This particular method has disadvantages, as pointed out in Section 6-9, because (1) environmental causes of resemblance (V_{EC}) may be important for a particular dietary regimen; and (2) the dominance variance (V_D) may be

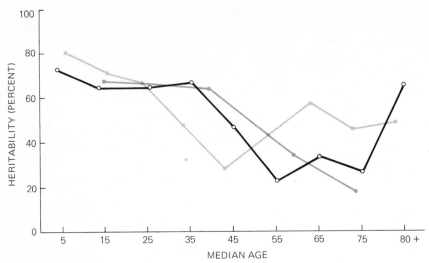

Figure 7-7 Changes in heritability of liability for diabetes mellitus with increasing age as estimated from sib correlations. *Dark gray line:* data on Canadian males; *solid line:* data on Canadian females; *light gray line:* data on both sexes in Birmingham, England. (*After Falconer, 1967.*)

relevant even though it is likely to be much smaller than the additive genetic variance (V_A). Therefore the heritability estimates have these limitations. Figure 7-7 shows the variation in heritability estimates for three samples. The heritability declines from between 60 and 80 percent for persons under 40 years of age to between 20 and 40 percent for persons between 40 and 70 years old. Therefore, as has been suggested in the literature, the genetic etiology of the disease for younger people may differ from that of older people whose vulnerability may be due to aging. Probably all that can be said at present is that the hereditary basis of the disease is multifactorial, with a threshold for its manifestation associated with the possibility that different genes may influence early and late onset. Some researchers are now tabulating the antibodies of diabetics in an attempt to connect viruses to diabetes; see Notkins (1979) for a description of this approach. Diabetes is discussed here in some detail because some of the diseases described in Chapter 11 reveal similar difficulties in analysis, for example, variations in diagnosis and in age of onset.

SUMMARY

We cannot define the environment of human beings as we can with experimental animals. This restricts all quantitative studies of human traits and has led us to great difficulties in the interpretation of data on traits such as IQ. In addition we cannot carry out breeding experiments, so the only data available are those collected from retrospective pedigrees.

In attempting to unravel the effects of genotype and environment, studies of twins play a major role. This is based upon the comparison of monozygotic (genetically identical) and dizygotic (genetically nonidentical) twins. Many have criticized the twin approach because of the apparent lack of environmental control. However, convincing evidence for the genetic similarity of twins comes from longitudinal studies where the conclusion is that individual differences in intelligence and related traits are determined at least partly genetically.

From the few studies of extended family groups a degree of genetic control of behavioral traits is found, and often quite large correlations between heredity and environment occur for traits such as IQ. Such studies, however, are extremely expensive and few have been carried out.

A further complication to studies of humans is the almost universal presence of positive assortative mating, which is a tendency for matings to occur between phenotypically similar individuals more frequently than would occur by random mating. One of the effects of assortative mating is to alter variance components compared with random mating. Other complications will be discussed in Chapter 12, such as variability among socioeconomic strata, and in environmental conditions accompanying racial differentiation. The analysis of quantitative data on humans is a complex and difficult task which should not be undertaken lightly.

GENERAL READINGS

Cavalli-Sforza, L. L., and W. F. Bodmer. 1971. *The Genetics of Human Populations*. San Francisco: Freeman. This excellent text includes an advanced chapter on the genetic analysis of quantitative traits in human beings. More detailed derivations of the equations given in this chapter are provided.

Mittler, P. 1971. *The Study of Twins*. London: Penguin. A very readable account of the place of twin studies in behavior-genetics research.

Shields, J. 1962. *Monozygotic Twins Brought Up Together and Apart*. London: Oxford University Press. One of the few classic analyses of twins brought up together and apart.

The Genetics of Behavior:
Drosophila

8-1 SINGLE-GENE EFFECTS

This chapter initiates the third section of this book (see Chapter 1 for its plan). To this point we have concentrated on principles, but in this and the next four chapters we look at behavior phylogenetically, beginning in the present chapter with *Drosophila,* followed by Chapter 9 on rodents and Chapter 10 on various other organisms ranging from bacteria and protozoa to assorted mammals. In Chapters 11 and 12, we proceed exclusively to *Homo sapiens.* The switch to specific organisms is made so that we can specify those behaviors that have been and can be studied in certain experimental subjects. As we saw in Chapter 1, the types of behavior that can be studied depend on the organism. In some species of *Drosophila* very sophisticated genetic techniques have been applied to many areas of study, including behavior. Furthermore, the ease of rearing large numbers of individuals from a single genetically controlled laboratory strain means that elegant statistical methods can be employed as an aid to drawing conclusions. Table 8-1 gives a full summary of the virtues of *Drosophila* as an experimental animal. (See also the ongoing *Genetics and Biology of Drosophila,* a most comprehensive anthological series on this genus listed in the General Readings at the end of the chapter. We call attention particularly to

Table 8-1 Reasons for *Drosophila's* Value as an Experimental Organism for Genetic Research (with Particular Reference to the Cosmopolitan Species, Especially *D. melanogaster*)

- *Short generation time.* Many species develop from egg to adult in less than 2 weeks

- *Ease of rearing.* Even novices can breed and rear fruit flies with rewarding success

- *Inexpensiveness.* Flies can be raised cheaply in large numbers; they eat fermenting fruit and yeasts

- *Small size.* Many flies can grow in a small space, often a few jars

- *Large numbers of progeny.* A single gravid *Drosophila* female can produce hundreds of offspring

- *Harmlessness. Drosophila* carry no known disease that affects human beings; adults possess no biting or piercing mouth parts

- *Sex ratio.* Most species produce approximately equal numbers of each sex among their progeny. Exceptions are usually situations of genetic interest

- *Parthenogenesis.* There are species with strains made up almost entirely of females, in particular *D. mercatorum* (Carson, 1973)

- *Many species.* There are more than 1,500 species described so far

- *Wide distribution. Drosophila* species are found throughout the world, from the arctic to the tropics

- *Ease of collection.* It is easy to collect *Drosophila* and to bring captured flies back to the laboratory in good condition

- *Low chromosome number. Drosophila* possess low numbers of easily identifiable chromosomes; some species have only six chromosomes or three pairs

- *Larval salivary gland chromosomes.* The large size of these giant, many-stranded chromosomes permits the researcher to identify even small sections of a single chromosome, much as one recognizes the face of a friend

- *Hybrids.* The large number of closely related member species permits the breeding of hybrid fruit flies in the laboratory

- *Races and/or subspecies.* A variety of races provides research material for those interested in the evolutionary process by which new species arise—speciation

- *Isolating mechanisms. Drosophila* exhibit several isolating mechanisms (such as sexual isolation and hybrid sterility) that prevent gene exchange between species

- *Mutants.* Because of their response to such mutagens as X rays and chemicals, *Drosophila* mutants can be produced easily in the laboratory; mutation can alter the size, color, number, and/or structure of almost all parts of the fly's body

- *Behavior.* Many behaviors lend themselves to genetic analysis and modification through selection

- *Symbionts.* Several *Drosophila* species carry a variety of microorganisms, allowing researchers to study symbiotic relations; since some of these microorganisms are "inherited," i.e., transmitted from parents to offspring; geneticists have a special interest in the process

- *Cytoplasmic inheritance. Drosophila* can in some instances transmit extrachromosomal units of heredity to progeny

Source: Modified from Ehrman, 1971.

volume 2b on behavior.) This chapter does not endeavor to be comprehensive—probably a hopeless goal in the face of the great mass of genetic studies on *Drosophila*. Many investigations appropriate to this chapter were discussed in earlier chapters and we will refer to them. Some of the evolutionary aspects are dealt with in Chapter 13. A review of the literature to mid-1971 appears in Parsons [1973].

Besides the well known *D. melanogaster,* there are now more than 1,500 formally recognized *Drosophila* species. Some eight of these are recognized as cosmopolitan: *D. melanogaster, D. simulans, D. ananassae, D. immigrans, D. hydei, D. funebris, D. repleta,* and *D. busckii.* They live in six life zones: Nearctic, Neotropical, Palearctic, Ethiopian, Oriental, and Australian, all of which contain a wide variety of habitats. These species can all be readily collected in the wild using fermented-fruit baits and are easily cultured in the laboratory. In addition, there are many widespread species occurring in fewer than six life zones, but including species such as *D. pseudoobscura* (see Parsons and Stanley 1980). Many of the discussions in this book are based upon cosmopolitan and widespread species, although some mention is made in Chapter 13 of more specialized species not attracted to fermented-fruit baits.

In deference to the "Queen Mother" of all *Drosophila* species, we begin with those *D. melanogaster* mutants that affect behavior (see Figure 2-3, a linkage map). Many diverse behaviors are controlled by single genes situated at loci scattered throughout the fly's chromosomes. *D. melanogaster* has by far the best known *genome* (one haploid set of genes and chromosomes) among the *eukaryotes* (organisms made up of cells with nuclei bounded by nuclear membranes, undergoing meiosis); this makes it particularly valuable in the analysis of new subdivisions of genetics as they arise. Behavior genetics is no exception to this rule.

As a specific example, Figure 8-1 presents data on the electroretinogram (ERG, a record of the mass electrical response of appropriate cells to illumination) of wild-type, normal-visioned *D. melanogaster* and of flies with photoreceptor mutants known to be sex-linked. Flies were dark-adapted for at least 15 minutes before electroretinography in response to half-second white light stimulus. Only intact, living flies were employed. The recording electrode was placed into the retina via a minute corneal hole. The independently occurring mutations either result in abnormal electroretinograms or preclude them altogether, producing diminished visual acuity in either instance. (See also Alawi et al., 1972, on mutation in *D. melanogaster* affecting phototransduction in insect vision, a mechanism by which the reception of sensory stimuli is associated with ionic events in receptor membranes.)

Benzer and collaborators have studied, among other point mutations affecting locomotor, visual, sexual, stress response, and neuromuscular behaviors, three single-gene mutations altering the normal circadian (about 24 hours) rhythm exhibited by *D. melanogaster* (Konopka and Benzer, 1971; Benzer, 1971, and references therein; Benzer, 1973). Flies of this species normally eclose (emerge) "around and about" dawn, when the presence of dew and conditions of temporarily higher humidity satisfy their omnipresent requirements for moisture; this indeed is the root of the noun *Drosophila*—"dew lover." Then, for most *Drosophila* species observed, a morning period of activity follows. It is over by midday and is followed by an evening spurt of activity.

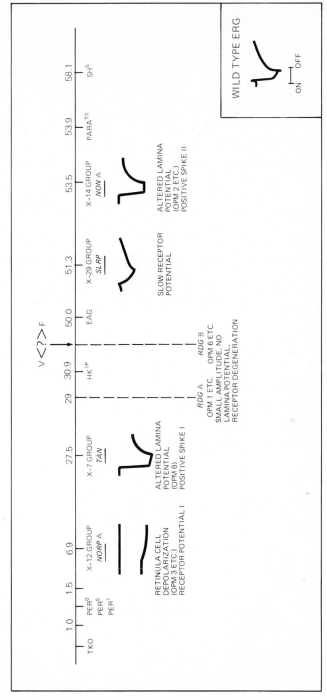

Figure 8-1 Neurological mutants on X chromosome of *D. melanogaster*. Map positions (not to scale) of mutants are indicated above line. Symbols for loci are shown below line together with a representative trace of the phenotypes of the electroretinogram (ERG) mutants. Two such traces are shown for X-12 group of mutants to demonstrate that some alleles at this locus have no response to a light stimulus, while other alleles show a small depolarization. An ERG on wild-type flies, with the duration of light stimulus indicated below, is presented for comparison. A phrase beneath each ERG provides a brief description of each mutant. Terms in parentheses show alternate designations of same lesions. (*From Grossfield, 1975.*)

167

Compelling evidence for the genetic control of this biological clock (Pitten-drigh, 1958) is presented by flies with the *arrythmic* mutant, which eclose ad lib all day long; flies with the *short-period* mutant, which display a consistent 19-(not 24-) hour cycle; and flies with the *long-period* mutant, which have a 28-hour cycle (Figure 8-2). Consider how effectively such different rhythms could establish isolation between and among flies with variant clocks (Section 5-5).

Genetic *Drosophila* mosaics can be generated—as opposed to passively awaited for exploitation after they occur spontaneously, usually at very low rates—by depending upon the loss of a reliably unstable ring X chromsome. Ring chromosomes do not as a rule successfully migrate to either pole in the terminal phases of either mitosis or meiosis, and so they are omitted or destroyed in telophase, the final separatory stage of these cycles; the extent of male tissue depends upon the time during development when the loss occurs. At that point, two embryonic cell clones are established, one with a single X and the other with two X chromosomes. Subsequently, the XO imago sections are male (unlike the XO sterile human Turner female discussed in Section 4-3), while the XX sections are female, and the imago (adult insect) sections can be readily identified by using sex-linked mutants for eye color, body color, and bristle shape. Given this basic technique plus the extreme genetic control extant with (only) *D. melanogaster*, two-dimensional embryonic *fate maps* can be constructed. These correlate precise anatomical sites with abnormalities affecting behavior (Hotta and Benzer, 1972; see Figure 8-3).

Consider one of several detailed fate maps that has been constructed for the sex-linked *Hyperkinetic* (*Hk*) gene (Ikeda and Kaplan, 1970a,b): Homozygous females and hemizygous males, and to a lesser degree heterozygous females, display arrhythmic leg-shaking action during ether anesthetization. Fig-

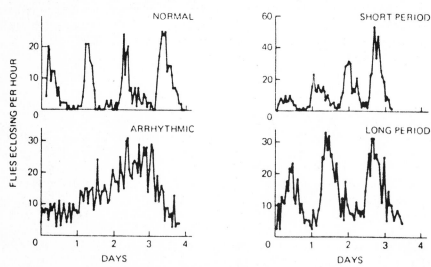

Figure 8-2 Rhythm of eclosion in *D. melanogaster* from populations of wild-type and mutant pupae kept in constant darkness. (*From Konopka and Benzer, 1971.*)

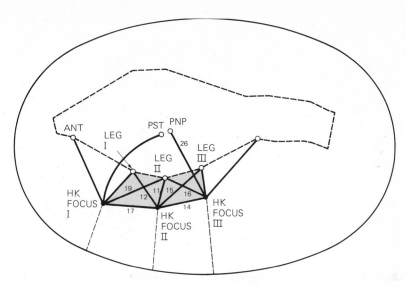

Figure 8-3 Fate map for *Hk* gene in *D. melanogaster*. Three foci related to mutation are shown on a background representing one side of the embryonic blastoderm. *ANT:* antennae; *PNP:* postnotopleural bristle; *PST:* presutural bristle. Mutation at these sites is related to mutant behavior. (*After Hotta and Benzer, 1972.*)

ure 8-4 presents nine different gynandromorphic (i.e., mosaic with cells of different sex) patterns, mosaics of female and male parts, with respect to this mutant. Six hundred such mosaics were collected, scored for surface landmarks (male surfaces have different colors, eye and body, and differently shaped bristles), and examined for shaking while anesthetized. Ikeda, Kaplan, Hotta, and Benzer all agree that the shaking or control of each of the six legs is independent of all other legs and that there is strong likelihood of identity between the shakiness of a leg and the genotype of its cuticle (noncellular, waterproof external covering). Figure 8-3, shaped like the ovoid *Drosophila* blastoderm (a layer of cells surrounding all the yolk in an insect's fertilized egg), depicts three separate foci, the structures in the fly that, when mutant, cause mutant behavior. There is one focus for each leg on each side of a fly. All foci lie within the blastodermal region associated with the origin of the insect's ventral nervous system—ventral to blastodermal cuticular regions. This last is so despite the often observed sameness of the leg genotype and the cuticular genotype. There exists electrophysiological evidence of abnormally behaving thoracic ganglia (clusters of nerves in insect's "chest") in *Hk* individuals; expression of *Hk* is genetically autonomous, and the left and right sides of each thoracic ganglion are independent of each other. We may then cogently speculate about the location of parts of the thoracic ganglia on this fate map, for example, perhaps in the shaded area imposed upon Figure 8-3.

The hyperkinetic mutants, *Hk*[1] and *Hk*[2], show another aberrant behavior—they jump and fall over when an object moves near them. This

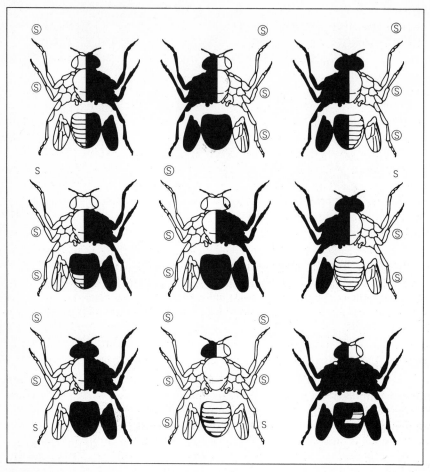

Figure 8-4 Hyperkinetic *D. melanogaster* genetic gynandromorphs. *Black:* female tissue; *white:* male tissue; *S:* shaking leg; *encircled S:* rhythmic bursts from motor regions associated with the leg. (*From Ikeda and Kaplan, 1970a,b.*)

behavior, the kinetogenic response, is measured by the experimenter moving his hand above a vial containing a single fly and scoring the number of responses in 50 trials (Kaplan and Trout, 1974). At the fifth day of age, Hk^1 responds on average 42 times; Hk^2 responds 28 times out of 50. The two genes are allelic since the Hk^1/Hk^2 heterozygote also responds to movements. Another shaker mutant showing a kinetogenic response is *Shaker*⁵ (*Sh*⁵), which has a distinctly different shaking pattern from that of the *Hk* mutants. In addition, Williamson, Kaplan, and Dagan (1974) investigated the sex-linked temperature-sensitive mutant, *para*ᵗˢ, which moves normally at 22°C but is paralyzed within 10 seconds at 29.5°C, while recovery occurs within 5 seconds when flies are returned to 22°C. Kaplan and his coworkers hope that studies on these mutants may possibly serve as models for pathological conditions in human beings. Even if this does not turn out to be directly so, it must surely provide incentives for further fundamental research on these curious mutants and their aberrant nervous systems.

8-2 QUANTITATIVE TRAITS INVOLVING SEXUAL BEHAVIOR

We have seen elsewhere in this book that in *Drosophila* many quantitative traits are under genetic control and can be studied using biometrical methods and selection experiments. A list of the types of traits studied is presented in Section 5-1. Geotaxis (movements with respect to gravity) is discussed in Section 5-2 to illustrate the selection experiment approach and its application in obtaining information on the genetic basis of a trait. Diallel crosses for mating speed and duration of copulation are considered in Section 6-5, and in Section 6-10 locomotor activity is discussed as an example of the regression approach in the analysis of quantitative traits. We are now ready to consider those aspects of quantitative behavior in *Drosophila* whose hereditary bases are known and which we have not discussed in detail previously. This section considers traits involving sexual behavior.

Manning (1961) selected for fast and slow mating speeds in *D. melanogaster,* basing his selection upon the performance of 50 pairs of flies collected before mating could occur and placed together in a mating chamber. The 10 fastest and 10 slowest pairs were identified, selected, and used to initiate fast and slow selection lines. The response to this selection rapidly produced two fast and two slow lines; an unselected control line was maintained. After 25 generations, the mean mating speed averaged 3 minutes in the fast lines and 80 minutes in the slow lines. The differences between the lines is shown in Figure 8-5. Considerable variations in speed attributable to environmental fluctuations occurred during selection, but the fluctuations were generally similar in all lines for a given generation. An approximate realized heritability of 0.30 was computed from the rate at which the selection lines diverged during the first few generations. Although no further genetic analyses were carried out, Manning analyzed in some detail how selection affected behavior. Hybridizing the fast and slow lines in both directions (reciprocal crosses) gave intermediate F_1 mating speeds, while intercrossing the two fast and slow lines themselves reciprocally gave fast and slow speeds, respectively. These results indicate that both sexes were affected by selection. Confirmation of this came from testing mating speeds against an unselected stock of flies: both sexes of the selected lines gave altered mating speeds in the expected directions. Activity differences between lines were measured by admitting flies to an arena where the number of squares entered by a fly in a given time period was scored. The slow lines exhibited much more of this type of activity than did the fast ones. Experiments using unselected females with selected males showed that the lag before courtship was much smaller for the fast than for the slow lines; similarly, the frequency of licking (contact between the male proboscis [insect's tubular mouthpart] and the female genitalia; see Section 3-2) was higher in the fast than in the slow

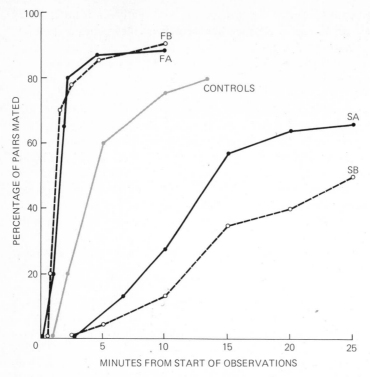

Figure 8-5 Mating speeds in *D. melanogaster* in two lines selected for fast speeds (FA, FB), two selected for slow speeds (SA, SB), and controls at the eighteenth generation of selection. (*From Manning, 1961.*)

lines. Thus the fast lines have a high level of "sexual activity" and a low level of "general activity," and the slow lines have the opposite. Under natural conditions these two components are presumably coordinated at an optimum, since overresponsiveness or underresponsiveness would obviously be undesirable.

Selection for this trait operates on both sexes, but there could be different genes controlling the response within the two sexes. Manning (1963) attempted to look into this by selecting for mating speed based on the behavior of one sex only. There was no response to selection for fast-mating males or slow-mating females, and a fast-mating female line was not set up. One wonders if natural selection had not already fixed these genes for rapid mating in males, since it so clearly enhances their fitness. Indeed, as discussed in Section 6-5, Fulker (1966) found just such evidence for directional selection for fast mating. The general importance of rapid mating speed as a component of fitness in *Drosophila* is further emphasized by Parsons (1974a). There was, however, a response in male lines selected for slow mating. In these lines, the mating speed of females was unaffected in early generations but somewhat reduced in later generations. Behaviorally, both sexes showed lower general activity and the males showed reduced courtship activity, which contrasts with other experiments. Manning was unable, however, to come to any definite conclusions concerning possible differences between sexes in the genetic control of mating behavior.

Genes affecting general activity and sexual activity are apparently wholly or partly controlled by separate genetic systems that can be modified independently. Ewing (1963), a student of Manning, selected for spontaneous activity and found that his inactive lines displayed greater sexual activity, as would be expected from Manning's observations. However, Ewing's technique involved placing 50 flies of one sex into the initial tube of a line of interconnected tubes and selecting the first 10 flies (active) and the last 10 (inactive) to reach the opposite end. This procedure separated flies that moved through rapidly from ones that did not, but when the two lines so formed were tested by introducing single flies into Manning's arena, there was no significant difference between them; thus the two types of behavior seem to differ. (Here a caveat: this is but one example of an often observed and often referred to *apparatus effect* on behavior.) Manning was measuring spontaneous activity; Ewing was measuring dispersal activity or the reactivity of flies with one another.

In a study of the control of receptivity, Manning (1966, 1967a) found female acceptance of a courting male to depend on two processes. The first determines whether a *D. melanogaster* female is accessible to the courtship of males. Young females are unresponsive for some 36 hours after eclosion; then quite suddenly they become receptive and accept a male after a few minutes of courtship. The evidence suggests that this rapid change of receptivity occurs when the concentration of circulating juvenile hormone rises with the activation of the corpus allatum (one of the insect's endocrine organs) and the ovaries show a growth cycle parallel to that of receptivity. The second process can be called *courtship summation* and involves the addition of all the heterogeneous stimuli provided by a courting male that induce a female to allow him to mount once a critical level of stimulation is reached.

The evidence advanced indicates that the two processes are distinct and that the change from the unreceptive state to the receptive is an all-or-nothing process. Females are completely unresponsive to courtship or they accept within the normal time range for receptive females (about 95 percent of females accepting males within 15 minutes of introduction to them). There is no evidence that females become gradually more receptive by requiring less and less courtship before accepting. Virgin females remain receptive for many days, but after the first week of adult life an increasing proportion become unreceptive, and the switch off seems to be a rapid all-or-nothing event like the switch on. Old females that have mated and used up their stored sperm (by depositing many fertilized eggs) are more often receptive than are virgins of the same age. It is suggested that this is because their corpora allata are more active and the juvenile hormone concentration is kept above the critical level longer.

Manning (1968) selected successfully for slow mating speed in *D. simulans,* a sibling species of *D. melanogaster,* in which the behavior of males was not affected but in which there were marked effects in females (Figure 8-6) in

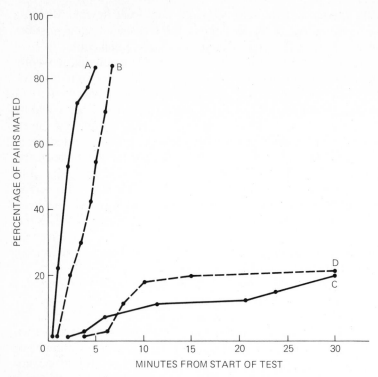

Figure 8-6 Mating speeds in *D. simulans* for (*a*) selected males × control females; (*b*) control males × control females; (*c*) selected males × selected females; (*d*) control males × selected females. (*From Manning, 1968.*)

contrast to the mating speed selection experiments in *D. melanogaster,* in which both sexes were affected. (*D. melanogaster* and *D. simulans* are very closely related and morphologically similar species—females more so than males; they are therefore referred to as sibling species. See Section 4-2.) Most of the slow-mating-line females failed to become receptive on the second day after eclosion, as do normal flies. The females in fact performed the most vigorous repelling movements, by extruding the ovipositor and twisting or lifting the abdomen beyond the reach of the courting male. These movements normally are performed in these species by elderly virgins that have become unreceptive or by fertilized females whose receptivity is inhibited by the presence of already stored sperm in their seminal receptacles. However, the females in Manning's slow-mating line had normal ovarian growth and a corpus allatum complex that, when implanted into normal hosts, was capable of producing precocious receptivity. The experiments suggest that the females have a normal supply of juvenile hormones and that the genetic change involves one or more links in the chain of neural "target organs" on which the juvenile hormone acts. As Manning (1968) pointed out, this situation is comparable to that in some mammals, such as guinea pigs (Valenstein, Riss, and Young, 1955).

Kessler (1968, 1969) selected for fast and slow mating speeds in *D. pseudoobscura,* crossing three wild-type strains (from British Columbia, California, and Guatemala) and using a technique essentially similar to Manning's. After the twelfth generation of selection, tests were made on all possible combinations of fast, slow, and control lines as a 3×3 diallel cross in which observations were carried out for 30 minutes using 50 pairs in one container, copulating pairs being removed with an aspirator. It was found that slow-mating females reduced mating whenever they were involved and that fast-mating females were not significantly faster than the controls. Fast-mating males speeded up all matings in which they were involved, but slow-mating males were not significantly different from the controls. An analysis of variance showed females to account for more of the total variance than did males. This contrasts with data of Kaul and Parsons (1965) for ST/ST, ST/CH, and CH/CH karyotypes (ST is the Standard and CH the Chiricahua third chromosome gene arrangements in *D. pseuboodscura*), in which male determination was very strong. Note two things, however: first, Kessler was dealing, not with known karyotypes but with selection lines derived by crossing three wild-type strains; second, he was dealing with a mating chamber with 50 pairs of flies, while Kaul and Parsons examined single-pair matings. Of these the first may be the most important since Spuhler et al. (1978) used AR, PP, CH, and TL lines of *D. pseudoobscura* scrambled together to promote a more coadapted population than that used by Kessler. Spuhler's flies came from stocks maintained since Dobzhansky collected these lines in 1959 at Mather, California, a frequently used collection site. The realized heritabilities for two fast and two slow selection lines over 19 generations, while being slightly positive, were insignificant. This suggests that even though the lines had been in the laboratory for some time before the experiment was commenced, there was virtually no additive genetic variance within the Mather population; this had presumably been exhausted by previous natural selection. The contrast with Kessler's experiment is of course that he did obtain a response to selection based on crosses between different populations where coadaptation in a particular population would be broken down, so that additive genetic variability becomes available.

We can, however, conclude that in two species of *Drosophila*—*D. pseudoobscura* and *D. melanogaster*—male mating speed is usually an important, perhaps the most important, component of fitness within populations (more details in Sections 4-2 and 13-1 on these and other species).

8-3 QUANTITATIVE TRAITS INVOLVING MOVEMENT

Hirsch and Boudreau (1958) studied phototaxis (movements with respect to light) in *D. melanogaster* in an apparatus consisting of a Y maze (part of Figure 8-7c), one arm of which was exposed to light during tests. Quite rapid responses to selection were found for both positive and negative phototaxis, which is reasonable as a heritability in the broad sense was estimated in excess of 0.5. Phototaxis has been studied with mazes having a design similar to those used

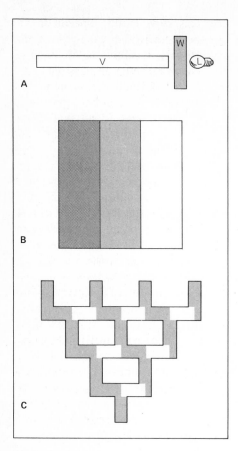

Figure 8-7 Experimental designs for phototaxis analysis in *Drosophila*. (A) Measurement of rate at which flies approach light source. *V:* vessel containing flies; *W:* water-filled heat absorber; *L:* light source. (B) Measurement of distribution of flies in lighted area. Shaded areas correspond to areas of different intensities of overhead light. (C) Measurement of movement of flies in partly illuminated maze. Unshaded areas represent possible light choices. Lateral movement across several intersections is prevented by one-way cone inserted in each arm. Illumination comes from above. (A) *after Carpenter, 1905;* (B) *after Rockwell and Sieger, 1973b;* (C) *after Spassky and Dobzhansky, 1967.*

for geotaxis. Responses to selection have been found in *D. melanogaster* (Hadler, 1964) and in *D. pseudoobscura* (Dobzhansky and Spassky, 1969). In both cases positive and negative selection lines were rapidly established. Upon relaxation of selection in *D. pseudoobscura*, convergence occurred almost as rapidly as the divergence under selection, showing that the photoneutrality of natural populations is a trait subject to genetic homeostasis, the capacity for self-regulation and adjustment (Lerner, 1968). In fact, Dobzhansky and Spassky (1962) made a similar observation about geotaxis. Spassky and Dobzhansky (1967) found that geographical strains of *D. pseudoobscura* and *D. persimilis* respond differently in tests of phototaxis. Later, Rockwell and Seiger (1973b), using laboratory stocks, demonstrated the existence of a large amount of variation for photobehavior within populations of each species. Rockwell, Cooke, and Harmsen (1975) corroborated these findings with freshly isolated samples of sympatric natural populations. They further demonstrated that the level of this genotypic variation is higher in a population of *D. persimilis* than in a sympatric population of *D. pseudoobscura*. A store of variability for phototaxis is apparently available in natural populations, as is the case for almost every quantitative trait (Section 6-7). In *D. melanogaster,* Médioni (1962) found variations between wild strains collected at different localities in the northern hemisphere; greater positive phototaxis was shown in flies of more northern origin. These variations are presumably under genetic control, but their adaptive significance is not known.

Phototaxis, as a behavioral trait, also illustrates a rather subtle environmental problem in that under the usual conditions of handling in the laboratory, *D. pseudoobscura* shows positive phototaxis. Pittendrigh (1958), however, found the flies to be negatively phototactic. Lewontin (1959) carried out a series of experiments showing that *D. pseudoobscura* is negatively phototactic under conditions of low excitement, but if flies are forced to walk rapidly, or fly, they lose their negative phototaxis and become strongly attracted to light. Rockwell, Cooke, and Harmsen (1975) have shown that the extent of these environmentally induced changes differs among the genotypes in natural populations of this species. Hadler (1964) did indeed list numerous environmental variables affecting phototaxis in addition to those just mentioned—temperature, hour of day when tested, time since anesthesia, rearing conditions, time since feeding, energy and wavelength of light, state of dark adaptation, number of observations or trials per individual, age, and sex. A phototactic response is, therefore, a product of particular stimulus-environment variables as well as particular genotypes. It is quite clear that any form of accurate genetic analysis must be based on precisely defined environments, as has been repeatedly stressed.

Another complication is the method used in studying phototaxis. Three different experimental designs (Figure 8-7) have been employed (Hadler, 1964; Rockwell and Seiger, 1973a): (1) measurement of the rate at which flies approach a light source at the far end of a tube (e.g., Carpenter, 1905; Scott, 1943); (2) recording, after a specific period, of the distribution of flies in a field with a directed or undirected light source (e.g., Wolken, Mellom, and Contis, 1957; Koch, 1967), and (3) analysis of movement in mazes (Rockwell and Seiger, 1973b). Hadler considers that one of the major difficulties in comparing studies of phototaxis originating from different laboratories results from interlaboratory differences in experimental technique. They may be measuring slightly different behaviors; for one example, the first method confounds phototaxis with photokinesis. Phototaxis is directed movement with respect to a light source; photokinesis is nondirected.

Rockwell and Seiger (1973a) concur that the measurement of phototaxis is operationally defined. They further discuss how the various general designs differentially accentuate components of the chain of events composing the totality measured as phototaxis and indicate that the designs thus differ in their utility for research directed at diverse aspects of the response. They warn, for those interested in the potential adaptive significance and evolution of the behavior, that the operational nature of the measurement must be taken into account in any generalizations, since there is no certainty that the response measured in the laboratory is necessarily the same as that occurring in nature.

This, of course, is a crucial problem relevant to every laboratory-analyzed behavior discussed in this book.

While it is clear that all environmental factors influencing either the sign or intensity of the response must be carefully controlled and taken into account in any comparisons, Rockwell and Seiger (1973a) consider that the flexibility of the response relative to environmental changes may be an important adaptive qualification of the response allowing increased survival in a heterogeneous environment. For this reason they, among others, propose that in studies concerning the adaptive significance and evolution of phototaxis and of other behaviors, responses relative to several values of various environmental parameters found in the behaving animal's habitat are as important as the response to a certain set value of each parameter (Perttunen, 1963). For example, Parsons (1973, 1974a) maintains that the only valid way of assessing the importance of mating behavior in natural populations is to study it in all the environments to which the population under surveillance is likely to be exposed—unfortunately, a difficult charge. See, however, Section 6-2 where an attempt is made for mating speed in relation to temperature, a critically important environmental variable, in the sibling species *D. melanogaster* and *D. simulans*.

Optomotor response (compensatory movements in response to visual stimuli) in *D. melanogaster* is another trait for which evidence of genetic determination has come from selection experiments (Siegel, 1967). The optomotor response to an illuminated moving striped plate was measured, and each fly was given 10 opportunities to respond. Scores ranged from 0 (no response) to 10. Selective breeding procedures based on low, middle, and high scores were instituted. This led to the development of three strains differing with respect to optomotor response.

Becker (1970) has made a start on the genetics of chemotaxis (movements with respect to the concentrations of chemical compounds) in *D. melanogaster* with a Y-maze of the type used for geotaxis and phototaxis. Selection over 12 generations yielded two lines insensitive to insect repellents; the appropriate crosses indicated that the genes responsible for insensitivity were at least in part dominant. Compared with geotaxis and phototaxis, chemotaxis seems to have the likely advantage of allowing researchers to relate the stimulating molecule to the specificity of the receptor. Further attempts to select genetic variants for chemotaxis are awaited with interest. This point applies not only to *Drosophila* but also to bacteria, protozoa, nematodes, and a variety of organisms, some of which are discussed in Chapter 10.

A number of other traits appear amenable to genetic analyses, especially with selection techniques, but they have not yet been fully exploited. One such trait is preening or cleaning behavior, which has been described in terms of a number of separate behavioral elements (Connolly, 1968). The various preening movements serve to keep the insect clean and to free the sensory surfaces from contaminants. The presence of other flies increases the amount of preening behavior, even if no additional physical contact between these flies is permitted. Further descriptions of this type of behavior are given by Szebenyi (1969), whose approach is analogous to Bastock's (1956) detailed study of mating behavior of flies bearing the *yellow* mutant and of wild-type flies (Section 3-2). Szebenyi divided preening or cleaning behavior into a series of component behaviors and judged preening to be a good trait for behavior-genetics analysis. Hay (1972) agrees and has indicated this to be so both for preening and for more general activity, using sophisticated biometrical methods.

8-4 FREQUENCY-DEPENDENT MATING

Random mating is discussed in Chapter 2 in relation to the Hardy-Weinberg equilibrium. Tests for the randomness of mating are treated there. Here we conside frequency-dependent mating, known best but not exclusively in *Drosophila.* It occurs when the proportion of matings among different genotypes depends on the proportion of genotypes present in the mating population. Experimental investigation of this potentially important phenomenon must therefore employ populations, albeit small workable ones. In such experimental populations, it has been found that the minority genotype (which is analogous to the "rare" type in nature) tends to be favored in mating at the expense of the common type. As we saw in Section 3-2 in a series of multiple-choice mating experiments with *D. melanogaster,* Petit (1958) showed that females are apparently influenced in their choice of mates by the proportions of courting males available to them.

A number of more recent reports suggest that frequency-dependent mating occurs quite often (see Petit and Ehrman, 1969 for a review). Spiess (appendix to Ehrman, 1966) studied *D. persimilis* flies that were homozygous for two autosomal inversions, Klamath and Whitney. By varying the ratio of Klamath to Whitney males, he showed a clear advantage in mating was awarded to the minority homokaryotype.

Using Elens-Wattiaux mating chambers (see Figure 3-3), Ehrman and co-workers (Ehrman et al., 1965; Ehrman, 1968; review in Ehrman, 1972) found that frequency-dependent mating in *D. pseudoobscura* (see Figure 8-8) occurred between strains with different karyotypes and different geographic origins, between mutants versus wild types, between positive versus negative geotaxis, and even between flies of the same genotype raised at different temperatures. (The lower the temperature, the larger the fly: Parsons, 1961.)

Rare-male mating advantages have so far been reported in seven *Drosophila* species—*melanogaster, pseudoobscura, persimilis, willistoni, tropicalis, equinoxialis,* and *funebris* (Spiess, 1968; Spiess and Spiess, 1969; Petit and Ehrman, 1969; Borisov, 1970)—as well as in the flour beetle *Tribolium* (Sinnock, 1970) and, to a lesser degree, in the wasp *Mormoniella* (Grant, Snyder, and Glessner, 1974; also see Neel and Schull, 1968; and Mayr, 1970). The preliminary work has revealed three discrete aspects of this fascinating phenomenon—chemistry, population genetics, and behavior.

Chemistry

Recognition implies discrimination. What cues are available to a *Drosophila* female? How does she use them? Can she be deceived into making erroneous

Figure 8-8 The female *Drosophila pseudoobscura* at the left, her abdomen distended with eggs, preens with both forelegs, apparently unaffected by the mating couple sharing the observation chamber with her. Both females are orange-eyed mutants. The male bears the Chiricahua gene arrangement and has the dark red, nonmutant eye color of wild-type *Drosophila*. The wings of the copulating female provide a base of support for her mate; she may even fly while carrying him. (*Photo courtesy of A. Heder.*)

choices? These are the questions that have concerned Ehrman in attempting to unravel the relationships between behavior and genetics in *D. pseudoobscura*.

The initial experiments were simple but effective (Ehrman 1969). Two Elens-Wattiaux chambers were stacked together, separated only by cheesecloth. Male flies of the Arrowhead (AR) strain were introduced into the bottom chamber, and an equal number of male and female pairs of AR and Chiricahua (CH)—another wild-type strain—were placed in the top. Studies of mate choices made under these circumstances clearly showed that the females were treating the males in the bottom chamber as part of the whole population, since they strongly preferred to mate with CH males. Analogous results were obtained when CH males were placed in the bottom chamber, when females above preferred to mate with AR males.

Because separatory space does sometimes make a difference, a special chamber was devised to ascertain the effects of distance by creating a "wind tunnel" effect (for a description of this sophisticated apparatus, see Ehrman 1969). The results obtained with these chambers clearly indicated that discrimination took place over some distance. Only vibrational (auditory) and odor (olfactory) cues would seem to be involved, because the distances and modes of separation precluded visual and tactile appraisals.

To determine which cue was operative, another series of experiments was run in which the AR males were killed and crushed on the floor of the mating chambers. An equal number of pairs of live AR and CH were then introduced, and again CH males clearly had a mating advantage. Thus auditory cues should be ruled out as being essential to the discrimination process. Further confirmation of the importance of odor cues was obtained when it was found that extracts of dead flies made by homogenizing the flies with organic solvents were also effective in bringing about a rare-male advantage in a population of equal numbers of the two strains of flies (Ehrman, 1972; Leonard, Ehrman, and Schorsch, 1974).

Therefore the answers to our three questions are: olfactory cues are sufficient cues for recognition; the female seems to employ them as a means of estimating the proportions of various strains of males in the population, rather than employing any more direct procedure; and the female can be easily deceived by chemical signals. Having these answers, we could then elaborate our questions further. What are the chemical cues? How specifically defined is the molecular structure? Can we duplicate the behavior with synthetic compounds or mixtures of compounds?

The chemical signals—or *pheromones*—are kinds of airborne "hormones" that serve to influence the behavior of conspecifics. Although the phenomenological consequences of using volatile substances as attractants had been noted by the great nineteenth-century naturalist Jean Henri Fabre, the isolation and identification of the first such substance, and indeed the name *pheromone* itself (which means "to carry a message"), dates from the work of Karlson and Butenandt in 1959 on the sex attractant of the silkworm moth. Pheromones were divided by Bossert and Wilson in their classic 1963 review into two kinds—releasers and primers. *Releasers* are compounds that stimulate an organism to produce an immediate, overt behavior response. *Primers* function by bringing about a change in the physiological state of the organism that will manifest itself overtly at some later time.

The chemical evaluation of these compounds requires three kinds of operations: an assay, to tell us whether or not we are actually working with the right compounds; procedures for purifying the extracts; and methods for identifying the chemical components of the purified extracts.

The bioassay is of first importance. Ever since the work of Karlson and Butenandt, almost all pheromones studied have been releasers, whose effects are easier to investigate. For example, in some cases the sex attractant of the silkworm moth was purified by gas chromatography and detected by watching how excited the males became when placed at the exit port of the instrument. This wingflaps-per-minute kind of assay is a chemist's dream: it is quick and quantifiable and requires a minimum of animal handling. Releaser pheromones that have been studied include sex attractants, alarm pheromones, and trail pheromones (see review by Law and Regnier, 1971). The primers are much more difficult to study. At the time Bossert and Wilson wrote (and to this day), the only good example of such substances was the "queen substance" of the honeybee, which is fed to the larvae in order to make them develop into queens (also see Barrows, Bell, and Michener, 1975).

The pheromones Ehrman investigated represent a third class, undefined by

Bossert and Wilson. These "recognition" pheromones had long been suspected among the social insects, since the insects clearly possess the ability to discriminate nest mates from aliens and to act upon such information. However, no chemical investigations of these substances had been published, nor had any effective bioassay been set up. The rare-male advantage gives us just such an assay, time-consuming though it is, since it involves watching an average of 96 individual matings for each data point (Ehrman and Probber, 1978).

Ehrman began by homogenizing the flies in organic solvent and then using thin-layer chromatography to effect a further purification of the active fraction. Bioassay shows that the active material is quite nonpolar and inert to mild $KMnO_4$ oxidation but susceptible to hydrolysis by both acid and base. On the basis of this and preliminary gas chromatographic data, Ehrman and colleagues investigated the methyl esters of the fatty acids as possible compounds. These compounds are of the general type

$$CH_3(CH_2)_n\overset{\displaystyle O}{\overset{\|}{C}}OCH_3$$

Compounds for $n = 4$ to 30 were tested. Two were found to be active in mimicking the behavior of the recognition pheromone in the Chiricahua strain, and two were unusual in that they could mimic either strain, depending on the concentration of active substance used. However, in no case was any single compound active at a level likely to be present in the intact organism. The quantities required were much larger than would be reasonably expected in an organism of this wee bulk (Leonard, Ehrman, and Pruzan, 1974; Leonard, Ehrman, and Schorsch, 1974; see also Jacobson, 1972).

However, Ehrman and colleagues found that they could use significantly less total material if they used mixtures of compounds, including even compounds that were inactive when tested alone. This synergistic effect was quite marked, leading to a reduction by a power of 10 in the amount of material required. Thus the pheromone system would appear to be multicomponent, and such multicomponent systems are now believed to be quite common (Silverstein, 1977). It is likely that the wing-flaps-per-minute assays were misleading, since they tend to emphasize the activity of purified compounds. However, many of these purified compounds are found to be inactive in the field unless supplemented by other compounds, and thus they are actually merely one component of a multicomponent system.

Recognition systems have a particular advantage in using a multicomponent pheromone, since this allows the use of fewer kinds of pheromones and recognition sites. If the organism can discriminate 10 compounds, each at 10 different levels of strength, then there are potentially 10^{10} different pheromonal signals possible. Clearly, it would be more efficient to evolve the capability to discriminate levels of strength among these few compounds than it would be to add an additional 10^{10} different compounds to production/recognition capabilities!

The exact nature of the *D. pseudoobscura* recognition pheromone in the CH strain (the one that Ehrman has explored most fully) has been put in doubt by recent studies using gas chromatographic–mass spectroscopic analysis. Although there is certainly a homologous series of compounds involved, all of a nonpolar character, it is now certain that there are no methyl esters of the fatty acids present in the active fraction (Ehrman and Probber, 1978).

Further investigation will be required to identify the compounds involved and to determine the relative importance of compound type and concentration in the recognition mechanism. Furthermore, it will be of interest to discover where these recognition substances are produced. Ehrman and Probber hypothesize at present that they are simply compounds that are normal metabolic products but that provide a characteristic "bouquet" to the males. This would not only account for their production but would also provide a reasonable basis for the evolution of the recognition system (since the compounds to be recognized are already in existence) and would demystify the question of the variety and environmental sensitivity of the recognition pheromone system.

Population Genetics

Consideration of the rare-male advantage from the viewpoint of population genetics leads to the hypothesis that an initially rare genotype will increase in frequency if there are no other selective forces operating against it. As the rare type becomes more common, its advantage diminishes, leading to equilibrium (see Table 8-2).

Table 8-2   Distribution of Matings in Mass Cultures of *D. pseudoobscura*, in Which Orange-eyed (*or*) and Purple-eyed (*pr*) Females Had a Choice of *or* and *pr* Males, Showing That as the Minority Male Becomes More Common, Its Advantage Diminishes, Leading to Equilibrium (Similar Results Were Obtained When a Reciprocal Experiment, in Which *or* Males Were Initially Predominant, was Performed—i.e., *or* Males Became Rarer)

Generation	Pairs		Matings observed with rare ♂♂	Matings expected with rare ♂♂
	or	*pr*		
1	20	80	20	14
2	29	71	24	19
3	38	62	25	27
4	35	65	26	22
5	41	59	20	16
6	50	50	31	30
7	52	48	42	44
8	50	50	37	42
9	44	56	36	34
10	47	53	15	16

Source: Ehrman, 1970a.

It seems that a number of gene and chromosomal polymorphisms in *Drosophila* are maintained by such frequency-dependent equilibria. Because these are polymorphisms for which minimal fitness differentials between competing component genotypes are expected at equilibrium, a different sort of selection would prevail from that of the heterozygote advantage model (Section 4-2). Therefore, frequency dependence may represent a way of maintaining a high level of genetic variability without obviously associated fitness differentials. This could be of considerable evolutionary significance, since it has been argued that there is a limit to the amount of variability a population can maintain under the classic heterozygote fitness advantage model (see Dobzhansky, 1970).

In the studies of frequency dependence cited, direct observation techniques were used to determine the number and nature of matings between flies confined to a small space. In other experiments (Ehrman, 1970a,b), two different types of *D. pseudoobscura* individuals of both sexes were allowed to mate in mass cultures. The proportions of the two types in each generation were decided by their mating success in the previous generation. From an initial 80 : 20 ratio, the proportion of the two types converged to approximate equality because of the advantage of the rare type in securing mates (Table 8-2).

Even more significant are experiments (summarized in Table 8-3) carried out in a room with a volume of about 75 m³ (Ehrman, 1970b). Two *D. pseudoobscura* strains were used—the wild type and flies homozygous for the previously wild recessive orange-eyed mutation (*or*), a vigorous and easily

Table 8-3   The Mating Behavior of *D. pseudoobscura* Released in a 75-m³ Room and Then Recaptured Revealing an Advantage for the Minority Male (*or* = Orange-eyed, + = Wild Type)

Released	% recaptured*	% ♂♂ accepted by *or* ♀♀	% ♂♂ accepted by + ♀♀
Experiment 1			
800 *or* ♀♀	33		
800 *or* ♂♂	27	40	24
200 + ♀♀	63		
200 + ♂♂	80	60	76
Experiment 2			
800 + ♀♀	40		
800 + ♂♂	38	79	60
200 *or* ♀♀	31		
200 *or* ♂♂	33	21	40

* Total number recaptured in experiment 1 = 761, in experiment 2 = 753

Source: Ehrman, 1970b.

identified mutation found in nature in the heterozygous condition. About 2,000 flies, in a 4:1 ratio, were used in each of the two experiments: one in which *or* was rare and one in which the wild type was less abundant. In both cases, an advantage was demonstrated for the rare type, although it varied in magnitude depending upon which genotype was the rare one. This is the closest approximation to a natural population yet studied (but see Borisov, 1970, on *D. funebris*) and suggests that this phenomenon, if widespread, may serve an important evolutionary role.

Until now, experimental studies of rare-male mating advantage have employed either genes with visible morphologic effects or inverted chromosomes carrying linked, adaptive complexes of genes as markers. Both of these have pronounced phenotypic effects, and thus it comes as no surprise that some of them affect mating behavior (Sections 4-2 and 8-2). What is surprising is that so many of those genotypes tested show the rare-male advantage, since this form of selection should cancel itself if it occurs simultaneously at very many loci. As Lewontin (1974) has noted, "A generalized intrapopulation advantage for rare males with respect to almost any genotype could not exist, if for no other reason than that every male is genotypically rare in a population with 10% heterozygosity and 40% polymorphism!"

As long ago as the 1920s, Chetverikov (1926) postulated that populations were storehouses of genetic variability. Subsequent studies of "lethals" and "visibles" in *Drosophila* have confirmed his hypothesis, and with the application of electrophoretic techniques (measuring movements in a charged field) to population genetics a decade ago (see Powell, 1975), extensive genetic variability has been demonstrated for a vast array of organisms. The genes assayed by electrophoresis code for enzymes or other proteins and need have no visible effect on the organism's phenotype. Alleles of the gene at such "enzyme loci" are called *allozymes*. It is only natural to inquire if the rare-male mating advantage extends to enzyme loci.

With this query in mind, matings between *D. pseudoobscura* strains differing at the *Amylase* locus were observed in Elens-Wattiaux chambers (Ehrman, Anderson, and Blatte, 1977). Males homozygous for either $Amy^{1.00}$ or $Amy^{0.84}$ alleles in the homozygous CH gene arrangement enjoyed a mating advantage when moderately rare, but none when quite rare. The minority-male advantage for strains differing at the *Amylase* locus, and other loci linked to it, was comparable in size to that observed between strains carrying the Standard (ST) or CH arrangements and either alike or different at the *Amylase* locus. Although some features of the results are puzzling, there is evidence that the *Amylase* locus, and others for which it serves as a marker, has an effect on mating behavior that includes some degree of rare-male mating advantage.

The frequencies reported in Table 8-4 were used to test several hypotheses about matings between the three genotypes studied. Perhaps it is best to begin with the tests we have not felt it necessary to report in the table. Randomness of mating was tested by χ^2 ("goodness-of-fit") tests between observed mating combinations and those expected on the basis of random crosses between females and males that mated. The fit was uniformly good, and there is no evidence of departure from random mating. The observed mating frequency for each type of female was also calculated and compared by means of the χ^2 test to the numbers of each type put into the chambers. In no case was there a statistically significant difference between observed and expected. This outcome was anticipated: since most females were in optimal condition and thus mated in the chambers, the observed mating frequencies had to be close to the number of females introduced into the chambers.

Unlike the females, males may mate repeatedly during the observation period, and observed male mating frequencies may vary widely among males put into the chamber. The observed and expected mating frequencies of males are listed in Table 8-4 along with the results of χ^2 tests for goodness-of-fit between them. If the difference is statistically significant and the observed number of matings by the rarer type is greater than expected, then of course there is a demonstrable mating advantage of the rarer male.

There was no indication of a mating advantage by any of the three male genotypes at equal frequency with one of the other genotypes. In the first set of

Table 8-4 The Number of Matings Between *D. pseudoobscura* Strains Having the Standard (ST) or Chiricahua (CH) Gene Arrangement of the Third Chromosome and Either Allele 0.84 or 1.00 at the *Amylase* Locus Showing a Degree of Frequency Dependence

Pairs per chamber		Male mating frequency				
		Observed		Expected		
A	B	A	B	A	B	χ_1^2
$Amy^{1.00}$ (ST)	$Amy^{0.84}$ (CH)					
2	18	31	81	11	101	38.88†
5	15	54	46	25	75	44.85†
10	10	57	52	50	50	0.23
15	5	79	22	76	25	0.57
18	2	95	14	98	11	0.98
$Amy^{1.00}$ (ST)	$Amy^{1.00}$ (CH)					
2*	18*	7	93	10	90	1.00
5	15	36	68	26	78	5.13‡
10	10	60	47	54	54	1.58
15	5	63	50	85	28	22.33†
18	2	96	6	92	10	1.92
$Amy^{0.84}$ (CH)	$Amy^{1.00}$ (CH)					
2	18	14	89	10	93	1.47
5	15	48	64	28	84	19.05†
10	10	91	109	100	100	1.62
15	5	63	39	77	26	9.53†
18	2	87	14	91	10	1.67

* The ratio of A to B females was probably 5:15 in error; the male ratio was 2:18
† $P < 0.005$
‡ $P < 0.05$
Source: Ehrman, Anderson, and Blatte, 1977.

matings, the $Amy^{1.00}$ (ST) males showed a strong mating advantage when in the minority, but the $Amy^{0.84}$ (CH) males did not. When $Amy^{1.00}$ (ST) and $Amy^{1.00}$ (CH) were tested together, both types of males showed a minority advantage at the 5 : 15 and 15 : 5 ratios, but not at the more extreme 2 : 18 or 18 : 2 ratios. It is puzzling that no male advantage occurred at the extreme ratios. The *Amylase* locus, and of course the genes linked with it that cannot be detected, did have an effect on mating, since the first two sets of matings gave somewhat different results.

The effect of the *Amylase* locus is clearer in the last set of matings, which is the most important here. These tests involved CH strains differing at the *Amylase* locus. Again there was an advantage of both types of minority males at the 5 : 15 or 15 : 5 ratio, but not at the more extreme 2 : 18 or 18 : 2 combinations. It is again puzzling that males at 25 percent frequency have a mating advantage, while males at 10 percent do not. If there is an effect at the lower frequency, it is so small that only a very large experiment could establish it.

The ST and CH strains used had been maintained in laboratory culture for many years. Any genetic differentiation among them could serve to obscure or at least to complicate behavioral effects of the *Amylase* locus or of chromosomal type. However, three of the CH strains with $Amy^{1.00}$ and three with $Amy^{0.84}$ were derived from three original CH stocks that were polymorphic at the *Amylase* locus. Thus we expect that many of the background differences should average out in the crucial comparison of $Amy^{1.00}$ (CH) and $Amy^{0.84}$ (CH). In addition, all the strains in this experiment, which were collected at Mather, California, began with very similar genetic backgrounds on chromosomes other than the third, since these gene-arrangement strains were isolated by crosses to the same "analyzer" stock that had a Mather background. Even with random differentiation in laboratory culture, these strains will be quite similar except for the third chromosomes, each of which was contributed by a different single ancestral fly collected at Mather.

On balance, we believe that the data show some degree of mating advantage associated with two variants at the *Amylase* locus and with whatever genes are associated by linkage with each of the allozymes. The mating advantage is neither as clear nor as consistent as we would have liked, but nature is not always kind enough to provide neat results. Only experiments with additional enzyme loci will show whether the effect reported here is general or not, and only if results are general do they possess any evolutionary significance; seriatim experiments are clearly suggested by these results.

Behavior

Recently, frequency-dependent sexual selection has been studied from a psychological viewpoint by Pruzan, who was interested in the effects of differing experiences on females. Exposure to copulating *Drosophila* couples, exposure to males only, actual mating experience, and the effect of age on subsequent choices of mates were the variables analyzed (Pruzan and Ehrman, 1974; Pruzan, 1976). Direct observations were made of *D. pseudoobscura* females of the AR and CH homokaryotypes. Four-day-old virgins conferred a rare-male advantage on all tested minority males—*or,* AR, and CH—which confirmed previously published results (see above).

A micromesh divider was used to allow the passage of light, airborne stimuli such as odor and vibration, and some touching, but to prevent copulation. Test females were exposed either to copulating couples or to males only. Results of these experiments were not clear-cut; in some cases, mere exposure shifted the mating disadvantage of particular males employed in a female-choice test toward random mating. When aged AR virgins (11 days old) underwent preference trials, they mated significantly more frequently than expected with minority males when those males were of their same karyotype; otherwise, matings were nearly random.

Exciting and consistently replicable results were obtained, however, when females were initially inseminated by males of one karyotype, were allowed to use up the stored sperm by depositing their eggs, and then were retested for male preferences. These experienced females awarded a rare-male advantage only when the rare male was of the same karyotype as their first mate; otherwise, mating was random. Such females, then, were showing a change in behavior as a result of previous experience, and, using a basic definition (Le Francois, 1972), one can make a case for the demonstration of learning. This conclusion is strengthened by recent evidence showing that cycloheximide, a protein synthesis inhibitor that disrupts the memory of learned tasks in mice, rats, and goldfish, also affects *Drosophila* mating choices altered by experience (Pruzan, Applewhite, and Bucci, 1977).

In another *Drosophila* species, the ubiquitous *D. melanogaster,* wild-type flies have been shown to avoid an array of related compounds dissolved in absolute ethanol, these serving as olfactory cues associated with mild electric shock. Flies avoid these odorants by shifting "rooms" in a two-part plastic observation chamber. An induced sex-linked mutant appropriately named *dunce* by Dudai et al. (1976) fails to avoid the irritant despite repeated opportunities to perform appropriately and despite apparent behavioral normality otherwise. Tests of just 12 or so odorants (a hexanol, an octanol, an aldehyde, etc.) show that *dunce* individuals do sense the test compounds but are unable to adjust their responses to such olfactory cues. There are five primary reports of learning via conditioning in *D. melanogaster* (Spatz, Emanns, and Reichart, 1974; Menne and Spatz, 1977; Quinn, Harris, and Benzer, 1974; Quinn and Dudai, 1976; Dudai et al., 1976) and two in *Phormia regina,* the blowfly (see Section 10-5). Conditioning, which involves reorientation as a result of training, has been notoriously difficult to demonstrate reliably in the Diptera. It has not been as problematical in rodents; discussions of conditioned avoidance responses in mice may be found in Section 9-3 and in rats in Sections 6-5 and 9-6.

Thus frequency-dependent mating itself seems to be dependent on age and experience and is produced by a complex interaction of these variables. To elaborate this point, a shift in focus is needed from sexual selection to sexual isolation (as in Section 3-2). That is, a "preference" so strong that it eliminates all but one possible type of mate. Both behaviors are different magnitudes of the same phenomenon: they occupy variant positions on a behavioral continuum (Petit and Ehrman, 1969). Most experiments dealing with sexual isolation have utilized young virgin females (as in Section 5-3), but recent work has been completed with aged, experienced females (O'Hara, Pruzan, and Ehrman, 1976; Pruzan, 1976; Pruzan et al., 1979). *D. paulistorum* provides rich material for a survey of the effects of early experience on later sexual selection, because it comprises six semispecies, or incipient species, among which many degrees of sexual isolation exist. These differences vary with allopatricity or sympatricity of the species (see Section 5-3). Also, females of this superspecies have been shown to mate repeatedly (Richmond and Ehrman, 1974).

In direct observation of the mating of *D. paulistorum* semispecies females with heterogamic (unlike) or homogamic (like) males, it was demonstrated that aged females' sexual selection did not significantly differ from that of young females. Previous heterogamic copulatory experience did not consistently change the degree of sexual isolation; however, females with homogamic copulatory experience showed a significantly higher preference for homogamic males. A test of the proportions of homogamic matings relative to total matings indicated significant differences between subjects with homogamic experiences and naive ones (9 days old) across all combinations.

It has been demonstrated that olfactory stimuli serve as one of the bases of species and strain discrimination in this system (e.g., Ehrman, 1969). Using specially constructed observation chambers, it was shown that if a gentle current of air first passes through a rear section that contains courting and copulating couples of the rare type, minority male advantages disappear in the front portion of the chamber. Since tactile cues are precluded by dead space between the sections, it would seem that the commonness of the olfactory stimuli derived from the rare males can override and obscure their actual rarity.

Work by Shorey and Bartell (1970) is suggestive in this context. They found that a volatile sex pheromone produced by *D. melanogaster* females stimulates and initiates courtship behavior in the male and increases the probability of his approach to a nearby female. Male courtship behavior is also stimulated by odors released by other males, but the male odor appears to have less than one-tenth the effect of the female's. More recently, the ongoing work of Averhoff and Richardson (1974, 1976) on pheromonal control of *D. melanogaster* mating is producing valuable information. They found that individuals were sexually unresponsive to their own pheromones and those of their close relatives. Indeed pheromones from different strains stimulated male courtship. Such a process based on a diversity of pheromones would help to prevent inbreeding in small populations.

Mating trials were conducted with 24 pairs of *D. pseudoobscura* in observation chambers (Table 8-5). Females were equally divided between AR and CH strains, while males were used in a variety of ratios. Four days before the mating trials, virgin flies of both sexes were collected within 3 hours of emerg-

Table 8-5   Tests in Which Varying Ratios of 24 Arrowhead (AR), Chiricahua (CH), or Orange-eyed (*or*) Males Were Presented to 12 AR and 12 CH Females Showing that Females Can Discriminate Among Males (Data Tabulated as the Natural Logarithm of the Ratio of the Observed Mating Frequency to the Expected Frequency)

Ratio of males* AR : CH : *or*	Odds for male mating success (observed matings/expected matings)†		
	AR	CH	*or*
1 : 1 : 1	−0.054	−0.130	0.310
4 : 1 : 1	**−0.378**	**0.701**	0.410
1 : 4 : 1	**0.644**	**−0.285**	0.173
1 : 1 : 4	0.060	0.134	−0.113
5 : 5 : 2	−0.108	0.160	**0.855**
5 : 2 : 5	−0.076	**0.362**	−0.237
2 : 5 : 5	**0.407**	**−0.261**	0.059

* Each ratio was tested 144 times.

† The expected matings have been corrected for differences in male vigor. A positive value indicates a mating advantage for that particular strain; a negative value indicates a disadvantage. Boldface entries are statistically significant when tested by χ_1^2 against the other two strains of males in the trials ($P < 0.05$).

Source: Leonard and Ehrman, 1976.

ing from the pupae and separated by sex under mild etherization. The AR and CH flies were marked by wing clipping to permit visual scoring of matings.

In these trials, males mated repeatedly, while females mated only once. On the basis of vigor alone, AR and CH males are equally acceptable as mates. However, an orange-eyed (autosomal recessive) male mutant is only 50 percent as likely to mate as either wild-type inversion strain under the same conditions. The expected mating frequency is corrected for this disadvantage.

Matings in which males are present in $5 : 5 : 2$ ratio show the classic rare-male advantage. However, the interpretation of the other four cases investigated is more complex. In the $1 : 1 : 1$ case there is a slight advantage for the *or* strain, while in the three $4 : 1 : 1$ cases there are two rare strains, but only one strain enjoys an advantage. Advantages in all trials are of two types: when the number of AR and CH are unequal, the minority strain is advantaged; and when the number of AR and CH males are equal, the *or* males enjoy an advantage if they are the minority strain relative to the sum of the two nonmutants.

We conclude that the characteristics of the *Drosophila* frequency-dependent pheromonal system include (1) specificity—different strains can be discriminated; (2) quantifiability—the chemical signals can be used as a measure of the proportion of strains in a population; (3) ability to respond to novel combinations—females can recognize a great variety of strains without prior experience, although there is an influence attributable to experience; and (4) hierarchical processing—there is an apparent neural ordering involved in the importance given to signals from different strains.

More than a dozen years have passed and many experiments have been conducted since Dobzhansky first suggested that an investigation of those odd events we have discussed might be fruitful. His concerns were conveyed in *Genetics of the Evolutionary Process* (1970, p. 174):

> Nothing is known about possible mating advantages of rare genotypes in natural environments. If they exist in the natural habitats of the flies, the resulting frequency-dependent selection may be a potent instrumentality for maintaining the polymorphic equilibria of gene alleles without heterosis. Even mildly deleterious alleles could be maintained in natural populations by these means. Rare alleles will grow in frequencies until the mating advantages of their carriers decrease and disappear. More research in this field is evidently needed.

8-5 LARVAL BEHAVIOR

Although the behavior of adult *Drosophila* has been the subject of extensive investigations, as we have seen in this chapter and in previous chapters, far less is known about larval behavior despite its certain importance in the development of the organism. Sewell, Burnet, and Connolly (1975) found that larvae of *D. melanogaster* feed continuously during their period of development and that the rate of feeding activity, measured as the number of cephalopharyngeal retractions per minute, varies with the physiologic age of the larva. Feeding rate responded rapidly to directional selection giving nonoverlapping populations with fast and slow feeding larval strains with realized heritabilities (Section 6-11) between 0.20 and 0.21 in different selection lines. Crosses between selected lines show significant dominance for fast feeding rate, which presumably is adaptive in nature in times of food shortage, when a high feeding rate would be favored. Locomotor behavior was studied in the selection lines as a correlated response, but little association was found between locomotor and feeding behavior. It is argued that under optimal environmental conditions an increased propensity for locomotor behavior would be maladaptive if it causes a reduction in feeding time. Consequently, it would be of advantage if these behaviors are regulated independently, as the results suggest.

The larval stage of the *Drosophila* life cycle is of course the stage of maximum resource utilization. It is known, for example, from release-recapture experiments in a winery that adult *D. melanogaster* are attracted to a fermentation tank during wine-making and that its sibling species *D. simulans* is not (McKenzie, 1974). This result is predictable given that *D. melanogaster* utilizes ethanol as a resource up to high concentrations (McKenzie and Parsons, 1972). In nature, larvae of the two species may therefore seek out discrete and different microhabitats in response to available resources. Larval preference data (Table 8-6) on strains from southern Australia (latitudes 37° to 37.5°S) support this hypothesis, while from Townsville, North Queensland (latitude 20°S), the difference between species is smaller but in the same direction. That is, if the two species are competing in the same medium, the larvae may occupy

Table 8-6   Means and Ranges of Isofemale Strains of *D. melanogaster* **and** *D. simulans* **for Number of Larvae Out of 10 Choosing Agar Containing 6 Percent Ethanol in Preference to Agar Containing No Ethanol After 15 Minutes on a Petri Dish**

		D. melanogaster		*D. simulans*	
	Latitude	Mean	Range	Mean	Range
Melbourne	37.5°S	7.8	8.8–7.0 = 1.8	5.5	6.5–5.0 = 1.5
Chateau Tahbilk	37°S	7.5	8.8–6.5 = 2.3	5.3	6.2–4.6 = 1.6
Townsville	20°S	6.4	8.8–3.6 = 5.2	5.8	6.8–5.0 = 1.8

Source: Parsons, 1977b.

different microniches. Field observations in support of this from winery grape residues are considered in Section 13-2.

It is not however, so simple as this. While the isofemale strains (see Section 6-7) from the southern Australian populations of *D. melanogaster* all show a uniformly high preference for alcohol, only some from Townsville show such high preferences and others show little or no preference. This explains the lower mean for the Townsville population, as well as its extremely high range compared with the two southern populations. No such interpopulation heterogeneity was found in *D. simulans.* The working hypothesis is that as *D. melanogaster* spread south, there was a premium on selection for alcohol resource exploitation, which can be regarded as a process of selection among isofemale strains. Indeed the decline in heterogeneity in southern *Drosophila* agrees with the general principle of declining biological diversity with increasing latitude. This is demonstrated here by assessment of a behavioral response to a chemically defined resource. The approach to the study of larval resource utilization with isofemale strain comparisons would appear to have considerable potential since many possible metabolites can be tested.

Other studies on chemotaxis in adults referred to in previous sections of this chapter may well have larval parallels. Begg and Hogben (1946) showed that acetic acid, ethyl acetate, and DL-lactic acid are attractants for *D. melanogaster* adults, and Fuyama (1976) found interpopulation differences to attractants showing the effects of natural selection within this species. Parsons (1979a) found that larvae of both *D. melanogaster* and *D. simulans* are attracted to these three compounds with small differences between species and populations, as compared with ethanol. Not unexpectedly, the differences between these two species of subgenus *Sophophora,* and the cosmopolitan species *D. immigrans* of subgenus *Drosophila* are much greater, given that the former two species almost exclusively use fruit resources and the latter fruit and vegetable resources (Atkinson and Shorrocks, 1975)—as will be further discussed in Section 13-2.

These recent results suggest that the study of larval behavior is going to play an increasing role in our understanding of organisms such as *Drosophila* in relation to habitats occupied in nature. It is an area needing further work, as shown by results demonstrating variability among *D. melanogaster* strains for digging behavior (Godoy-Herrera, 1977) and between sibling species by Barker (1971). This is further emphasized by Pruzan and Bush (1977), who found that significantly more larvae went to the arm of a maze containing their own strain (as larvae or biotic residues) as a stimulus, suggesting again that larval strains have the capacity for olfactory discrimination and hence may differentially select habitats.

SUMMARY

In many areas of genetics, *Drosophila melanogaster* studies have made fundamental advances, simply because of the thorough knowledge of the genome of this species, and the array of special genetic stocks available for sophisticated genetic manipulations. For example, utilizing genetic *Drosophila* mosaics, precise anatomical sites can be correlated with abnormalities affecting behavior using "shaker" mutants. It follows too, that this becomes one species for research in neurogenetics.

Furthermore, the genetic basis of quantitative traits, especially sexual behavior, phototaxis, and chemotaxis, can be readily investigated in *Drosophila*. A phototactic response is a product of particular stimulus–environment variables as well as particular genotypes, so that any form of accurate genetic analysis must be based upon precisely defined environments. Indeed responses relative to several values of the environmental parameters found in the animals' habitat are of great importance. Studies on chemotaxis have the likely advantage of allowing attempts at relating the stimulating molecule to the receptor.

Far less is known about larval behavior despite its importance in the development of the organism, being the stage of maximum resource utilization. There may well be larval parallels with adult chemotaxis, considering metabolites such as ethanol and acetic acid. Because larval responses to metabolites vary among populations of *D. melanogaster* and between closely related *Drosophila* species, evaluations of larval behavior are likely to play an increasing role in understanding habitat selection in nature.

Finally, in various *Drosophila* species, especially in *D. pseudoobscura*, the population phenomenon of frequency-dependent mating in which rare genotypes mate more frequently than would be predicted under random mating has been demonstrated many times. This fascinating phenomenon appears to have its basis in characteristic "bouquets" of males of different genotypes, which are normal metabolic products. If frequency-dependent selection is common, it becomes a potent force in maintaining large amounts of genetic heterogeneity in natural populations. Even so, frequency-dependent mating depends upon age and experience; there is a learned component.

GENERAL READINGS

Ashburner, M., and E. Novitski (eds.). 1976. *Genetics and Biology of* Drosophila, Vols. 1a, 1b, 1c. New York: Academic Press.

Ashburner, M., and T. Wright (eds.). 1978. *Genetics and Biology of* Drosophila, Vols. 2a, 2b. New York: Academic Press. Volume 2b is on behavior. Most subsequent volumes in this series will have information of importance and are worth checking as they appear.

King, R. C. (ed.). 1974–1976. *Handbook of Genetics,* Vols. 1–5. New York: Plenum.

Parsons, P. A. 1973. *Behavioural and Ecological Genetics: A Study in* Drosophila. Oxford: Oxford University Press.

The Genetics of Behavior: Rodents

This chapter considers the behavior of rodents, with reference particularly to mice and to a lesser extent to rats and guinea pigs. The aspects of rodent behavior discussed in previous chapters can be summarized as follows:

- The behavior-affecting mutant gene *fidget* is considered early in Chapter 2 in connection with the basic principles of mendelian genetics. In Section 2-4 comments are made on the behavioral effects of the mutant gene *yellow* in mice.
- In Section 3-3 mating success as determined by single genes is discussed as an example of sexual selection.
- In Chapter 4, where sex chromosome aberrations are considered, mention is made of the need to study chromosomally aberrant mice in greater detail.
- Chapter 5 begins detailed considerations of polygenes and behavior. Commonly studied rodent behaviors under such genetic control are listed in Section 5-1. In Section 5-5 one of the classic behavior-genetic selection experiments concerning defecation in rats is described.
- In Chapter 6 the genetic analysis of quantitative traits is reviewed in experimental animals, including rodents (Sections 6-4 and 6-5). An example of the effect of early experience on the time required to reach food is presented as illustrative of the complex genotype-environment (GE) interactions often occurring in mammals (Section 6-2).

- In Section 7-4 maze learning in rats is used as an illustrative example of GE interaction when considering such issues in *Homo sapiens*.

Tallying all examples, we note that much information on the genetics of behavior in rodents has been presented prior to this chapter. The discussion to follow derives from our selection of certain examples from which conclusions supplemental to those already presented emerge.

9-1 MICE: SINGLE-GENE EFFECTS

A diversity of single genes affect behavior in rodents. Many genes with obvious behavioral effects have been mapped in mice. In addition, a number of studies have shown that genes identified via morphology, usually coat color, display subtle behavioral effects (Thiessen, Owen, and Whitsett, 1970). The work on mice exemplifies the point made in Section 8-1 in relation to *Drosophila:* no genes affect behavior per se. Rather, the behavioral changes result from genetic effects on enzyme level, hormone level, tissue sensitivity, membrane permeability, and other functions. The influence of heredity on behavior is indirect.

Over 300 mutant genes (some allelic) occupying over 250 loci were listed for the mouse by Green (1966). By mid-1965 92 mutant genes associated with neurological defects were known (Sidman, Appel, and Fuller, 1965), affecting almost every conceivable biological function. The effects of these mutant genes were classified into defects of regional development, structural defects of individual cells that fail to make some special product, and functional defects requiring biochemical study. Most of the known mutant disorders affect the nervous system during its development; several mutant genes are known specifically to affect the cerebellum. A number of mutants affect the inner ear and related structures (see Deol, 1975), leading to a primary disorder in the early embryonic central nervous system and interfering with the subsequent induction of peripheral structures. Some representative genes that affect the nervous system that are considered worthy of detailed behavioral study are listed in Table 9-1; they range from those that produce severe defects of the central nervous system to those responsible for minor disturbances. Behavioral changes are associated with various morphological, physiological, or biochemical effects, just as in *Drosophila* (Chapter 8). Detailed studies on various mutants associated with behavioral effects are summarized by Wilcock (1969). Included are the mutant genes *short ear, hairless, furless, pintail, looptail, tailless, waltzer, wobbler lethal, jerker, quaking, ducky, twirler, dystrophic, reeler*, and *jumpy*. For example, jerkers, waltzing mice, and quaking mice are unable to walk normally and cannot balance well. Wobblers develop more slowly than normals in open-field activity level, ability to rear up on the hind legs, and ability to climb an inclined plane. Looptail mice perform fewer face-cleaning acts than normals. On histological examination they are found to have enlarged ventricles and deformation of extrapyramidal motor systems in the forebrain.

Table 9-1 Mutant Genes Affecting Nervous System, and Consequently Behavior, in the Mouse

Gene name	Gene symbol	Linkage group	Biological* phenotype	Behavioral phenotype
Absent corpus callosum	ac		Absence of all or part of corpus callosum	None now known
Cerebral degeneration	cb		Degeneration of cerebral hemispheres and olfactory lobes	Diffuse; progressive overall degeneration of behavior
Dancer	Dc		Absence of macula of utriculus	Circular movements; failure to swim
Deafness	dn		Degeneration of Deiters' cell in organ of Corti	Accentuation of sniffing
Dilute lethal	d	II	Myelin degeneration; lowered phenylalanine hydroxylase activity	Convulsions
Eyeless	ey		Absence of eye and optic tract	Visual incapacity
Muted	mu		Absence of otoliths in one or both ears	Postural defects
Pirouette	pi	XVII	Degeneration of organ of Corti	Circular movements
Wobbler lethal	wl	III	Myelin degeneration; elevated succinic dehydrogenase levels	Locomotor difficulties

* By *biological* is meant morphological, physiological, or biochemical effects produced directly by the mutant gene and in turn responsible for the abnormal behavior.
Source: Thiessen, Owen, and Whitsett (1970).

Quaking (autosomal recessive) and *jumpy* (sex-linked recessive) are two examples of neurological mutants in which enzyme activities that are absent in the brain are present in the peripheral nervous system and some other tissues. Both mutations result in defective myelination in the brain and spinal cord (Mandel et al., 1973). In contrast, the peripheral nerves of all mutants are normally myelinated. Peripheral-nerve myelin is not chemically identical to central myelin, nor is it formed by oligodendrocytes but by the morphologically distinct Schwann cells. The enzyme absent in the mutants is 2', 3'-cyclic AMP 3'-phosphohydrolase (CNP). Here then we have an enzyme, CNP, affecting myelination of the oligodendrocytes of the central nervous system resulting in the observed behavioral abnormality. Thus the behavioral change in this case has a known molecular basis. Such detailed studies of gene expression in higher organisms could well assist in our understanding of the genetic bases of human development, function, and behavior.

Assessment of the work on nervous-system-affecting mutants is difficult because the effects of these mutant genes on behavior are gross. To date these studies have contributed little information about the relation of genes to normal behavior, since the behavior associated with the mutant is so abnormal as to be outside the usual variation found in behavioral phenotypes. However, as Wilcock (1969) has suggested, developmental studies of such mutants could have important clinical implications.

Thiessen, Owen, and Whitsett (1970) have pointed out that only a few single-gene substitutions with more "normal" functions have been studied in terms of their behavioral effects (Table 9-2). These involve mainly alterations in coat color. Failure to observe behavioral effects due to single genes has been reported, but this does not constitute proof that behavioral effects are lacking, since in no case can the complete spectrum of possible behaviors exhibited by an organism ever have been tested. Consider, for example, the *albino* allele. The double recessive *cc* at the C locus on linkage group I blocks the enzymatic

Table 9-2   Influence on Behavior of Single-Gene Substitutions in the Mouse

Gene name	Gene symbol	Linkage group	Biological phenotype	Behavioral phenotype
Albino	*c*	I	Absence of pigment in fur and eyes	Decreased nibbling; delayed audiogenic seizure; decreased water-escape performance; decreased active avoidance, increased passive avoidance; decreased activity; competitive advantage in sex partners; decreased alcohol preference; decreased open-field activity; poor black-white discrimination
Brown	*b*	VIII	Brown fur instead of black pigment	Increased grooming
Dilute	*d*	II	Blue-gray fur	Decreased activity
Misty	*m*	VIII	Dilute coat color, tail and belly spots	Decreased nibbling
Pink eye dilution	*p*	I	Pink eyes, reduced or brown pigment	Decreased staring at examiner, less paw lifting, more grooming and shaking
Pintail	*Pt*	VIII	Short tail	Fast avoidance extinction
Short ear	*se*	II	Reduced cartilaginous skeleton	Impaired avoidance learning to sound
Yellow	A^y	V	Yellow or orange fur and black eyes	Decreased short and long-term activity; failure of male to synchronize mating in groups of females

Source: Thiessen, Owen, and Whitsett (1970).

synthesis of tyrosinase, which is necessary for the conversion of tyrosine to dopa and finally to melanin:

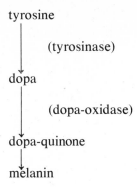

tyrosine

      (tyrosinase)

dopa

      (dopa-oxidase)

dopa-quinone

melanin

Extensive behavioral changes are noted in *cc* mice; the problem is to elucidate the pathway from gene to behavior. There is some indication (DeFries, Hegmann, and Weir, 1966) that photophobic reactions related to a loss of eye pigment mediate the extreme hesitancy exhibited by albino mice.

To assess the generality of behavioral effects associated with traits lacking obvious behavioral effects, a number of coat-color mutants were investigated (Thiessen, Owen, and Whitsett, 1970). The pigment in rodent fur is melanin of two kinds, pheomelanin and eumelanin. The former is always yellow; the latter may be brown or black. Therefore, the wide variety of coat colors in mice is the result of genetic influences on only two kinds of pigments. For reference, the most common inbred strains of mice encountered in behavior-genetics research are listed in Table 9-3; the genetic determinants of their coat colors are specified. The primary coat-color loci are *agouti, black, albino,* and *dilute,* which specify the distribution of black, brown, and yellow pigments in rodent hairs; *pink eye* and *piebald* are secondary loci that govern not only the amount of different pigments but also the shape, size, and distribution of individual

Table 9-3 Genetic Determinants of Mouse Coat Color at Common Loci

| Strain | Locus | | | | | |
	Agouti	Black	Albino	Dilute	Pink eye	Piebald
C57BL	aa	BB	CC	DD	PP	SS
C3H/2	AA	BB	CC	DD	PP	SS
DBA/2	aa	bb	CC	dd	PP	SS
I	aa	bb	CC	dd	pp	ss
BALB/c	AA	bb	cc	DD	PP	SS
A	aa	bb	cc	DD	PP	SS
R III	AA	BB	cc	DD	PP	SS
Linkage group	V	VIII	I	II	I	III

Double recessive phenotypes are aa nonagouti, bb black, cc albino, dd dilute, pp pink eye, ss piebald.

pigment granules. Generally, at each locus three or more alleles are possible. Dominance and epistatic relations are often complex.

In an F_2 derived from intercrossing F_1s from the inbred strains AKR/J (*aaBBccDD*) and DBA/2J (*aabbCCdd*) (see Table 9-3), albinos were compared with nonalbinos using a battery of 11 tasks with the apparatus illustrated in Figure 9-1. The tests are described briefly below to illustrate the type of test battery the behavior geneticist interested in rodents can use. It is important to follow detailed experimental procedures in any test to minimize the degree of subjectivity. Reference to the original publications involved is needed to assess and appreciate this.

The testing equipment illustrated in Figure 9-1 includes:

- An open field (*a*), used to assess open-field activity, a measure of general exploration, described by the number of lines crossed. This apparatus used as an inclined plane allows an estimate of geotaxis.
- A brightness plane (*b*), used to measure the degree to which an animal prefers a light or a dark environment.
- A tactual plane (*c*), with floor divided into smooth and rough halves, used to measure tactual preference.
- A visual cliff (*d*), used to measure depth perception. The tendency of the animal to go to the shallow or the deep side is assessed after placing the animal on the centerboard.
- An arena (*e*), used to measure general activity as assessed by the number of lines crossed in a 2-minute period.
- An olfactory alley (*f*), in which olfactory sensitivity to a noxious stimulus is assessed by placing ammonia at one end of the runway and water at the other. After placing an animal in the middle of the alley, the experimenter records the amount of time spent on the side of the alley containing water during a 5-minute test.
- An activity wheel (*g*), whereby the number of rotations can be counted over long periods to provide measures of long-term activity.
- A water-escape apparatus (*h*), in which learning and escape performances are assessed. It consists of an aquarium filled with water at a temperature of approximately 25°C. A small restraining box with a trapdoor is positioned above the water at one end; at the opposite end a wire-mesh ramp extending into the water provides an exit. A trial consists of placing a mouse in the restraining box and orienting it toward the exit ramp; the trapdoor is opened and the mouse dropped into the water. Latency or tendency to swim to the ramp over five trials is measured.
- A temperature gradient (*i*), ranging from 10°C to 51°C, used to determine mouse temperature preferences.
- An auditory alley (*j*), in which preference for or aversion to an auditory stimulus is measured. At the end of each arm of the apparatus is a speaker into which a 14,000 Hz sine wave is fed by an audiogenerator and an audiostimulator. An animal is placed in the center of the alley and allowed to explore for 5 minutes. Then sound is delivered at one end of the alley and the time spent by the animal at either end is recorded. Scores range from primarily positive, indicating a preference for the sound, to primarily negative, indicating aversion to the sound.

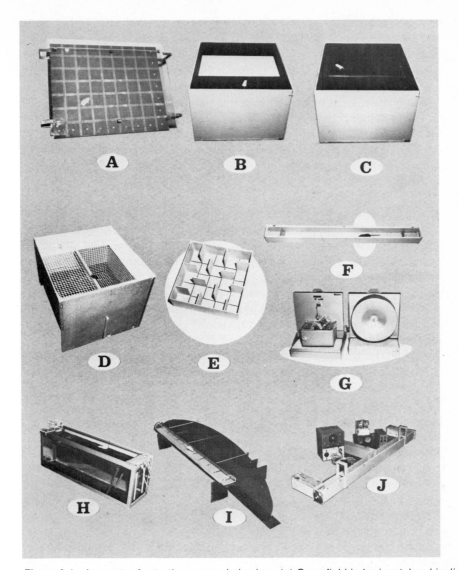

Figure 9-1 Apparatus for testing mouse behaviors. (*a*) Open field in horizontal and inclined plane. (*b*) Brightness plane. (*c*) Tactual plane. (*d*) Visual cliff. (*e*) Arena. (*f*) Olfactory alley. (*g*) Activity wheel. (*h*) Water-escape apparatus. (*i*) Temperature gradient. (*j*) Auditory alley. (*From Thiessen, Owen, and Whitsett, 1970.*)

Because the albino allele was studied in F_2 animals, results may not be entirely free of linkage effects, although Thiessen, Owen, and Whitsett (1970) consider such effects relatively unimportant. A considerable array of behaviors was found to differ between albino and nonalbino mice. The *albino* gene lowered the normal sensitivity to change in environmental incline (the inclined plane), reduced activity under white-light conditions (arena, water-escape apparatus, and visual cliff) but not red-light conditions for the open field or when activity was measured primarily at night on an activity wheel. Albinos, reflecting their generally lower level of activity, stayed in a lighted environment or on a rough surface longer than pigmented mice. Both types tended to avoid light and remain on a rough surface. Albinos responded toward a sound source whereas nonalbinos moved away. The albinos tended to avoid the olfactory stimuli more. Photophobia, mentioned above, seems the basis of the light reaction, but explaining the whole complex of behaviors in an integrated way is clearly difficult. The one generalization that can be made is that the albinos hesitate in their reactions to environmental changes; further evidence in agreement with this is presented in Section 9-3.

Fourteen other genotypes involving coat-color variations (but not albinos) were studied in the C57BL/6J strain. Since for practical purposes the only gene allowed to vary was the one of interest, any behavioral effects can be ascribed to that gene. Four tests were used: open-field activity, geotaxis, water-escape behavior, and activity wheel revolutions. The results showed that of the 14 genotypes tested, 71 percent modified some aspect of behavior; in other words, it is not difficult to find effects. Further, the larger the number of behaviors sampled, the more likely it is that a gene substitution affecting behavior will be recognized. Thus 14 percent of genotypes affected open-field behavior alone; 36 percent open-field behavior and/or geotaxis; 57 percent open-field behavior, geotaxis, and/or water-escape behavior; and 71 percent one or more of the four behaviors. Clearly, a number of behavioral effects are associated with the ordinary coat-color alleles, suggesting that almost any mutant may have behavioral effects if the battery of tests used is adequately comprehensive.

9-2 MICE: QUANTITATIVE TRAITS INVOLVING OPEN-FIELD BEHAVIOR

Much work has been done on quantitative traits, usually without the identification of the specific loci or chromosomes involved. Studies of open-field behavior date from Hall's (1951) and Broadhurst's (1960, 1967) work in rats. An enclosure is used that provides a strange open-field situation. In this enclosure both a measure of emotionality, as assessed by defecation and urination, and a measure of activity, as assessed by the number of squares crossed in a given time, can be obtained. As shown in Section 5-5, selective breeding affects emotionality. Negative correlations between ambulation and defecation in the open field have been found (Hall, 1951; Broadhurst, 1967). Such negative correlations seem fairly general, although the association is affected somewhat by environmental variables such as light and noise (Archer, 1973). The relation also depends to some extent on the species, strain, sex, sample size, and early experience of the subjects.

A number of mouse studies indicate heritable differences in activity. Thompson (1953) found striking differences among 15 strains when individuals were tested in a square arena (30 × 30 inches). The floor was divided into 25 squares, and barriers were placed at the base of every other square. The number of squares traversed by a subject during a 10-minute test was used as the activity score. Of the 15 strains found to differ widely in this behavior, 5 were tested later for activity level in a Y-maze and the arena (Thompson, 1956). For arena activity, the same rank order of the strains was observed; with one exception, the rank order of the strains in the Y-maze corresponded to that in the arena. These parallel results reflect the degree of situational generality in these activity data—an important consideration.

These and many other studies (see Fuller and Thompson, 1960) show clearly that the observed individual differences in level of activity in an unfamiliar situation are a function of heritable differences. The same conclusion applies to defecation. Open-field behavior itself can be objectively and efficiently measured, permitting the large sample size that is essential for detailed genetic analyses. DeFries and Hegmann (1970) carried out a detailed analysis of the differences in open-field behavior for two inbred strains of mice and their derived generations. The two parental inbred strains, BALB/cJ and C57BL/6J, were known to differ widely in open-field behavior. The field used was a square (36×36 inches) of white-painted Plexiglas divided into 36 squares, each 6×6 inches. Testing was carried out on mice 40 ± 5 days old. Activity was measured as the total number of light beams (used to demarcate the squares) that were interrupted during two 3-minute tests. The total number of fecal boluses dropped was also recorded. Data were obtained on the inbred strains P_1 and P_2, their F_1, backcross, F_2, and F_3 generations, and five generations of selection in both directions, that is, for high and low open-field activity.

Heritabilities were estimated based on the data for the parental, F_1, BC_1 ($P_1 \times F_1$), BC_2 ($P_2 \times F_1$), and F_2 generations after applying the square root transformation to both activity and defecation scores. These transformations were carried out in an attempt to fulfill scaling criteria—an issue discussed in Section 6-3. The heritabilities so estimated were appreciable—of the order of 0.4 (Table 9-4). Comparing sections A and D of Table 9-4, the heritabilities in the narrow sense h_N^2 were generally a little lower than the heritabilities in the broad sense, h_B^2, indicating that most of the genetic variability is due to the additive genetic variance. Heritabilities were also estimated from the regression of offspring on midparent (Section 6-9) and from half-sib correlations. Immediately there is a problem, since although heritabilities are generally greater than zero (except for females in half-sib correlations), they vary highly according to the breeding procedure. This could argue for the effects of inadequate scale, genotype-environment interactions, or other difficulties in the data. The same problem occurs for defecation. Heritabilities for defecation are rather lower than for activity, indicating a higher environmental component in determination of this behavior. The heritability for the half-sib correlation in females is negative (but not significant); the remaining heritabilities are positive, some being significant. For both traits we can thus ask: Which value of heritability forms the best estimate? A possible way of approaching this is to assess the realized heritability (h_N^2) from a directional selection experiment (Section 6-11).

Such an experiment was done for open-field activity beginning with a foundation population of 40 litters chosen at random from the F_3 generation of the

Table 9-4 Heritabilities in the Narrow Sense h_N^2 and in the Broad Sense h_B^2 of Activity and Defecation Scores of Mice in Open-field Behavior

A. h_N^2 from parental, F_1, BC_1, BC_2, and F_2 generations

	Males	Females
Activity	0.58 ± 0.06	0.28 ± 0.04
Defecation	0.42 ± 0.07	0.36 ± 0.06

B. h_N^2 from regression of offspring on midparent

	Males on midparent	Females on midparent
Activity	0.24 ± 0.12	0.19 ± 0.12
Defecation	0.04 ± 0.09	0.17 ± 0.08

C. h_N^2 from half-sib correlations

	Males	Females
Activity	0.50 ± 0.32	-0.25 ± 0.31
Defecation	0.30 ± 0.32	-0.29 ± 0.31

D. h_B^2 from parental, F_1, BC_1, BC_2, and F_2 generations

	Males	Females
Activity	0.63 ± 0.06	0.49 ± 0.06
Defecation	0.39 ± 0.06	0.38 ± 0.06

Source: DeFries and Hegmann (1970).

animals used in the preliminary analyses (with the restriction that each litter must contain at least two males and two females). The most active male and female and the least active male and female from each of 10 litters were selected. The resulting 10 high-active males and 10 high-active females were then mated at random at approximately 60 days of age so as to produce progeny representing the first selected generation (S_1) of one high-active line (H_1). Similarly, the 10 low-active males and females were mated at random to produce the S_1 generation of a low-active line (L_1). (The parents of lines H_1 and L_1 were littermates.) In addition, high-active and low-active males and females were selected from each of 10 other litters and mated at random within their level of activity; their offspring represented generation S_1 of lines H_2 and L_2. Two unselected control lines, C_1 and C_2, were set up. (See DeFries and Hegmann, 1970, for further details.)

Over five generations of selection there were clear and consistent responses to selection, from which realized heritabilities can be computed. After five generations of selection, the realized heritability, h_N^2 as measured by the response to selection (R) divided by the selection differential (S) (Section 6-11) was 0.31 ± 0.04 for H_1 versus L_1, and 0.19 ± 0.07 for H_2 versus L_2, with a pooled value of 0.26 ± 0.03. This is in good agreement with the single-generation regression of offspring on midparent (Table 9-4). Therefore, under the conditions of this study, the regression of offspring on midparent can be argued to be a good predictor, but there is no theoretical reason why this should always be true, in view of the complications and assumptions that have to be accommodated in these various estimates. In addition, see Section 6-9 for sources of possible bias using this particular method. This difficult-to-interpret example is presented as an instance of the problem of interpretive assessment in behavior-genetics analysis; we chronicle it here precisely because of these difficulties.

This selection experiment has now been continued for 30 generations (Figure 9-2), with more than a thirtyfold difference in mean activity between the high and low lines (DeFries, Gervais, and Thomas, 1978). Even at 30 generations there is no evidence for an approach to a selection limit. Open-field defecation scores of low-active lines were about seven times higher than those of the high-active lines, substantiating earlier reports (see Section 6-11) of a high negative genetic correlation between these characters.

One remaining facet of this study merits mention. The above analysis

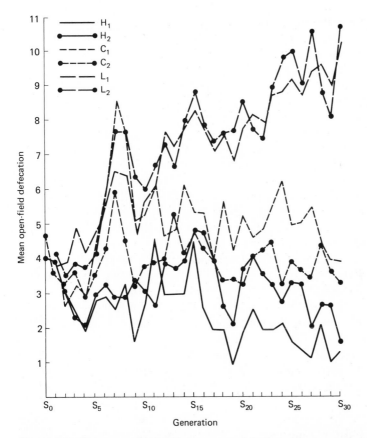

Figure 9-2 Mean open-field activity scores of six lines of mice, two selected for high open-field activity (H_1 and H_2), two selected for low open-field activity (L_1 and L_2), and two randomly mated within lines to serve as controls (C_1 and C_2). (*From DeFries, Gervais, and Thomas, 1978.*)

shows that open-field activity is a quantitative trait, presumably under multifactorial additive control. However, a major gene effect on activity has been found, as discussed in the previous section, since albino mice have lower activity and higher defecation scores than do pigmented animals. The relative importance of this single-gene effect was measured by assessing its contribution to the additive genetic variance associated with these behaviors. Segregation at the C locus was estimated to account for 12 percent of the additive genetic variance of open-field activity and 26 percent of the additive genetic variance of defecation. Therefore, even though there is an important major gene effect, a relatively large proportion of the genetic variance of the genotypes tested remains and is due to segregation at an unknown number of unidentified loci.

9-3 MICE: QUANTITATIVE TRAITS: VARIOUS BEHAVIORAL PHENOTYPES

The observation from several sets of data in mice (and in rats) of a negative correlation between activity and emotionality in the open field under a variety of genetic situations (such as comparisons between major genes, different strains, and in lines selected for high and low activity and for high and low defecation scores) indicates the complexity of the total behavioral phenotype. Furthermore, as mentioned in Section 6-11, in Broadhurst's (1960) rat lines selected for high and low defecation scores, correlated responses for many traits, some behavioral and some physiological, were found to agree with what is expected from the dichotomy of emotionality occurring in the reactive and nonreactive lines. In addition, Blizard (1971) found that reactive rats had higher heart rates after handling than did nonreactive rats. This leads us to ask whether there is generally a *behavioral phenotype* corresponding to a particular genotype. In other words, does a particular genotype lead to a set of behaviors, as suggested for the *albino* locus? It is difficult to answer this completely, but the evidence favors it as a working hypothesis, although in many cases inadequate breeding experiments have been carried out.

The proposal of a complex behavioral phenotype corresponding to a particular genotype is suggested by Parsons (1972a, 1974b) for activity (open-field and exploratory), emotionality, and weight in studies on three inbred strains of mice: C57BL, BALB/c, and C3H (Table 9-5). C57BL, the lightest strain, was the most active, had the highest exploratory activity, and was the least emotional, while BALB/c was the complete reverse, with C3H intermediate but often quite close to C57BL. It must be stressed here that an analysis of these traits in the F_2 and backcross generations is necessary to assess the degree to which traits remain together in inheritance. Hence comments on the behavioral phenotype are often tentative.

By the use of inbred strains and mutant stocks, it can be shown that much of the variation in skeletal morphology among strains is genetic (Grüneberg, 1963), and in fact Grüneberg and others have suggested that many if not most minor skeletal variants are expressions of generalized or localized size varia-

Table 9-5   Order of Three Inbred Strains of Mice for Morphological, Physiological, and Behavioral Traits

Trait	Order*
Open-field activity	C57 > C3H > BA
Exploratory activity	C57 > C3H > BA
Open-field emotionality	BA > C3H > C57
Percentage of no-shock jumps†	C3H > C57 > BA
Weight	BA > C3H > C57
Skeletal divergence	BA > C3H > C57
Temperature preference‡	C57 > C3H ≫ BA
Body temperature‡	BA ≫ C3H ≃ C57
Abdominal fur density‡	C57 > C3H ≫ BA
Tail length‡	
55–58 days at measurement	BA ≃ C3H ≃ C57

* C57 = C57BL; BA = BALB/c.
† See Table 9-6
‡ See Table 9-7
Source: Data of Howe and Parsons (1967), Rose and Parsons (1970), Silcock and Parsons (1973).

tions. For this reason, Howe and Parsons (1967) classified skeletons of mice in the three strains for the presence or absence of 25 minor skeletal variants—15 affecting the skull, 8 the vertebral column, and 2 the appendicular skeleton. From the percentage of incidences of each variant in the strains, a mean measure of divergence among strains can be obtained, as given in Berry (1963). Data on skeletal divergence show BALB/c > C3H > C57BL (Table 9-5). Weight differences and the pattern of skeletal divergence are thus associated, and the incidence of many skeletal variants may indeed be associated with body weight, as suggested by Grüneberg (1963). Even though the number of strains examined is limited, these results allow one to argue for a possible association between genotype, skeletal morphology, weight, and various behavioral parameters. This argument is intuitively reasonable since skeletal variants are presumably associated with variants of the muscular, nervous, and vascular systems, and such variants presumably have consequences at the behavioral level. Therefore, even if the argument cannot be generalized fully, it seems worth considering in the study of any quantitative behavioral trait. Generalization or otherwise would be assisted by a consideration of other strains. In this regard the tables of Staats (1966) and others on the origin of the various inbred strains are essential. As is clear from the papers in Lindzey and Thiessen (1970), there is a vast amount of information on the various strains, much of it involving behavioral traits. Unfortunately, although many strains and behaviors have been studied, comprehensive studies of many simultaneous behaviors in many strains are rarer.

In the same three strains of mice, a measure of learning was assessed by testing conditioned avoidance (Rose and Parsons, 1970). The apparatus used consisted of a see-through Perspex box with a grid floor. The floor was divided into two equal parts by a low central barrier. Shock could be applied to either side and to the central barrier. The barrier was shocked to prevent the mouse from "sitting on the fence." The mouse was placed into this apparatus for 1 minute; then a light source above the apparatus was switched on. Two seconds later a shock was administered to the mouse's feet through the grid floor. Times were recorded from the commencement of the light signal until the mouse had jumped the central barrier to the safe side. The times recorded for the first jump in the apparatus were used as a measure of "initial reaction to shock." The mouse was then removed from the apparatus, allowed to rest for 1 minute, and returned for another shock test. Altogether, the mouse was given 10 shock trials in the following sequence:

- Four trials, each 1 minute apart
- 1 hour rest
- Three additional trials, each 1 minute apart
- 24 hours rest
- A final three trials, each 1 minute apart

Table 9-6 shows the percentage of no-shock jumps (jumps to the safe side of the apparatus after the light signal was switched on but before shock was applied). Trials 2 to 10 were used to determine this parameter. The highest percentages of no-shock jumps occurred for trials 4, 7, and 10 at the end of each set of trials, and low percentages occurred for the first trial in each series after rest, as expected. The order of superiority of the inbred strains is C3H $\gg$ C57BL > BALB/c, which does not correspond to the sequence obtained for activity and emotionality. Therefore, the association between morphology and the behavioral phenotype does not hold. In this case, the link between genes and the behavioral output can be regarded as less direct than for the various simpler forms of behavior discussed earlier because of the assumed importance of the learned component. The data show some variability accord-

Table 9-6   Percentage of No-Shock Jumps in Trials 2 to 10 Recorded for Male Mice of Three Inbred Strains

Trial Number	BALB/c	C3H	C57BL	BALB/c × C3H	BALB/c × C57BL	C3H × C57BL
2						1.3
3	0.7	14.8		3.1	2.2	9.3
4	2.8	22.2	2.0	10.2	9.6	22.7
5	1.4	16.0	1.0	2.0	7.4	12.0
6	0.7	20.0	3.2	19.4	14.8	17.3
7	3.6	20.0	7.4	20.4	29.6	32.0
8	1.6	4.5	6.9	12.7	18.9	12.0
9	4.0	13.6	9.7	11.4	23.6	24.0
10	9.5	31.8	12.5	16.5	37.8	29.3
All trials	2.6	15.8	4.3	10.4	15.8	17.8

The rules in the body of the table represent the rest periods.
Source: Rose and Parsons (1970).

ing to trial number, since up to trial 5, BALB/c > C57BL, but thereafter the situation is reversed (C57BL > BALB/c). This represents a genotype–environment interaction among trials in that C57BL mice take longer for a score to be registered, but even so, C57BL mice end up with a higher score in later trials. The experiments were repeated using a similar technique but with a buzzer rather than a light as the signal (Rose and Parsons, unpublished data). Sequences of 10 trials were used. After the first sequence C57BL > BALB/c > C3H, but after the second sequence, C3H and C57BL were very similar but superior to BALB/c. Differences occur according to the technique of assessment—a result that is not surprising given the known photophobia of BALB/c mice.

Of more ecological significance are the remaining traits listed in Table 9-5. Temperature preference was assessed in a cage with a temperature gradient along the floor ranging from 23°C to 43°C over a distance of about 120 cm (Silcock and Parsons, 1973). Mice could be seen actually choosing a preferred temperature. The behavioral process consisted of a mouse lowering its abdomen onto the floor of the cage as it moved about in a certain section, eventually settling in the position that presumably was the most comfortable temperature for it. The C3H and BALB/c mice often slept in these positions. The BALB/c strain preferred the lowest temperature (Table 9-7). Associated with this was a high body temperature (as measured anally within 30 seconds of death with a quick-read thermometer) and low abdominal fur density (the abdomen was

Table 9-7 Temperature Preference, Mean Weight, Body Temperature, Abdominal Fur Density, and Tail Length of Three Inbred Strains of Mice and Their Hybrids, Tested at Age 55 to 58 Days

Strain	Temperature preference (°C)		Mean weight (g)	
	Males	Females	Males	Females
1. BALB/c	25.67	26.30	24.0	20.8
2. C3H	36.78	35.92	21.5	19.0
3. C57BL	34.30	37.47	21.1	18.1
4. C57BL × C3H	30.94	37.95	22.5	19.8
5. BALB/c × C3H	30.00	30.65	22.8	19.3
6. BALB/c × C57BL	33.10	37.25	23.6	20.6

Strain	Body temperature °C	Abdominal fur density hairs/mm²	Tail length (cm)	
			Males	Females
1. BALB/c	38.03	31	8.3	8.1
2. C3H	35.95	64	8.4	8.2
3. C57BL	35.55	59	8.0	8.0
4. C57BL × C3H	37.05	63	9.1	8.9
5. BALB/c × C3H	37.29	64	9.0	9.0
6. BALB/c × C57BL	36.98	64	8.9	8.8

Source: Modified from Silcock and Parsons (1973).

chosen because of its apparent importance in selecting optimal temperature). There is a positive association between fur density and temperature selected by the various strains, associated with a negative relation between these variables and body weight and body temperature. Because the larger BALB/c mice have a low surface area/volume ratio, they would be favored in a cooler environment and so may be expected to select it. In a cooler environment the high metabolic rate may be adaptive; this may be indicated in these mice by the higher body temperature. Based on the assumption of these adaptations, high fur density may not be of great importance to them. Conversely, C57BL and C3H mice have a lower body temperature and a higher fur density and are lighter in weight. In a warm environment, heat is clearly of less significance than in a cold one; hence, under these circumstances little selection for high body temperature is expected. Thus, temperature preference is apparently associated with morphological and physiological traits, namely weight, skeletal divergence, body temperature, and fur density. The behavior patterns observed during the process of choosing a preferred temperature show a direct selection indicating that the behavior is essentially innate.

There are discussions in the literature on the role of the tail as a thermoregulatory organ. In Table 9-7 data for the inbred strains show no significant differences in tail length among strains. Even though the tail is regarded as having a thermoregulatory function (Harrison, Morton, and Weiner, 1959), the data of Silcock and Parsons (1973) and other published data suggest the conclusion that tail length may be less important in temperature preference selection than the other variables here considered. There are, however, tail-length differences in nature such that mice in cool environments often, but not always, have longer tails than do mice in warm environments (Berry, 1970).

The value of the learning and tail-length data is reinforced when data from hybrids are considered (Tables 9-6 and 9-7). Heterosis occurs for the learning measure between the two hybrid pairs, BALB/c × C57BL and C3H × C57BL, and for all hybrids for tail length. None of the other traits listed in Table 9-5 shows heterosis. Thus the two exceptional traits based on the ordering of the three inbred strains are noteworthy for showing heterosis (and inbreeding depression). It has been argued that such traits are subject partly or wholly to directional selection in the direction of the hybrids (Mather, 1966) and are traits with a relatively direct relation to fitness. The adaptive significance of directional selection for rapid learning ability is obvious, although not obvious for long tails. Perhaps tail length is subject simultaneously to directional and stabilizing selection, a possibility indicated in certain plant populations (Allard, Jain, and Workman, 1968). Traits not showing heterosis or inbreeding depression have been argued to be relatively peripheral to overall fitness and to be subject to stablizing selection. For these latter traits there is apparently a relatively direct connection from genotype to morphology and physiology and then to behavior, and hence a complex behavioral phenotype associated with a given genotype may occur—although as stressed, more crosses are needed to confirm (or modify) this suggestion.

Other data showing heterosis for traits with a learning component include those by Collins (1964) on conditioned avoidance responses, Wahlsten (1972) for a review of many experiments, and Anisman (1975) on a number of tasks ranging from simple to complex. Quoting from Anisman (1975):

> In the relatively simple inhibitory task and in the activity and reactivity situations, the pattern of inheritance was primarily one of an intermediate nature. With increasing task complexity, in terms of the cue-shock and response-shock contingencies, the patterns of inheritance were altered considerably. In the oneway avoidance situation, complete dominance was typically observed, whereas in the shuttle task the rule seemed to be one of overdominance.

Presumably the shuttle-avoidance task consists of several phenotypes that must be well-coordinated as compared with simpler tasks. For these complex tasks involving learning, inbreeding depression with associated heterosis appears to be the rule. Another manifestation of this is that for those traits showing heterosis, there is a tendency for the variances of the hybrids to be lower than for the inbred strains—an observation apparently true for a number of behavioral traits as discussed by Parsons (1967b). This means that the hybrids are better buffered than the inbred strains themselves against systematic or unsystematic environmental variables. Such buffering is referred to as *behavioral homeostasis* in Section 6-2.

Considerable other published data compare different strains of mice. Rodgers and McClearn (1962) found when giving mice a choice of alcohol of various concentrations that the ranking of three strains on the basis of mean daily consumption was C57BL > C3H > BALB/c. For learning performance, McClearn (1972) concluded from a literature survey that C3H mice are generally poorer performers than C57BL and BALB/c in various experiments, but they performed relatively well in a water-escape situation and a shuttle-box avoidance apparatus. So as already mentioned, variable results may be obtained depending upon how learning is assessed.

Erlenmeyer-Kimling (1972) surveyed the literature on a number of strains in relation to early experience. The C57BL strain responded to all treatments more frequently than other strains, verifying observations made by Ginsburg (1967), Henderson (1968), and others on the particular lability of the C57BL strain to environmental variables. BALB/c showed generally low responsiveness to the treatments considered, and C3H seemed intermediate. The BALB/c result is not surprising, in view of the known hesitancy of albinos in reactions to environmental changes. For strains C57BL and C3H, general background variables such as isolation, environmental enrichment, or cage illumination may be less critical than more specific and possibly traumatic events, such as handling, shock, and noxious noise. The opposite may be true for BALB/c. Therefore, for lability to early experience effects we can probably write C57BL > C3H > BALB/c. However, further work paying particular attention to those traits in which learning is involved seems necessary.

With respect to learning, Henderson (1970) carried out a biometric analysis of performance using a food-seeking task with a 6 × 6 diallel cross (six inbred strains of mice and their 30 F_1 crosses). He used two environments: standard laboratory cages and an enriched environment where, for 4 days prior to the onset of the test, mice were permitted more perceptual motor activity than standard cages provide though far less than is normally available in the wild state. An examination of the hybrid performance scores in Figure 9-3 shows that animals reared in the natural cages had a relatively uniform low performance, whereas enriched animals showed both better performance and greater genetic variability. Apparently the restrictive conditions reduced the performance of all genetic groups to some lowest common denominator below that of all enriched groups. The figure also shows heterosis in that hybrid scores were more often than not significantly higher than the average means of the higher-scoring parents. A genetic analysis of the diallel cross revealed strong unidirectional dominance, as expected for a task involving learning, and a considerably larger additive genetic effect among enriched as compared to standard-caged animals. Taking into account environmental variation, estimates of additive genetic variation of 0.13 and 0.29 and dominance variation of 0.13 and 0.35 for standard and enriched cage groups, respectively, were obtained. Therefore, the various genetic analyses show that the primary effect of the restrictive environment is to reduce the overall magnitude of genetic effects. Later work shows that this enrichment effect is detectable over periods of time as short as 6 hours (Henderson, 1976). There are associated brain weight changes, since brain sizes are larger in enriched environments. A genetic analysis indicates significant directional dominance toward larger brains in enriched animals only—showing, among other things, a complex association of behavioral with morphological change (Henderson, 1973). We will return to these complex types of genotype-environment interactions in later chapters. Let it suffice to remark here that if we assume the enriched environment to approximate the situation in nature

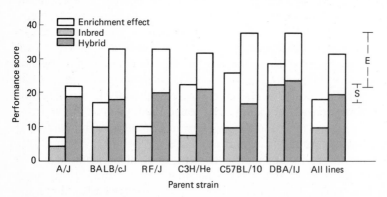

Figure 9-3 Mean performance scores for six inbred strains and their F_1 hybrid crosses reared in standard (grey bars) and enriched cages (grey bars plus white bars). Dashed lines show the range of responses for hybrid genotypes reared in standard (S) and enriched (E) cages. (*From Henderson, 1970.*)

more than does the standard environment, then such interactions may be important in natural populations (including those of human beings).

This result points out one difficulty with most behavior-genetic research on rodents, namely, that interpretations of evolutionary significance arise *following* the accumulation and analysis of experimental data. It would be more convincing to begin with a prediction. Henderson (1978) therefore argued that although most studies on locomotor behavior had demonstrated intermediate inheritance or some dominance toward high activity, this should not be the case in very young animals. Infant mice less than 1-week-old have extremely limited visual and auditory sensitivity associated with poor locomotor coordination. Therefore if animals are removed from the nest, a high rate of locomotor activity should be maladaptive; activity is then as likely to result in an animal moving farther from the nest as toward it. The former would increase its chances of being attacked by predators, including mice from other nests. Low activity with a highly efficient maternal retrieval process would be more adaptive. Therefore the genetic pattern of locomotor behavior in very young mice would be predicted to be one of genetic dominance favoring low activity associated with low heritability, thus differing from the genetic pattern of older animals. Using 4-day-old mice, a triple test cross (TTC) analysis (Section 6-5) gave $V_A = 0.0077$, $V_D = 0.0034$, $V_E = 0.0920$, so that $h_N^2 = 0.075$. The low heritability is therefore associated with a very high dominance component, which is in fact in the direction of low activity as predicted, thus confirming that low infantile activity is apparently adaptive. In addition, Henderson (1978) tested some wild mice and found that infantile activity was even lower than in hybrids, which suggests that selection for low infantile activity occurs in nature. During laboratory domestication this selection has presumably been reduced. The results are therefore in agreement with what one might expect for a behavioral character having adaptive significance in nature but less in the laboratory.

Henderson (1979) followed these experiments with another prediction based on 10- to 11-day-old mice. If such mice are moved 10 to 15 cm from the maternal nest, they will usually turn and crawl back to the nest on their own if not immediately retrieved. Intuitively, it appears adaptive for mice to return rapidly to the home nest. Presumably olfactory, thermal, and other cues can guide the return of mice, so that locomotor activity directed toward the nest would be expected to have a significant dominance component. In contrast, for mice removed from the nest and placed in a totally new environment, the relation of locomotor activity to fitness is unclear. As Henderson comments, this is an artificial "nonsense" test rarely if ever existing under situations in which natural selection would take place. An activity measure would then be predicted to produce a smaller ratio of dominance to additive genetic variance than the above more natural situation.

Appropriate biometric (TTC and diallel crosses) experiments revealed (1) directional dominance associated with overdominance for the fast nest-return times when mice were placed 15 cm from the nest and (2) significant additive genetic variance and no dominance variance for mice placed in the "nonsense" test situation in the new environment. The predictions are therefore confirmed and stress the importance of measuring variables in reasonably meaningful situations for the species. It therefore appears that it is now possible to make moderately accurate predictions about genetic changes that occur as a function of varying environmental parameters, some of which may be directly relatable to nature.

These results begin to suggest that the mouse must be a candidate for behavior-genetic studies in wild populations that exist in a wide variety of habitats. The evolutionary questions posed in this section must be extrapolated to nature; Henderson's experiments are an excellent start. Bruell (1970) drew attention to the enormous array of races and subspecies of *Mus musculus* on a worldwide basis that await our attention. That such an approach has potential is indicated by observations of the deer mouse *Peromyscus* discussed in Section 13-3 where it is shown that various phenotypes can be directly related to the habitats selected. It makes sense to devote more attention to those traits that are of apparent significance in such habitat selection, as for instance, temperature preference and the associated variables shown in Table 9-7, and to subject such traits to more sophisticated genetic analyses than have so far been reported.

9-4 MICE: MALE SEXUAL BEHAVIOR

This section is concerned with a genetic analysis of sexual behavior in the mouse. McGill (1970) describes the behavior of the males after the females are brought into estrus by hormone injections:

> In a typical series a male first encountering an estrus female carefully investigates her, concentrating his attention on the anogenital regions. If sufficient sexual-arousal occurs, the male mounts the female palpating her sides with his forepaws while simultaneously executing a series of rapid, pelvic thrusts. Quite frequently the first attempt at gaining intromission fails and the male dismounts and engages in genital cleaning. When a male is successful in gaining intromission, the rate of pelvic thrusting is greatly reduced while the amplitude is increased. The thrusts of intromission average about one-half second and are easily counted. During intromission, the male keeps one hindfoot on the floor and rests the other on the hindquarters of the female. The number of thrusts in each intromission varies from only a few to 300 or more. After an intromission, both animals generally engage in genital cleaning. This behavioral sequence of mounts, intromissions, and genital cleaning usually continues until the male ejaculates. During the ejaculatory intromission, the speed of pelvic thrusting increases, and, finally, the male quivers strongly while maintaining deep penetration of the female. At this point, he raises the hindfoot which has been resting on the floor and clutches the female with all four limbs. Most frequently, this results in both animals falling to one side. Following the male's ejaculation, both male and female again engage in genital cleaning.

Preliminary data on male sexual behavior using the inbred mouse strains C57BL/6J and DBA/2J and the F_1 are given in Table 9-8 for the 14 measures

Table 9-8 Median Scores and Significance Levels of the Three Possible Comparisons for 14 Measures of Male Sexual Behavior in Three Inbred Strains of Mice

Measure*	Median scores			Significance levels		
	C57BL/6J	DBA/2J	F_1	C57 vs. DBA	C57 vs. F_1	DBA vs. F_1
1	42	85	42	0.02		0.002
2	400	129	546	0.002		0.002
3	17	5	18	0.02		0.02
4	0	20	0	0.02		0.001
5	23	17	19	0.02	0.02	
6	15	20	19	0.02	0.01	
7	2	0.5	0	0.01	0.001	
8	28	137	42	0.002	0.002	0.002
9	2	7	3	0.002	0.001	0.02
10	1	4	2	0.002	0.02	0.002
11	18	16	7		0.02	0.05
12	16	20	25		0.02	
13	107	179	93			0.02
14	1252	1376	1091			

* Definitions of measures

1. Mount latency (number of seconds from introduction of female until male mounts).
2. Total number of thrusts with intromission preceding ejaculation.
3. Number of intromissions preceding ejaculation.
4. Percentage of times male bites female following ejaculation.
5. Ejaculation duration (number of seconds male spends clutching female and maintaining vaginal contact, following ejaculation).
6. Time of intromission (number of seconds from beginning of mount with intromission until male dismounts).
7. Number of head mounts during a series.
8. Interintromission interval (number of seconds from end of one mount with intromission until beginning of next).
9. Time of mount (length of mount without intromission in seconds).
10. Preintromission mount duration (number of seconds from beginning of mount with intromission until male's penis is inserted into female's vagina and first thrust of intromission occurs).
11. Number of mounts without intromission per series.
12. Number of thrusts making up each intromission.
13. Intromission latency (number of seconds from introduction of female until male gains intromission).
14. Ejaculation latency (number of seconds from beginning of first intromission until beginning of ejaculation).
Source: McGill (1970).

defined therein. Considerable differences among inbred strains are shown for the various components of male sexual behavior. The inheritance of sexual behavior is clearly not simple, since the data suggest three different modes of inheritance: (1) dominance of one parental genotype or the other (measures 1 to 4 where C57BL is dominant and 5 to 7 where DBA is dominant); (2) absence of dominance, where the F_1 is between the parents (measures 8 to 10); and (3) overdominance or heterosis, where the F_1 is superior to both parents (measures 11 to 14). In other words, for this spectrum of behaviors, all associated with male mating, there is a complete spectrum of modes of inheritance. Furthermore, when a different cross (DBA/2J × AKR/J) was carried out, differing results were obtained for many of the traits. Therefore, the modes of inheritance found are specific to the particular strains studied. More generality would be possible if many strains were studied, as can be done for a diallel cross or the simplified triple test cross.

In discussing his results, McGill (1970) commented on the importance of environment, since the matings described in Table 9-8 were observed and scored while pairs were housed in plastic cylinders under normal room illumination. Under these circumstances, a C57BL/6J male mated with three females during 10 nights of testing. Using a cage placed in the dark, which would approximate more closely the situation normally experienced by the animals, this figure increased above five. This demonstrates a point we have repeatedly made—the results of any experiment are specific to the environment in which it is carried out. Generalizations on inheritance should be made only for experiments carried out under a wide range of conditions, preferably including some which are relatable to habitats in nature.

A full biometrical analysis was carried out for the two parental strains C57BL(P_1) and DBA(P_2), and the F_1, F_2, BC_1 ($P_1 \times F_1$), and BC_2 ($P_2 \times F_1$) generations from which components of variance can be estimated. It was found from reciprocal crosses that sex linkage or maternal effects could be largely ruled out.

As an example, consider intromission latency (measure 13, Table 9-8). The mean values in seconds were:

C57BL(P_1)	DBA(P_2)	F_1	F_2	BC_1	BC_2
151.91	171.02	115.40	123.48	127.87	136.03

Some heterosis is evident, as in Table 9-8. Biometrical analysis showed that a logarithmic transformation provided the best scale. The components of variance then came out to be: $V_E = 0.045$, $V_A = 0.008$, and $V_D = -0.002$, giving $h_N^2 = 0.154$. Thus in the F_2, about 15 percent of the variance is due to additive genetic causes and the dominance component is unimportant. McGill concluded that this trait is controlled by additive genes plus a large environmental variance. He concluded from repeated testing that the large environmental variance was due primarily to specialized nonlocalized variation occurring from test to test within individual animals.

For ejaculation latency (measure 14), the mean times in seconds were:

C57BL(P_1)	DBA(P_2)	F_1	F_2	BC_1	BC_2
1368.91	1977.27	1189.82	1204.73	1354.35	1316.94

Again some heterosis is evident, as in Table 9-8. An attempt to find a suitable scale was unsuccessful; therefore the analysis was carried out on the raw data. This gave h_N^2 in the region of 0.15 to 0.25, which is similar to that for intromission latency. Again there was major variation due to within-animal variation from test to test. For both intromission and ejaculation latencies it may be reasonable to attribute much of the variation to differences in the behavior of the females, even though efforts were made to control this. Possible environmental factors include barometric pressure, recency of feeding or drinking, time of day at test (circadian rhythms), and interaction with other males prior to test—in addition to differences between females. In both the examples discussed, $V_A > V_D$; therefore, if directional selection were carried out, a positive response would be expected. The response would probably be slow because of the large environmental variances.

9-5 MICE: TRAITS WITH DETECTABLE PHYSIOLOGICAL BASIS

Audiogenic seizures are a series of psychomotor reactions to acoustic stimuli of the intensity of an electric doorbell (intensity, 90+ decibels), 12 to 18 inches distant. One complete syndrome (Schlesinger and Griek, 1970) consists of (1) a latency period of variable duration during which the mouse may huddle while appearing to be attending to the stimulus or appear to ignore the stimulus while washing and grooming excessively; (2) a wild-running phase characterized by frenzied running along the boundary of the container; (3) a clonic convulsion during which the animal falls on its side while drawing up its rear legs toward its chin; (4) a tonic seizure during which all four legs are extended caudally; and (5) death due to respiratory failure. Variations of this pattern have been observed. The duration of the latency period ranges considerably. The wild-running phase, which clearly differentiates audiogenic seizures from other convulsive patterns, may be accompanied by a change of gait appearing as a series of stiff-legged bounds. It may then terminate without a subsequent clonic seizure. Tonic seizures may or may not be fatal, and death may or may not be prevented by artificial resuscitation. Even so, the five discrete phases itemized above are characteristic of an audiogenic seizure.

Interest in audiogenic seizures has resulted in the publication of an enormous research literature over the last 40 years, perhaps because of the possibility that the seizures are an appropriate model for human disorders such as epilepsy. This remains to be proved. Information of interest to the behavior geneticist is reviewed in Fuller and Thompson (1960, 1978) and Schlesinger and Griek (1970). Prior to 1947 most of the literature was in psychological journals, but this emphasis has now disappeared. The other concomitant change, as mentioned in Chapter 1, is a shift away from an almost total reliance on the laboratory rat as an experimental animal. Indeed much of the recent work is with mice. Schlesinger and Griek (1970) consider that studies of audiogenic seizures have now become truly interdisciplinary, involving biochemistry, genetics, pharmacology, psychiatry, and psychology.

Coleman (1960) studied *dilute* mice (*dd* and *d'd'*), which have a light coat color provided the background genotype is such that the phenotypic effect can be seen. They have a lower audiogenic seizure threshold than normal mice. Furthermore, *dilute* mice have only 14 to 50 percent as much phenylalanine hydroxylase as nondilute wild-type mice. The particular interest in this enzyme is due to the fact that it converts the essential amino acid phenylalanine to tyrosine and is the enzyme missing in human phenylketonuria. However, there are some differences between the enzyme defect in *dilute* mice and in humans. First, in *dilute* mice the enzyme deficiency is only partial and the remaining activity is sufficient to metabolize dietary phenylalanine adequately. No excess phenylalanine can be demonstrated in *dilute* mice maintained on standard laboratory diets. Second, the defect in *dilute* mice does not appear to be due to a failure to produce the enzyme, since enzyme activity is normal in the supernatant fraction of liver homogenates after centrifugation. The implication is that in *dilute* mice an endogenous inhibitor of phenylalanine hydroxylase is associated with the mitochondria (subcellular organelles). When fed a diet containing excess phenylalanine, *dilute* mice excrete this amino acid more slowly than do nondilute mice; under normal dietary conditions, they excrete certain abnormal phenylalanine metabolites such as phenylacetic acid. Phenylacetic acid has been shown to inhibit decarboxylating reactions in a number of tissues, and it is possible that *dilute* mice are deficient in certain products of decarboxylating reactions. Specifically, it is possible that these animals are deficient in the neurotransmitter substances GABA (γ-aminobutyric acid), NE (norepinephrine), and 5HT (5-hydroxytryptamine, also called serotonin) in the brain, and that the deficiencies in these brain amines in turn account for the high proportion of seizures in *dilute* mice. (*Neurotransmitters* are chemical substances that mediate the transmission of nerve impulses.) Of these, serotonin is considered by Fuller and Thompson (1978) to be the critical substance. As pointed out by Schlesinger and Griek (1970), several assumptions are needed if the above suggestions are to form a working model. It must be assumed that GABA, NE, and 5HT are inhibitory in their action on the central nervous system and that phenylacetic acid in the amount present in *dilute* mice inhibits decarboxylation. Problems remain to be dealt with, but Schlesinger and Griek regard the working model as reasonable. Irrespective of the final situation, there is likely to be a fairly intimate association of genes, biochemical and physiological processes, and behavior.

The inbred strain DBA/2J is genetically *dilute*. Predictably, DBA mice are sensitive to audiogenic seizures; C57BL/6J mice are not. F_1 mice are intermediate, although phenotypically closer to the nonsusceptible parent. Age is a major factor in seizure susceptibility. In one survey (Schlesinger and Griek, 1970) DBA mice had a 90 percent risk at 21 days (as assessed by full clonic-tonic seizures) and 13 percent at 14 and 28 days, while C57BL mice were resistant to clonic-tonic seizures at all ages, and F_1 mice had a developmental pattern similar to that of DBA mice. These results parallel those reported by other investigators (e.g., Fuller and Thompson, 1960, 1978). The exact age of onset varies in different laboratories, indicating the importance of as yet unspecified environmental factors such as diet, housing conditions, temperature, and/or diurnal rhythms, all of which may interact with genetic factors to give slightly different developmental patterns. Acoustic priming is the most fascinating of the environmental effects; Henry (1967) found that resistant C57BL mice could be made highly susceptible by exposure to bell-ringing at 15 to 24 days of age and testing 3 days later. Fuller and Collins (1968) among others extended the observation to other mouse strains. Apart from acoustic priming studies mice are conventionally tested at 30 days of age, which happens to be optimal for demonstrating the C57BL, DBA strain differences.

These same two strains have been tested for seizures resulting from the drug Metrazol and for electroconvulsive seizures. In both cases the DBA strain was more susceptible, suggesting that this particular strain is simply more susceptible to seizures regardless of the inducing agent. In agreement with the metabolic hypothesis stated above was the finding of lower endogenous 5HT and NE levels in DBA mice. Furthermore, when 5HT or NE was depleted by drugs depleting brain amines, susceptibility to audiogenic, Metrazol-induced, and electrically-induced seizures was increased. Conversely, increased levels of 5HT, NE, or GABA protected the animals. The general conclusion is that DBA mice have extremely excitable nervous systems. (It should be noted, however, that in a different set of strains Castellion, Swingard, and Goodman, 1965, found no association between electroshock threshold and audiogenic seizure susceptibility.) Additional work at the pharmacogenetic level is reported by Maxson, Cowen, and Sze (1977), suggesting the importance of corticosteroids for the development of seizures. In conclusion however, we must assume that we have evidence for a complex behavioral phenotype with physiological correlates of some complexity. For similar considerations in human beings see Section 11-8 on the epilepsies.

The genetic situation needs further investigation, since Schlesinger and Griek (1970) have argued that the *dilute* locus may not be directly implicated but just closely linked. The evidence comes from Schlesinger, Elston, and Boggan (1966), who obtained single-gene mutations to full coat color in DBA/2J strains. In these mice, genetically *Dd* or *DD* on a DBA background, the dilute locus did not contribute importantly to seizure susceptibility. But Lindzey et al. (1971) consider that no single genetic hypothesis can account for these data. Evidently we must await further work. Indeed Fuller has experimental data on a cross between the two inbred strains, followed by repeated backcrossing to the C57BL strain, that clearly argue for polygenic control (Fuller and Thompson, 1978).

As another example of a trait having a physiological basis, we consider alcohol preference and aversion. A good recent review of this appears in Lindzey et al. (1971) with particular reference to mice, although studies have been carried out on rats and human beings. Like all the traits discussed, differences have been reported among strains of mice. Given a choice of fluids to drink, some strains (e.g., C57BL/6J) prefer alcohol while others (e.g., DBA/2J) do not. In one series of experiments (Rodgers and McClearn, 1962) four inbred strains of mice were offered simultaneously ad lib choices of water and six alcohol solutions from 2.5 to 15 percent. The proportion of the liquid consumed weekly that was alcohol is given in Table 9-9; this provides a single index representative of the alcohol preference of each strain on a weekly basis. In each week the rate of alcohol consumption for the four strains was in the order C57BL > C3H/2 > BALB/c > A/3. For strains C57BL and C3H/2 the percentage of alcohol consumed increased over a 3-week period, a marked preference for

Table 9-9 **Proportion of Absolute Alcohol to Total Liquid Consumed Weekly for 3 Weeks in Four Inbred Strains of Mice**

Week	Strain			
	C57BL	C3H/2	BALB/c	A/3
1	0.085	0.065	0.024	0.021
2	0.093	0.066	0.019	0.016
3	0.104	0.075	0.018	0.015

Source: Summarized from Rodgers and McClearn (1962).

10 percent alcohol developing by the third week. In strains BALB/c and A/3, however, there was a progressive reduction in alcohol consumption and the development of an increasing preference for water. Thus in the strains tested, the tendency for alcohol preference to increase is positively correlated with initial preference. An analysis of variance of the proportion of liquid consumed that was alcohol revealed a highly significant effect due to different strains (genotypes). Clearly, alcohol preference is under genetic control, but also depends on the environment, in that alcohol preference varies according to the period of previous consumption. Nachman, Larue, and LeMagnen (1971) showed that the removal of the olfactory bulbs eliminated the aversion to alcohol in BALB/c but did not abolish the preference for alcohol in C57BL mice. This, along with the observation that BALB/c mice appear to avoid alcohol immediately, without prior experience, led to the hypothesis that BALB/c mice are more responsive than are C57BL mice to alcohol as a sensory stimulus.

From the physiological point of view, there seems to be some association between differences in the liver enzyme alcohol dehydrogenase (ADH) and alcohol preference, although the correlation broke down in the F_2 offspring of a cross between high preference C57BL and low preference DBA mice (McClearn and DeFries, 1973). ADH is involved in the first step in the metabolism of ethanol to acetaldehyde and so could be of critical importance. Most of the earlier research concentrated on ADH (Lindzey et al., 1971), but attention is now turning to the enzyme that oxidizes acetaldehyde—aldehyde dehydrogenase (ALDH). In combination, the two enzymes are responsible for the breakdown of ethanol through the citric acid cycle. ALDH may be of considerable importance, since alcohol-drinking and nondrinking strains differ by as much as 300 percent for this enzyme. Another point of potential significance is the observation that acetaldehyde has a strong inhibitory effect on brain metabolism because of interference with enzymes related to catecholamines (a specific neurotransmitter—an aromatic amine) (Eriksson, 1973). Certainly audiogenic reactivity appears to be associated with ethanol consumption; indeed ordinarily resistant mice can be made susceptible to audiogenic seizures by prolonged exposure to ethanol early in life (Yanai, Sze, and Ginsburg, 1975). We await with interest the further linking of the genetic, biochemical, physiological, and behavioral components of both alcohol preferences and audiogenic seizures, especially as there may be some links between both behavioral phenotypes through brain amines.

9-6 OTHER RODENTS

Quite an amount of early work was carried out by psychologists on rats, in particular the Norway rat, *Rattus norvegicus*. In fact, as Beach (1950) pointed out (Chapter 1), the trend in American journals specializing in comparative psychology was to reduce the number of species studied over the period 1911 to 1948, with mammals displacing the invertebrates; and of the mammals, the Norway rat was by far predominant. Slightly more than 50 percent of the articles dealt with conditioning and learning, and about 15 percent to 20 percent with reflexes, simple reaction patterns, and sensory capacities. Other forms of behavior, such as reproductive behavior, emotional reactions, social behavior, feeding, and general habits were much less frequently studied. Therefore, it is not surprising that the proportion of this literature directly relevant to behavior genetics is minimal, since the behavior geneticist depends on comparisons within and among species and strains. In addition, the concentration on conditioning and learning is rather restrictive.

With these comments in mind, we can turn to some of the relevant experiments. Tolman (1924) reported the results of what was the first selection experiment for maze learning by rats. The foundation population consisted of 82 white rats of heterogeneous ancestry. From this population, nine "bright" and nine "dull" pairs were mated to obtain the first selected generation. A second selected generation was produced by further selection among the brights and dulls. Selection was successful in the first generation, but less so in the second, and Tolman suggested the discrepancy may have been due to extraneous environmental factors. The general problem, however, was not abandoned. Tryon (1942) published the results of eight generations of selection for maze-learning ability. He bred rats selectively on the basis of their error scores in learning in a multiple T-maze. The foundation population was again a heterogeneous sample of rats. By the eighth generation the maze-brights and maze-dulls did not overlap and there was negligible response thereafter. Detailed biometrical genetic analyses were not carried out, although Broadhurst and Jinks (1963) later calculated a heritability in the broad sense at about 40 percent using the F_1 and F_2 crosses between the selection lines. The data are somewhat unsatisfactory for biometrical analyses as there was a significant genotype-environment interaction that could not be scaled out. Further discussions of other maze-learning selection experiments differing in techniques are presented by Fuller and Thompson (1960), including some experiments with negative results.

Another form of behavior (already dealt with in mice), the conditioned avoidance response, which has a learning component, has been studied less in rats. Bignami (1965) carried out successful selection for the trait, indicating genetic control. Satinder (1971) found genotypic differences among four selectively bred strains of rats for escape-avoidance. Of particular interest was the finding that the strains responded differentially to the drugs d-amphetamine sulfate and caffeine—genotype-environment interactions of a special kind. This area should be explored fully with the aim of testing the range of effects of drugs on behaviors of different genotypes (see next section).

In a few other traits studied in rats, there is evidence for genetic control either from crosses between inbred strains or from selection experiments. Broadhurst's work on defecation in rats as a measure of emotionality and its associated variables, especially activity score, has already been discussed. Directional selection for activity has also been carried out, since Rundquist (1933) successfully selected for active and inactive strains based on activity in a revolving drum. Generally the data favor a multifactorial interpretation for the genetic basis of these traits. Finally, just as in mice, a number of major gene effects have been described in rats (reviewed by Wilcock, 1969), but no new principles emerge. Much of this work involves major pigmentation genes. Since the behaviors are similar to those described for mice, a detailed discussion is not given here.

An experiment in guinea pigs is relevant. Goy and Jakway (1959) and Jakway (1959) studied sexual behavior in two inbred strains and in the F_1, F_2, BC_1, and BC_2 generations. For females the response to the injection of female hormone was assessed in terms of four behavioral measures elicited experimentally by tests for lordosis (arching of the back) prior to copulation. For the males, females in heat were used as the stimulus and six behavioral measures were taken. Broadhurst and Jinks (1963) found difficulties in scaling for all but two variables—the number of malelike mountings by females during estrus and the number of penile intromissions by males. Estimates of heritability fell between about 0.5 and 0.6 for these two traits, and there was a marked dominance component for both traits. Heterosis was found for the highest measures of activity in males—the intromission rate and number of ejaculations. In rats, Dewsbury (1975) carried out a 4×4 diallel cross for components of copulatory behavior. He found directional dominance for rapid copulation, and for ejaculation after a relatively few mounts and intromissions. That is, the data are consistent with the reasonable expectation that it is adaptive to maximize the transmission of sperm to females.

The basic conclusion is that, parallel with results in the mouse, most quantitative behavioral traits in other rodents are controlled multifactorially, as can be shown by differences among strains and the results of selection experiments. For this reason, a more detailed review of these results is not necessary. Mice have an advantage as experimental animals because they have a short generation time and a much better known chromosome map than other rodents. It is likely that behavior-genetics analyses of mice will develop rapidly even though they were initiated later than those with rats.

9-7 PSYCHOPHARMACOGENETICS

The discussions in the last two sections suggest that one approach to the physiological basis of behavior is via the effects of drugs and their differential effects on different genotypes—a field referred to as *psychopharmacogenetics* (Eleftheriou, 1975). If it is assumed that a drug acts by affecting some metabolic pathway, then presumably we are dealing with some sort of modification of this pathway leading to a behavioral effect. The degree of modification depends partly on the genotype. Large individual differences in responses to drugs are found in human beings and in animals (Meier, 1963). Different inbred strains of mice show large variations in sleeping time in response to a given dose of hexobarbitone. Related to these results are experiments of Nicholls and Hsiao (1967) showing that rats can be selectively bred for susceptibility to morphine addiction, i.e., for an "addictive personality." In mice heritability estimates of susceptibility to morphine addiction were very high (Oliverio, 1975), based on a biometric analysis between the two inbred strains CBA/Ca and C57BL. In some of the lines derived by selection for behavioral traits in rats, especially those of Tryon (1942) for maze brightness and dullness, and of Broadhurst (1960) for defecation scores in rats, various drugs have been shown to give drug-strain interactions (Broadhurst and Watson, 1964).

Not unexpectedly genotype-drug interactions are frequently documented. Taking C57BL and DBA mice, Oliverio (1974) reported on the effects of the adrenergic drugs amphetamine and scopalamine, and the cholinergic drug physostigmine, on exploratory activity (Figure 9-4). Amphetamine decreased the activities of C57BL and C57BL × DBA mice so that DBA was dominant over C57BL. For scopalamine, C57BL was recessive to DBA. Physostigmine diminished exploratory behavior in all strains. This is consistent with other evidence showing that C57BL and DBA mice differ in activity and may respond to the same centrally active drugs in opposite directions. It is likely

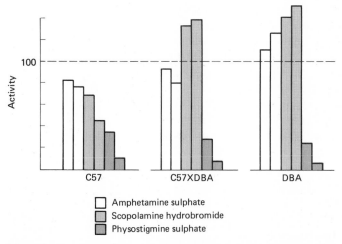

Figure 9-4 Modifications of exploratory activity (expressed as percentage of control level) in two strains of mice and the F_1 hybrid. Amphetamine sulphate (0.5 and 1.0 mg/kg, i.p.), scopolamine hydrobromide (2.5 and 5.0 mg/kg, i.p.), and physostigmine sulphate (0.15 and 0.30 mg/kg, i.p.) were injected 30 minutes before the test. The lower and the higher dose of each drug are represented by the left and right columns, respectively. Each column represents the mean of crossings in a tilt-box for 15 mice. (*After Oliverio, 1975.*)

that the different reactivities of these mouse strains are related to the known neurochemical diversity among them, as is also suggested by the discussion on audiogenic seizures and alcohol preferences for the same strains in Section 9-5.

Finally, extending the definition of psychopharmacogenetics a little, differential responses for detecting sweet, salty, and bitter substances in mammalian populations can be considered. In mice Ramirez and Fuller (1976) found moderate to high heritabilities for saccharin and sucrose consumption, and they refer to studies showing large individual differences in response to sweet taste in mice, rats, cattle, and pigs. The human PTC tasting polymorphism (Section 2-3) is well-known and occurs in nonhuman primates (Section 11-5). In addition there are clear differences in bitter taste sensitivity of inbred strains of mice (Klein and DeFries, 1970) which are probably controlled by an autosomal locus. More recently Tobach, Bellin, and Das (1974) showed differences in bitter taste perception in three strains of rats for PTC which is relatively toxic and for the antibiotic cycloheximide (CH)—a powerful inhibitor of ribosomal amino acid incorporation. They showed that Wistar and Long Evans rats are able to perceive millimolar concentrations of PTC and 0.2 μM CH, while Fawn-hooded rats are unable to detect millimolar PTC and first recognize CH at 1.5 μM concentration. In other words, the Fawn-hooded rats are unusually deficient in tasting abilities.

The few examples given here and elsewhere in this chapter should indicate the potential power of the psychopharmacogenetic approach for unraveling the path from genes to physiology to behavior. It is an approach of promising value, especially for behaviors with a learning component, because of the possibilities of careful extrapolation to human beings with regard to the effects of drugs on learning. Although extrapolation across organisms can only be done with caution, basic metabolic systems have considerable similarities in rodents and human beings. Many behavioral conditions in people are now treated with drugs, and, in addition, drug addiction itself is a problem of increasing concern. There are difficulties in prescribing drugs, since although a drug may affect one specific chemical pathway, the physiological effects may be complex. Also, a compound may well be greatly modified before reaching its target organ. Because of the blood-brain barrier, there are problems of introducing drugs to the brain. In spite of these complications, which cannot be regarded as insurmountable, the field of psychopharmacogenetics can be expected to develop considerably.

SUMMARY

Rodents, especially mice, have played a major role in behavior genetics. Many mutant genes are associated with neurological defects. In some cases, the behavioral changes can be related to underlying developmental, cellular, and even molecular changes. But almost any mutant (e.g., those affecting coat color) can be linked to behavioral effects if the battery of tests used is sufficiently comprehensive.

Many quantitative traits of mice have been analyzed including activity, emotionality, and sexual behavior, but there is now increasing emphasis on traits with learning components. A genetic architecture of directional dominance for rapid learning is usual, whereas the additive component is of greater importance for most other traits. Traits of obvious significance in nature have, however, been largely neglected.

Considering alcohol preferences and audiogenic seizures, the mouse appears to be a model organism for the linking of genetic, biochemical, physiological, and behavioral components in differing phenotypes. Furthermore, recent work in mice and rats shows that one approach to the physiological basis of behavior is via the effects of drugs and their differential effects on different genotypes. This psychopharmacological approach is of potential value, especially for behaviors with a learning component, because of possibilities of careful extrapolation to human beings.

GENERAL READINGS

Eleftheriou, B. 1975. *Psychopharmacogenetics*. New York: Plenum. A first integrated account of this hybrid field.

Lindzey, G., and D. D. Thiessen. 1970. *Contributions to Behavior-Genetic Analysis: The Mouse as a Prototype*. New York: Appleton. A collection of papers on various aspects of mouse behavior, considering genetic analysis, gene-environmental interplay, single-gene effects, gene-physiological determination, and evolutionary aspects.

The Genetics of Behavior:
Other Creatures

In 1962, in an eminently recommendable anthology entitled *Roots of Behavior* (Bliss, 1962), Dilger comments that "direct evidence of the genetic control of behavior in vertebrates is, unfortunately, even scarcer [than in invertebrates] and the precise identification of responsible genes is almost nonexistent."

Our purpose in this chapter is to correct the erroneous impression that the genetics of behavior can only be approached via the manipulation of genes and behavior in *Drosophila,* rodents, or human beings. We offer a survey of approaches to studying other organisms. These creatures are not generally so easy to tabulate as *Drosophila,* rodents, and human beings (about whom genetic knowledge has been accumulated despite the absence of planned crosses) because knowledge of their genomes is not so extensive. This, however, makes them all the more enticing as subjects, or so we believe and hope to convince our readers. Our examples are not all confined to this chapter: the trisomic monkey and its retarded behavior, Dilger's lovebirds, the barkless dogs, and Rothenbuhler's tidy bees have been utilized as examples of principles in the genetic analysis of behavior (Chapters 3, 4, and 5).

This chapter presents its material in the form of examples that indicate the range of organisms investigated and the scope of these often technically arduous investigations. Each organism is considered individually because of the specific limitations presented by the experimental subjects. Predictably, we commence with bacteria and conclude with vertebrates.

10-1 BACTERIA

"What is the repertoire of behavioral responses in an organism?" asked Adler, Hazelbauer, and Dahl (1973). When the organism is a peritrichous (possessing many hairlike extensions) bacterium such as the ubiquitous *Escherichia coli*, one may inquire about the orientation of the flagella in the presence of specific substances. Motile bacteria are attracted to a variety of chemicals (see Pérez-Miravete, 1973, for a survey). Chemoreceptors detect specific chemicals without participating in their metabolism. How is the clustering depicted in Figure 10-1 accomplished? Utilizing mutagens, Adler and coworkers isolated dozens of true-breeding mutant clones that do not display positive chemotaxis toward a variety of substances, such as sugars, amino acids, and oxygen, attractive to the wild-type, nonmutant *E. coli* (Mesibov and Adler, 1972; Adler, 1969, 1976 for review). This is so even though the mutant bacteria are perfectly motile and have full complements of normal flagella. They can and do respond normally to all but the particular attractant to which they no longer possess the capacity to respond. This situation is schematically presented in Figure 10-2. How many different chemoreceptors are involved? Nine exist just for strong sugar attractants (N-acetyl glucosamine, fructose, galactose, glucose, maltose, mannitol, ribose, sorbitol, and trehalose). Two serve to detect amino acids (aspartate and serine). Genes have been located for motility and chemotaxis on the *E. coli* genetic map, itemizing loci such as *curly* (altered flagellin protein, wavelength approximately half of normal, only rotational movement possible); *motile* (pos-

Figure 10-1 Attraction of *E. coli* to aspartate contained in a capillary tube. (*From Adler, 1969. Copyright 1969 by the American Association for the Advancement of Science.*)

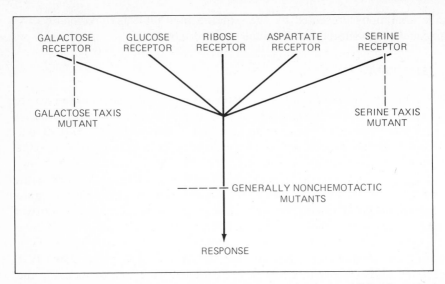

Figure 10-2   Chemotaxis in *E. coli*. Possible explanation of defects exhibited by mutants with impaired attraction toward specific amino acids or sugars. (*From Adler, 1969. Copyright 1969 by the American Association for the Advancement of Science.*)

sess normal looking flagella but cannot move); *flagella* (no flagella present, nonmotile); *chemotaxis* (nonchemotactic, fully motile, three genes involved). Major contributions to the science of genetics can be anticipated from the study of the behavior of *E. coli* and other bacteria. For instance, normal *E. coli* are repelled by acetate, benzoate, and indole. Muskavitch et al. (1978) have isolated reversed-taxis mutants that are attracted to these compounds and that are defective in one of a set of two methyl-accepting chemotaxis proteins that form a central processing component in the flow of information from chemoreceptors to flagella.

10-2   PARAMECIA

The new behavior genetics of *Paramecium aurelia* is aimed at a genetic dissection of the organism's outer, limiting, excitable membrane. The locomotor behavior of *Paramecium* is under the control of this surface structure (Eckert, 1972). The direction and beating frequencies of the *cilia* (fine membrane-enclosed cytoplasmic hairlike threads projecting en masse outward from the surface of a cell) are correlated with cross-membrane electrical events, that is, changes in the voltage-sensitive calcium conductance. Cilial reversal resulting in an alteration of swimming direction is called *avoiding*. Various stimuli elicit avoiding behavior in ciliated protozoa. These have been known since 1906, when Jennings described a paramecial avoiding reaction as an interruption in forward swimming due to the temporary reversal of the ciliary beat. This produces a short backward jerk or a sudden halt before forward propulsions begin anew in an altered direction. Recently *Pawn* mutants of *P. aurelia* have been isolated and subjected to behavioral, genetic, and electrophysiological analyses by Kung and collaborators (Chang and Kung, 1973; Satow and Kung, 1974; Chang et al., 1974, and references therein; Pérez-Miravete, 1973). *Pawns* cannot swim backward as the wild type do; they are named after the chess piece that functions under similar limitations (Figure 10-3). *Pawns* may be temperature-sensitive or temperature-independent; one of the former behaves normally at 23°C, moving backward, but at 35°C it fails to avoid strong stimuli (an assortment of toxic salts in a solution). Some of these mutations are induced with chemical mutagens such as versions of nitrosoguanidine (see Vogel and Röhrborn, 1970, and Hollaender, 1971, for general references on mutagenesis). More than 100 independently arising *Pawn* mutants are now available for study.

Specifically, the *Pawn* mutation is a single-gene defect involving the voltage-sensitive membrane that bears the animal's cilia. The earliest *Pawns* were found to have problems with conductance charges traceable to calcium cations. More recently discovered *Pawns* are defective in potassium ions; they are identified as K^+Pawns (Kung, 1978). *Pawns* with detergent-disrupted membranes can swim backward when sufficient Ca^{++} and adenosine triphosphate are added to the medium. Thus, only a membrane conduction defect causes the absence of backing movements; the ciliary motile apparatus is intact in *Pawns*. The temperature-sensitive *Pawns* are even more valuable in that they, as conditional mutants, allow the turning on and off of certain membrane processes at will. Most known independently arising, temperature-sensitive *Pawns* were found to be allelic at a locus known also to control temperature-independent *Pawn* mutations, and there are now a total of three *Pawn* loci, at least one unlinked with the other two. *Paranoiacs,* another mutation, violently overreact to Na^+ but not to other stimuli (Satow and Kung, 1974). Other mutants such as "fast" and "tetraethylammonium-insensitive" are reviewed by Kung et al. (1975) and Kung (1976). The maintenance of all these strains is facilitated by the facultative sexuality of *P. aurelia;* the organism can be easily cloned and/or crossed (Sonneborn, 1970). As with *E. coli,* we then have ample reason

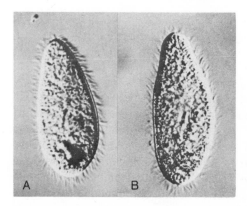

Figure 10-3 Locomotor behavior of *P. aurelia. Left: Pawn* depicted during its resting position at the end of a stroke; its cilia are static and pointing toward the rear. *Right:* Wild-type depicted swimming backward because salts were added to the medium. Photograph made with Nomarski interference-contrast illumination ×200. (*From Kung and Naitoh, 1973. Copyright 1973 by the American Association for the Advancement of Science.*)

to recommend this organism and its genre to behavior geneticists as a potentially fertile source material.

10-3 NEMATODES

The work of Brenner and of Ward (Ward, 1973, and references therein; Brenner, 1973) in studying the behavior of a nematode (an unsegmented round-worm), *Caenorhabditis elegans,* must be judged pioneer, not because of the behavior investigated (chemotaxis among other behaviors) but because of the organism studied. Brenner's and Ward's efforts are unique in that this invertebrate has heretofore not been investigated genetically. Yet the organism possesses inherent advantages for a geneticist. It is a self-fertilizing hermaphrodite wherein sperm is first produced and stored. Developing eggs, some 300 of them per organism, are subsequently deposited. The adult-to-adult life cycle spans 3 to 4 days at 20°C. Such endogeny fosters homozygosity, but different induced mutations can be coupled in the same individual because a small number of males (0.1 percent) are routinely produced by meiotic nondisjunction. (Where A = one set of autosomes: $\frac{4X}{4A}$ = low-frequency male producer; $\frac{3X}{4A}$ = high-frequency male producer; and $\frac{2X}{4A}$ and $\frac{2X}{3A}$ = male.) These may then be crossed with hermaphrodites, carrying genetic markers with them (Riddle, 1978).

   C. elegans has a cylindrical, threadlike, unsegmented body with tracks that can be seen quite easily in an agar-coated petri dish, thus leaving a record that can be analyzed. These visible grooves in agar can be made upon a gradient consisting of attractants such as chemical compounds (cyclic nucleotides), cations (Na^+, Li^+, K^+, Mg^{++}), or alkaline pH values. The animal's patterns of movement may reflect:

   • Orientation. Movement up a concentration gradient, involving "lateral" motion of the worm's head.
   • Accumulation. Persistent location of many nematodes at a specific point in a gradient.
   • Habituation. Always occurring last since the subject has by then become accustomed to both the container and its contents. This involves a familiarity with both the gradient and the attractant. The behavior of the worm changes after it remains in a region of maximum attraction; that is, it swims away only to repeat its cycle later.

   The chemotactic behavior of wild-type, nonmutant nematodes was recorded and compared with tracks made by mutants with head blisters or with tail blisters on their cuticles, with heads bent ventrally or dorsally, with defective head muscles, or with shortened heads. From such comparisons, we can conclude that the sensory receptors located in the head mediate orientation in a chemical gradient. Worms with distal tail blisters orient normally, but head blisters preclude this behavior. Bent-headed animals track complex spirals with the head bent toward the center. Defective head muscles or a shortened head reduces the efficiency of orientation compared with that of wild types. But why do these animals orient toward the common biologically active cyclic nucleotides such as cyclic adenosine monophosphate at all? Perhaps it is because *C. elegans* normally eats soil bacteria that release such compounds into their environment. Juvenile and adult worms, as well as *dauer* (lasting) larvae (which accumulate when cultures are starved or when bearing genes to endure regardless of the availability of their bacterial food), all respond similarly to these attractants. The role of attraction to ions or to pH in this nematode's natural environment is not known at present.

Besides the possession of unique cuticles (detergent and anesthetic inactivation resistant, among other features), these *dauer* larvae behave in a characteristic manner (Riddle, 1977, for an excellent historic and current survey). They remain motionless except in response to mechanical perturbation from which they hustle away. Pharyngeal pumping, a standard larval method of ingestion, is suppressed. And if the surface upon which they are located has a projection that disturbs them, "dauers" stand on their tails and wave their heads in the air. In their natural soil environment, this may serve to get them attached to vectors (animal hosts) for transportation elsewhere. Thermotactically their responses are opposite those of ordinary *C. elegans* larvae to which they may revert later. When they do, they swiftly catch up developmentally via accelerated cell divisions, so that the achievement of sexual maturity is concurrent with worms that never entered the *dauer* state. The genetics of all this—even initial, though sparse chromosome maps—is presented by Riddle (1977, 1978).

On many levels then, here is a splendid, easily laboratory-reared organism for exploitation by the behavior geneticist. It has a haploid number of six chromosomes corresponding to six linkage groups. One more tempting fact—this worm has less than 300 neurons in its entire nervous system. This figure should be evaluated in light of the estimated 612 million to 9.2 billion neurons in the cortex of just one human cerebral hemisphere (Blinkov and Glezer, 1968), and of the 70,000 to 80,000 "brain neurons" of the crayfish *Procambus clarkii* (Wiersma, 1967). This last range represents the only count available on an arthropod. Note though that the tarsal (distal part of arthropod limb) contact chemoreceptors of the Diptera are known to function with but one sensory neuron.

10-4 EPHESTIA

In the flour moth *Ephestia kühniella,* a tropical and temperate zone cereal-consuming pest in flour mills, spinning behavior just prior to pupation has been found to have both genetic and environmental bases (Cotter, 1951; Caspari, 1958). These may interact to lengthen the time from termination of larval feeding with emergence from the food mass to the initiation of pupation. The opposite, an abbreviated interval, results in what Caspari and Cotter have called *nonspinning.* This is a relative term as small amounts of "silk" are produced even by "nonspinners" for minimal cocoons. Normally, full-grown larvae in their final instar (after the final molt) stop eating, leave their food by crawling up culture dish walls or stay on the surface of the food mass depending upon degrees of crowding, spin a cocoon, and pupate. The normal wild-type cocoon is a closed tunnel of which the tip on the upside is less firmly constructed to serve as the site of eclosion. Full-grown larvae pupate in the laboratory in 7 to 10 days.

The lengthened interval, terminating in pupation up to a month later, results in sheet or mat spinning, with the very occasional omission of a spun cocoon altogether (Figure 10-4). But *Ephestia* able to spin mats will not do so if kept in the light: perhaps the negative phototaxis consistently displayed by these insects prevents them from leaving the food when cultured in light.

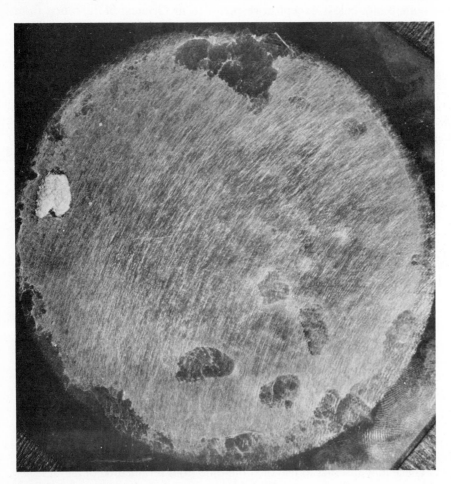

Figure 10-4 Mat silk spinning in E. *kühniella,* the flour moth. Two cocoons are located outside the food mass. (*Courtesy of William Cotter.*)

Caspari (1951) points out the difficulty of analyzing the genetics of spinning behavior in these subjects. Mats are spun by *populations,* not by *individuals,* and Cotter (personal communication) points out that there has been no adequate testing of differential spinning capacities; heavy spinners might be those individuals in the population with the greatest number of "spinning" alleles. In an F_1 generation produced by crossing the variant forms, nonspinning seems to be dominant, if not perfectly so. The F_1 spins little silk. The F_2 progeny produce somewhat more silk than their immediate F_1 ancestors. This observation is interpreted as indicative of genetic segregation. Furthermore, the backcross ($F_1 \times$ spinning strain) yields an amount of silk intermediate between amounts produced by the F_1 and by the spinning strain. The best explanation is mendelian inheritance and segregation for two or more pairs of unlinked genes (Cotter, 1951; Caspari, 1955; Caspari and Gottlieb, 1959).

In addition, a nutritional factor is vital to the manifestation of mat-spinning behavior since sheets are spun *only* if three conditions—two environmental and one genetic—are met: if the food is cornmeal, not whole corn; if light is absent; and, of course, if the appropriate genes are present. Larval density within individual culture dishes also must be considered as an environmental parameter. Keeping in mind that the fabrication of sheets lengthens the entire developmental time and the generation intervals, Caspari (1958) argues:

> It could be imagined that under some circumstances the spinning of sheets rather than of tunnels would constitute a selective advantage for the species. That this possibility exists is suggested by the fact that some close relatives of *Ephestia kühniella* actually spin mats rather than tunnels. This advantage would result in selective pressure in favor of the genes necessary for the spinning of sheets. At the same time, it would cause the organism to be adversely affected by whole-corn food, but favored by cornmeal. The choice of foods for the resulting organisms would therefore be severely restricted. This may serve to suggest the origin of food specialism in insects, where it is frequently found that a certain species is restricted to a specific kind of food plant.

This argument is cogent and is generally true of animals other than *Ephestia,* as will be discussed in Section 13-2 where evidence is presented indicating that food and habitat selection may be of considerable evolutionary significance in *Drosophila* and other insects. The very act of spinning silk outside a feeding tunnel in the food mass implies larval exposure to surface-feeding predators during that extended period of time. But if the food is additionally manipulated after harvesting, the subsequent mode of storage—whether in bags, boxes, or granaries—tends to exclude a very high percentage of potential predators. The mat itself presumably acts (1) to protect the surface-feeding larvae (the larger percentage of the population) from common parasitic wasps such as *Habrobracon juglandis,* and (2), perhaps less importantly, to function as a site for the deposition of eggs by the adult female.

10-5 HOUSEFLIES AND BLOWFLIES

Female houseflies *Musca domestica* produce (z)-9-tricosene, a pheromone that attracts conspecific males and induces courtship and mating behavior in them (Voaden et al., 1972; Rogoff et al., 1973). The compound has been appropriately called *muscalure*.

Pseudoflies were constructed of knots of black shoelaces either impregnated with benzene extracts of female flies (containing their pheromone) or with benzene alone (controls). The quantity and quality of the pheromone in solvent, the female flies from which the chemical was extracted, light, and temperature were all controlled. Three-hundred-forty-seven males were individually scored for their responses. The responses are sketched in Figure 10-5. Differences between flies exposed to treated and to control pseudoflies were due to two components of behavior: (1) pheromone-mediated attraction to treated pseudoflies and (2) responsiveness of individual flies, that is, number of mating strikes (from flight to mount) per male.

This latter category, responsiveness, was found to be a heritable characteristic. Selective breeding (breeding from the two males that each made the most or the fewest mating strikes during two separate tests with virgin females) for high (most mating strikes) and for low (fewest mating strikes) male responsiveness was practiced through the F_4 generation with two high and two low lines. Interestingly and unfortunately, one low line was lost due to failure to breed. The remaining low line averaged 6.24 ± 3.6 mating strikes per hour with a range of 0 to 15.6 strikes per hour at the F_4 generation. The two F_4 high lines averaged 21.72 ± 8.7 and 20.34 ± 9.8 mating strikes per hour with ranges of 0 to 90.0 and 0 to 41.7, respectively. Such results resemble those acquired when selection is practiced for fast and slow mating speeds in *Drosophila* (the work of Manning, 1961, 1963, discussed in Chapter 8).

Houseflies and blowflies are the subject of reasonably extensive research

Figure 10-5 Mating behavior in *M. domestica*. Transition from flight to mating position (a strike) in male housefly with a conspecific female and with a pseudofly (black material soaked in pheromonal extract). (*From Cowan and Rogoff, 1968.*)

simply because of their economic importance as pests. It is not therefore surprising that there are a few reports of significance in the field of behavior genetics. In Section 8-3 the problems of studying phototaxis under numerous environmental conditions and experimental designs in *Drosophila* were stressed. In houseflies, Kessler and Chabora (1977) found a genotype-environment interaction causing yellow-eyed mutant flies to show a reversal toward photonegativity under high compared with low light intensities (17222 lux vs. 86 lux), whereas wild types and hybrids show no such change. This result can be related to the pigment deficiency in yellow flies that leads to their increased visual sensitivity at higher light intensities, resulting in the photonegative response toward the darker exit tube of the apparatus used, and their subsequent emigration from it. In addition, this superstimulus was also responsible for reduced locomotor activity of yellow flies under high light intensities so that they made slower, more deliberate, movements (Chabora and Kessler, 1977). Here then we have a fairly direct association between a behavioral change and a physiological change; to what extent this may be of importance in the wild is difficult to say, but it is known that high-light intensities are important in the biologies of some insect species.

Turning to the blowfly, *Phormia regina,* and using proboscis extension as an unconditioned response to sucrose as a stimulus (Figure 10-6), individuals were classically conditioned to the artificial stimulus, saline water (McGuire and Hirsch, 1977). This demonstrated learning with reliably measured individual differences. Such work was built upon the careful analyses of responses and

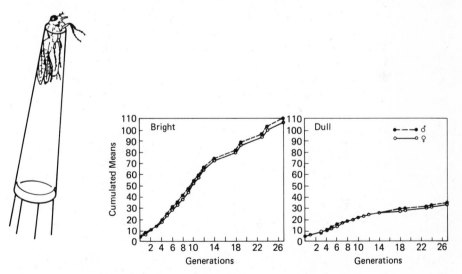

Figure 10-6 *Left:* McGuire and Hirsch's (1977) gentle fly restraining technique, confinement in a micropipette-tip body sheath. *Right:* Cumulative means over generations of conditioned responses by blowflies to distilled H_2O plus $1.0\,M$ NaCl. The unconditioned stimulus was $0.5\,M$ sucrose applied to the fly's mouthparts. A single conditioned response involved proboscis extension when the distal segments of the blowfly's forelegs were immersed in the salt solution.

conditioning provided by Nelson (1971), Dethier (1976), and others. Given such individual differences, McGuire and Hirsch (1977) theorized that artificial selection should be successful. Indeed high and low performance lines were obtained (Figure 10-6) that were markedly different from an unselected control line. These results mean that behavior-genetic analyses can now be carried out in this species for a trait with a learned component. The current experiments were begun with a free-mating wild-type population, so that natural variability was exploited. This is an approach recommended in some instances. See Section 6-7 where the use of isofemale strains in analyzing such variability is discussed. Learning from conditioning has also been reported for *Drosophila* (Section 8-4).

10-6 MOSQUITOES

Knowledge of the genetics of mosquitoes is beginning to accrue (Craig, 1965; Wright and Pal, 1967). The bulk of the information now available derives from formal genetics—the location of loci and knowledge of the chromosomes involved (Craig and VandeHey, 1962). *Aedes atropalpus* breeds in stream bed rock pools and occurs in two behavioral forms, autogenous and anautogenous. Autogenous reproduction involves egg maturation without an exogenous source of protein, such as a blood meal. Females fed only sugar and raisins deposit eggs that hatch. The resulting F_1 progeny can again be reared without blood meals. This is not true of anautogenous forms, the common ones, which must feed at least once on blood before egg maturation. Autogeny is controlled by an autosomal dominant gene.

Gwadz (1970) carefully picked two *A. atropalpus* strains for his study of the genetics of this mosquito behavior. One was homozygous for the dominant gene for autogeny and was designated GP (for Gunpowder Falls, Maryland, its site of origin). The other was homozygous for the recessive gene for anautogeny and was denoted TEX (for Austin, Texas). These subjects were raised under the optimal conditions for each strain, including the avoidance of crowding. Among other details, these conditions involved the control of temperature (27° ± 1°C), relative humidity (80 ± 5 percent), day length (16 hours of light), and age (females hatched within 30 minutes of each other). Males came from the same population as the females being tested and were older than these females. Figure 10-7 represents data acquired from dissections in physiological saline followed by microscopic examination for the presence or absence of stored sperm in the spermathecae of the females—evidence of mating and insemination.

The results are clear: GP mosquitoes mate sooner after eclosion than do TEX mosquitoes. Their hybrids in either direction are intermediate but closer to the GP parent. The details on mean insemination time after exposure to males are: GP, 38 hours; F_1GP/TEX or F_1TEX/GP, 54 hours; and TEX, 120 hours (5 days). The long refractory period TEX displays is not unexpected, since TEX females must fly and search and locate a blood meal before they can

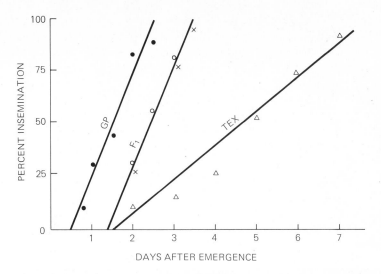

Figure 10-7 Mating behavior in *A. atropalpus,* the mosquito. Onset of receptivity to insemination for two parental and two hybrid populations. *Circles:* GP/TEX (GP female × TEX male)
hybrid; *crosses:* reciprocal TEX/GP (TEX female × GP male) hybrid. Each point represents a
minimum of 200 females. (*From Gwadz, 1970.*)

get down to the serious business of maturing hatchable eggs, sexual activity
aside.

The times of onset of sexual receptivity for the four (all the possible)
backcross generations were then ascertained and are graphed in Figure 10-8. If
one assumes genetic control exercised by means of a semidominant autosomal
gene, then the broken line represents expected rates of insemination.

Put another way, if R is the (autosomal) gene for rapid receptivity and R' is
its semidominant allele for delayed receptivity to insemination, then the
GP/TEX hybrid is RR', intermediate. (This intermediacy of the heterozygote is
referred to as semidominance at a locus.) Then

$$
\begin{aligned}
\text{GPTEX/GP} &= \text{TEXGP/GP} \\
&= RR'(\text{F}_1 \text{ hybrid}) \times RR(\text{GP male}) \\
&= 1RR \text{ (early)} : 1RR' \text{ (intermediate)} \\
\text{GPTEX/TEX} &= \text{TEXGP/TEX} \\
&= RR'(\text{F}_1 \text{ hybrid}) \times R'R'(\text{TEX male}) \\
&= 1R'R'(\text{late}) : 1RR' \text{ (intermediate)}
\end{aligned}
$$

And progeny of both backcrosses to GP males should, on the average, allow
insemination before either backcross to TEX males. They do so with appropriate overlap between the upper end of the two GP male backcross times and
the lower end of the two TEX male backcross times. The performance of the
backcross progeny fits expected results well. Note the impressive numbers of
individuals observed and scored in every category.

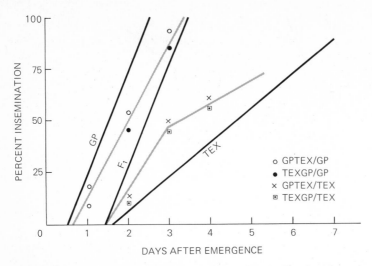

Figure 10-8 Mating behavior in *A. atropalpus*. Onset of receptivity to insemination for four backcrosses of F_1 hybrid females to parental males. *Gray lines:* expected percentages if a single-gene hypothesis is postulated. Each point represents a minimum of 200 females. The four backcrosses are GPTEX/GP (GP female × TEX male) F_1 female × GP male; TEXGP/GP (TEX female × GP male) F_1 female × GP male; GPTEX/TEX (GP female × TEX male) F_1 female × TEX male; and TEXGP/TEX (TEX female × GP male) F_1 female × TEX male. (*From Gwadz, 1970.*)

10-7 A PARASITIC WASP

During the course of his lifelong investigations into the genetics of wasps, Whiting (1934, 1939) obtained a number of gynandromorphs of the parasitic (on caterpillars) wasp *Habrobracon juglandis*. Figure 10-9 depicts one type of gynandromorph (the female-male form, a sexual mosaic) and the normal female and male of the species.

Like other members of the order Hymenoptera (bees, ants, hornets, etc.), wasp males emanate from unfertilized eggs. Unmated females can parthenogenetically (without a male partner) produce all male broods. A mated female, however, still produces fatherless sons but biparental daughters. If a female is homozygous for a recessive trait and is mated to a homozygous dominant male, her female offspring are heterozygous dominant and her male offspring hemizygous, since they are haploid, and recessive. (See Section 8-1 for a discussion of gynandromorphs in *Drosophila*.)

In wasps, gynandromorphs come from abnormal eggs with two nuclei, only one of which is fertilized. Female parts develop from the diploid part, male parts from the unfertilized portion. Consider, as one example, an individual in which the gynandromorphic parts could be easily identified as to sex because it was produced by a cross between a female recessive for orange eyes (*oo*) and defective wing venation (*dd*) and a wild-type male. Such a cross produced 82 wild-type (*OoDd*) females, 17 orange-eyed males with defective wing veins (*od*),

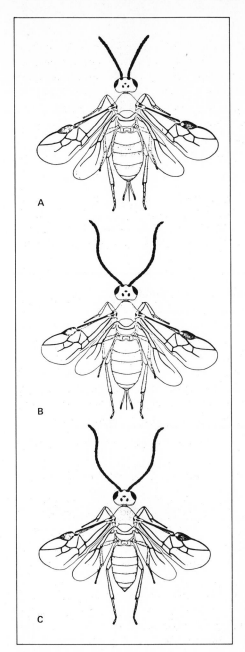

Figure 10-9 *H. juglandis,* a parasitic wasp. (*a*) Normal female. Note the relatively longer wings, shorter antennae, and sting sheathed by a pair of sensory appendages, the gonapophyses, at the tip of the abdomen. (*b*) Gynandromorph. (*c*) Normal male. Note the relatively shorter wings, longer antennae, and sexually dimorphic abdominal tip. (*From Whiting, 1932.*)

and 1 gynandromorph. These mosaics are rare, having an incidence ranging from 1 in 1000 to 1 in 10,000. Figure 10-10 shows the head of the gynandromorph resulting from the cross described. It had an orange (*o*) right eye, which was male; part of its left eye was orange, the rest black (*Oo*) and female.

Figure 10-10 Head of *H. juglandis* gynandromorph. Lateral view. This left eye is part male (lighter sections) and part female (darker section). (*From Whiting, 1932.*)

The right antenna was longer (male) than the left (female), and the two right wings had defective veins (*d*) and were shorter (male) than the left ones (female). Additional secondary sexual dimorphisms and pigmentation make it possible to identify other bodily parts. In this individual, the left side was female; the right, male.

Table 10-1 summarizes data on the behavior of 50 sexual mosaics; reactions, for the most part, are normal for one or the other sex. Though each body

Table 10-1 Behavior of *H. juglandis* Gynandromorphs Sexually (Toward Females) and Parasitically (Toward Caterpillars), According to Sex of Head and Abdomen

Head	Abdomen		Reactions toward females		Reactions toward caterpillars	
			Positive	**Indifferent**	**Positive**	**Indifferent**
Male	Mixed	9	9			9
	Female	20	20			15
Female	Male	1		1	1	
	Mixed	3		3	3	
Mixed	Male	2		2	2	
		1	1			
	Mixed	3	3			3
		3	3			
		1		1	1	
		1			1	
	Female	2	2			2
		1	1			
		3			3	
Total		50	39	7	11	29

Source: Whiting (1932).

is physically a sexual mixture, behavior is not. Clearly, it is the sex of the head upon which responses of the insect depend. For example, normal female responses to caterpillars (here, those of the Mediterranean flour moth, *E. kühniella;* see Section 10-4) involve thrusting the abdomen forward and downward to bring the sting into the protruding position and the antennae straight forward. Then she advances slowly and inserts her sting, with no specific region on her victim preferred for insertion. Antennae are passed over the caterpillar's body surfaces during stinging. After wriggling ceases, the sting is withdrawn and the female applies her mouth to the now quiet caterpillar to suck fluids from it. Afterward, a fold in the skin is selected for egg laying.

Males ignore caterpillars, even avoid them. Upon introduction to wasp females, they become agitated and attempt to mount as soon as a female is located. A male may copulate with a single female more than once or with several successively, beating wings and antennae while copulating. The male grasps and shoves aside the female's wings while mounting. He may try to mount other males. Copulation lasts up to 2 minutes.

The behavior of other insect gynandromorphs has been observed—e.g., the wasp *Habrobracon brevicornis* and the bee *Megachile gemula* (Mitchell, 1929)—but more recently and with greater precision, *D. melanogaster* mosaics have been created and analyzed by Hotta and Benzer (1973), discussed in Chapter 8. This work continues, with greater genetic control and varieties and incorporating the deliberate production of mosaics via chemical mutagenesis, the research of Sturtevant, Morgan, and Bridges (Morgan and Bridges, 1919).

For a report on point mutations in *H. juglandis,* see Petters, Grosch, and Olson, 1978. Of interest to behavior geneticists is a recessive variant that results in flightless wasps. These ectoparasites (living on the outer surface of the host) will not fly even when warmed, prodded with a brush, or dropped from a 6 inch (15.24 cm) height—even if any deviation from perpendicular fall is scored as flight. Such individuals have normal flight muscle ultrastructure.

10-8   SOME ACOUSTICAL INSECTS

> I think it will be admitted by naturalists, without my entering on details, that secondary sexual characters are highly variable. It will also be admitted that species of the same group differ from each other more widely in their secondary sexual characters, than in other parts of their organisation; . . . The cause of the original variability of these characters is not manifest; but we can see why they should not have been rendered as constant and uniform as others, for they are accumulated by sexual selection, which is less rigid in its action than ordinary selection, as it does not entail death, but only gives fewer offspring to the less favoured males. What ever the cause may be of the variability of secondary sexual characters, as they are highly variable, sexual selection will have had a wide scope for action, and may thus have succeeded in giving to the species of the same group a greater amount of difference in these than in other respects.
> Charles Darwin, 1859

Part of Darwin's *The Descent of Man and Selection in Relation to Sex* (1871) deals with species other than our own. Apparently Darwin found this discussion necessary to explain and to defend his newly introduced theory of behavioral sexual selection. One chapter (Chapter 10 of Volume 1) deals with the secondary sexual characteristics of insects. Darwin included the illustration reproduced here as Figure 10-11, showing the stridulatory (sound producing) apparatus of a male field cricket. He commented upon the prodigious volume and duration of nocturnal cricket choruses and the fact that "all observers agree that the sounds serve either to call or excite the mute females."

This section is mainly concerned with insects of the family Gryllidae, the common field cricket. Reproductive pairing is brought about by long-range acoustical signals. A sexually mature male produces sound pulses by moving a wing that bears friction mechanisms. Each closing stroke of the fore wings produces a sound pulse, and the wings are reset for another cycle by a silent opening of these structures. In this way, sound is produced by a cyclic opening and closing of the wings in which sound occurs only during the closing stroke. Each movement of the wings is precisely timed by the concerted contraction of two sets of wing muscles that operate antagonistically. Contraction is initiated by neural discharge in the motor neurons of each set of antagonists. Thus the wing movements that produce sound are said to be neurogenically controlled: contraction and motor neuron discharge are related causally in a one-to-one way (Bentley and Kutsch, 1966). This means that observing or recording the behavior (a procession of sound pulses) also provides a direct way to monitor precisely the activity of the motor components of the nervous system that underlies the behavior. This provides the neurobiologist with a simplified behavior for neural analysis and makes the system appealing for neurogeneticists, one of whose goals is to relate neural activity to genotype.

Let us now return to the behavior and its function. The male produces a calling song that attracts conspecific females to him, presumably over distances ranging from several meters to tens of meters. The effective radius of calling

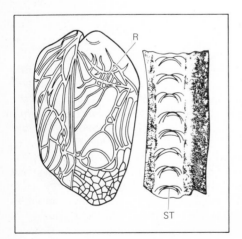

Figure 10-11 Stridulatory apparatus of male *Gryllus campestris,* the field cricket. *Right:* Highly magnified underside of part of the wing nervure showing the teeth (ST). *Left:* Upper surface of the wing cover, with the projecting smooth nervure (R) across which the teeth (ST) are scraped. (*From Darwin, 1871.*)

has not been carefully studied and is likely to be a complex matter affected by temperature, humidity, terrain, and wind conditions. Whatever the true effective limits of acoustical communication, it is certain that acoustical signals play the major, if not sole, role in attracting females. Most crickets are acoustically active at night. Chemical factors may play a minor part (this issue has not been investigated), but dating from the experiments of Regen in 1914, it has been known that acoustical signals are sufficient to attract females to the source of the sound in the "absence" of visual, chemical, or tactile environmental stimuli. Once a male has attracted a female to within a few centimeters of him, and tactual interaction between them occurs, he usually sings a courtship song that is obviously (even to human ears) different from the calling song. One might wonder about conferring species specificity on courtship song rather than on calling song. That this has not been found can be rationalized in terms of saving the female's time and energy (two factors known to be important in an animal's reproductive success). If many different species of crickets sang in close sympatry, a female would risk making many "wasted trips" if she did not know whether the caller was a conspecific male until she was within a few centimeters of him. In terms of a female's time and energy this would be a wasteful way of forming conspecific pairs for mating.

How does a species encode specificity into a calling song given the mechanism described above? The "song" can be varied in the physical domains of time and frequency. There is evidence that frequency, in this case the fundamental frequency, of the song plays some, but not the major, role in species specificity. Cricket calls are relatively pure in harmonic content, consisting of a fundamental frequency and a variable number of harmonics. Experiments with artificial songs have demonstrated that the fundamental frequency coupled with the temporal pattern of the species are sufficient to mediate phonotactic (movement with respect to sound) behavior. Although sympatric cricket species may differ in the value of the fundamental frequency, the range of variance is not great within the sympatric group. Furthermore, frequency is not modulated in cricket songs as it is in bird songs. The most striking difference in the songs of different species is in the temporal pattern of the procession of pulses that constitutes the calling song. These parameters of rhythm are remarkably stereotyped among individuals of a local population or species and markedly different from species to species. It is a simple matter to tape-record a cricket's calling song and subsequently display the pulse trains on an oscilloscope. Then one can film the pulse train displayed on the scope, measure precisely the interval (in millimeters) between pulses, and translate the measurement back to time (in seconds). This provides exact information about the temporal structure of the call. (For the most recent evaluation, see Pollack and Hoy, 1979.) As mentioned above, it is possible to infer the activity of the motor neurons involved in song generation by looking at the pulse train. Thus, knowledge of the temporal structure of the song not only allows characterization of the behavior but also provides a convenient "window" into the nervous system as it produces the behavior being studied. For purposes of providing species specificity, the rhythmic structure of song is the most characteristic aspect of the acoustical signal; most workers in the field feel that the species specificity resides in the call rhythm.

Genetic Control of Calling Song in Male Crickets

Earlier studies (Bigelow, 1960; Leroy, 1964) established the fact that hybrid F_1 crickets can be produced in the laboratory, even though hybrids are not commonly found in the field (Alexander, 1968; Hill, Loftus-Hills, and Gartside, 1972). In these earlier studies the focus was on measuring the pulse rate of hybrid calls compared with parental calls, or comparing the "phrase structure" of hybrid and parental calls. Bentley and Hoy (1972) analyzed the F_1 *Teleogryllus* hybrids (studied earlier by Leroy, 1964) with the specific aim of making exhaustive measurements of interval classes in the song by sampling calling songs from numerous individuals. Hybrids from the Australian field crickets *T. commodus* and *T. oceanicus* were produced in the laboratory. The calling songs of these species are complex because each phrase is composed of two pulse types (Figure 10-12). This gives rise to numerous interval classes whose stereotype provides behavioral units that can be followed in genetic experiments. It was also possible to produce reciprocal cross F_1 hybrids. These were denoted T-1 and T-2 by these researchers: T-1 was the hybrid *T. oceanicus* female × *T. commodus* male; T-2 was the hybrid from *T. commodus* female × *T. oceanicus* male.

Figure 10-13 shows oscillograms (produced by playing a section of a tape-recorded calling song and displaying it on an oscilloscope, where it can be filmed) of the calling songs of *T. commodus*, *T. oceanicus*, and the hybrids. Although visual examination of the oscillograms reveals differences in the F_1 song compared with the parental song, many interpulse intervals must be compiled before definite statements can be made about the genetic control of the

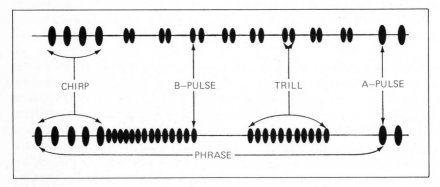

Figure 10-12 Phrase structure of calling song of *Teleogryllus*, a cricket. Each phrase is composed of two pulse types: A-pulses contained in the chirp portion of the phrase and B-pulses contained in the trill (warble). *T. oceanicus* is shown above and *T. commodus* below. (*From Bentley and Hoy, 1972.*)

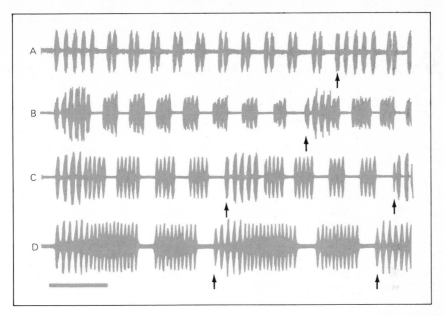

Figure 10-13 Oscillograms of calling songs of *T. oceanicus* (a), *T. commodus* (d), and their reciprocal hybrids, *T. oceanicus* female × *T. commodus* male (b) and *T. commodus* female × *T. oceanicus* male (c). *Arrow* marks the beginning of the next phrase. *Solid bar* is a 0.5-second time mark. (*From Bentley and Hoy, 1972.*)

calling song. From such measurements it is possible to construct interpulse interval frequency histograms, as shown in Figure 10-14.

Calling songs of the F_1 hybrids are distinctly different from either parental call. The intrachirp and intratrill intervals in the hybrid call are intermediate between corresponding parental intervals. This excludes a simple dominant-recessive genetic control of these rhythmic parameters. In fact, there is no evidence for simple dominance in any parameter of song rhythm. Intermediate inheritance is usually interpreted in terms of polygenic inheritance, a mechanism that has been invoked in earlier studies of song genetics (including *Teleogryllus* by Leroy, 1964) in crickets. Intermediate inheritance can also be explained by monofactorial control in which there is incomplete penetrance. Backcross hybrids can be examined to determine which mechanism is actually reponsible. In *Teleogryllus* four backcross classes can be produced. The backcross hybrids themselves reveal intermediate inheritance, so that the polygenic hypothesis is supported (Bentley, 1971). This finding confirms the same conclusion reached by Leroy (1964) in her studies of *Teleogryllus*.

The study of Bentley and Hoy demonstrated that sex-linked factors affect the calling song. The two reciprocal hybrids, T-1 and T-2, differed from each other in the duration and variability of the intertrill interval and in phrase repetition rate. The arrows in the frequency histograms of interpulse intervals

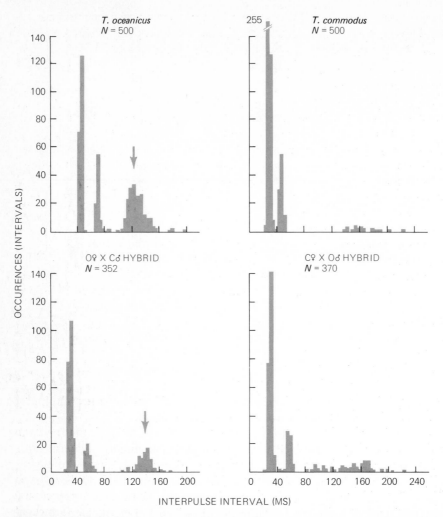

Figure 10-14 Interpulse interval frequency histograms of calling songs of *Teleogryllus* species and their F₁ hybrids. Each histogram represents the analysis of one individual cricket song. N = number of intervals measured. The intervals form three modes: intratrill, intrachirp, and intertrill intervals. *Arrows* denote well-defined intertrill intervals. (*From Bentley and Hoy, 1972.*)

(Figure 10-14) point to the intertrill interval that is prominent in the song of *T. oceanicus* and absent from the song of *T. commodus*. The intertrill interval is also present in hybrid T-1 songs and is absent from the songs of T-2. Crickets have XO sex determination; the male receives his single X chromosome from his mother. Thus, all T-1 males receive *T. oceanicus* X chromosomes; similarly T-2 males receive the X chromosome from *T. commodus*. The presence or absence of an intertrill interval in the calling song is associated with the origin of the X chromosome. We can, therefore, infer the presence of X-chromosome-linked factors that affect the rhythmic structure of calling song.

The genetic control of song production can be summarized as follows:

- Polygenic inheritance of the call rhythm is supported by the finding that intrachirp and intratrill intervals are intermediate in the F_1 between the parental values and are similar among individuals of both reciprocal hybrid classes. Hybrid calls are distinctly different from either parental call.
- A few aspects of call rhythm, such as the intertrill interval, are influenced by sex-linked factors.
- Genetic control is distributed among an unknown number of autosomes and the X chromosome. Thus, genetic control of song is multichromosomally distributed as well as polygenic.

Genetic Control of Female Response to Calling Song

The male produces a stereotyped calling song that serves to identify both his location and his species; but the intended recipient of his call, a conspecific female, cannot call back. Females are mute and respond to the call by walking to its source. It is obviously important that a female cricket be somehow attuned to the call of her species, presumably as a result of genetic programming. This can be tested by measuring the ability of a female to discriminate the call of her species from heterospecific calls. Females can and must discriminate and be attracted to the conspecific males' calling song.

We previously have described the genetics of calling song in the cricket *Teleogryllus*. Hybrid males were found to sing their own calls, quite distinctive from either parental call. What is the responsiveness of the female hybrid to the call of its parental species and particularly to that of its male siblings? The relative attractiveness of parental and hybrid calling songs for a female cricket can be measured by placing a tethered female on a Y-maze in a directional sound field. The tethered female walks on a Styrofoam maze in midair; when she walks, the maze is propelled backward beneath her (Figure 10-15). The maze consists of three straight paths, 10.5 cm long, interconnected by two Y-shaped choice points (at 120° angles). At each Y, the female must choose either the right or the left arm of the maze. Once having made the choice, she is back on one of the straight paths leading to another Y junction, and so on. Analysis of this behavior enabled Hoy and Paul (1973) to measure a female's responsiveness to calling songs under controlled conditions. Virgin females from each species as well as the hybrid T-1 were placed on the Styrofoam Y-maze. They were then presented the recorded mating calls from one of two loudspeakers, placed on the right and left of the maze. Choice behavior in a directional sound field (termed the *phonomotor response*) provided a measure of the relative attractiveness of a given song. The phonomotor response confirmed the locomotor behavior of a free-walking cricket in a directional sound field. Hill, Loftus-Hills, and Gartside (1972) found that free-walking *T. oceanicus* and *T. commodus* make species-specific discriminations. Hoy and Paul found that on the Y-maze, *T. commodus*

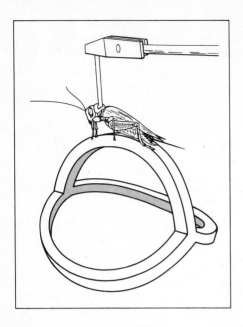

Figure 10-15 Tethered female field cricket on a Y-maze. Her movements on the maze reflect her preferences for different male calling songs. (*Courtesy of R. Hoy.*)

and *T. oceanicus* also perform species-specific discrimination. The surprising result involved the hybrids. T-1 hybrid females preferred the song of sibling T-1 males over the song of either parental type (Table 10-2). Hoy and Paul (1973) arbitrarily defined a phonomotor response as meeting a "strong" criterion if at least 15 of 20 choices were made in the direction of the sound source in *each*

Table 10-2   Phonomotor Responses of 147 Female Crickets to Conspecific and Heterospecific Calling Songs

Calling song	Females at criterion/ total tested	Females at criterion (%)
T. oceanicus on Maze		
T. oceanicus	14/22	63.6
T. commodus	4/22	18.0
T. commodus on Maze		
T. oceanicus	3/15	20.0
T. commodus	21/28	75.0
Hybrid (T-1) on Maze		
T. oceanicus	3/11	27.3
T. commodus	8/21	38.0
Hybrid (T-1)	21/28	75.0

Criterion was 15/20 correct choices in direction both for sound played from the right and for sound played from the left.
Source: Hoy and Paul (1973). Copyright 1973 by the American Association for the Advancement of Science.

direction. Calling sound was played through only one speaker at a time; when 20 choices were made, the same song was played through the other speaker for another 20 choices. With *T. oceanicus* on the maze, 14 out of 22 females met the established criterion when they were played the song of *T. oceanicus*. Similarly, *T. commodus* females on the maze preferred their homospecific calling song to that of *T. oceanicus*. This is precisely the behavior found by Hill, Loftus-Hills, and Gartside (1972) in the free-walking situation. Finally, Hoy and Paul found that 75 percent of hybrid T-1 females met the criterion when played the hybrid calling song, but only 38 percent met the criterion for the *T. commodus* song and only 27 percent met the criterion for the *T. oceanicus* song.

Hybrid male *Teleogryllus* produce unique calling songs that are easily distinguishable from either parental song. Hybrid female *Teleogryllus* find the calling song of their hybrid siblings more attractive than the songs of either parental species (Hoy et al., 1976). Presumably, both phenomena have a genetic basis. The implication is that the production of the signal in the male and its reception and translation into locomotor behavior in the female are somehow genetically coupled. Perhaps the receptive apparatus of the hybrid female is somehow attuned to the song of the hybrid male. It is tempting to propose neurological models by which this genetic coupling might lead to neurophysiological coupling, but it is too early to speculate on precise mechanisms. Clearly, however, this would be an appealing way to "design" a communication system. Evolutionary biologists have pointed out the attractiveness of coupling transmitters and receivers in species-specific communication (Alexander, 1962).

Conclusion

The study of behavior genetics in crickets offers a novel approach to the study of polygenically controlled behavior. The calling song provides measurable units of behavior that can be followed in hybridization experiments. The fact that acoustical behavior is the basis of species-specific communication provides the opportunity for applying genetic analyses that have implications beyond traits in an individual animal. Even though very little is known about the formal genetics of crickets (e.g., no chromosomal mapping has been done), a genetic approach to acoustical behavior promises to illuminate more general problems of behavior genetics and the evolution of behavior (Hoy, 1974).

10-9 FISH

Records of sexual behavior in the platyfish, *Xiphophorus* (*Platypoecilus*) *maculatus,* and in the swordtail, *X. helleri,* and in the males of their F_1, F_2, and backcross hybrids have been taken from a series of over 1,000 ten-minute-long observations made during experiments designed to elucidate the mechanisms of insemination in these freshwater fishes. (Our account follows closely that of the authors, Clark, Aronson, and Gordon, 1954.) Some features are:

- Gonopodial (appendages used for reproductive purposes) thrusting that can be distinguished behaviorally from copulation. As determined by a gonaductal smear technique (the observation of living sperm extracted with a pipette under a microscope), thrusts alone never result in insemination of the female.

• The gonopodium is a holdfast organ in which the tip is modified to form an effective device for attachment. In the absence of this holdfast mechanism, copulations do not occur and males so mutilated do not inseminate females.

• The pelvic fin on the side to which the gonopodium is swung also moves forward, and this is an integral part of the copulatory mechanism. In the absence of both pelvic fins, the ability to transfer sperm to the female is greatly reduced.

• Although courtship patterns are similar in platyfish and swordtail, a number of qualitative and quantitative differences are revealed. Several behaviors—swinging, sidling, quivering, nipping, thrusting, and copulation—are observed during courtship in both platyfish and swordtails. Two behavior patterns typically shown during courtship by male platyfish—pecking and retiring—are not observed during swordtail courtship. Male swordtails, on the other hand, perform types of courtship behavior referred to as exaggerated backing and nibbling that are not seen in platyfish.

The most striking quantitative differences in the sexual behavior among platyfish, swordtails, and their various hybrid combinations were associated with copulation. The mean duration of copulation was longer for swordtails (2.39 seconds) than for platyfish (1.36 seconds), and swordtail pairs mated sooner (averaging 1 minute) during the 10-minute observation period than did platyfish (averaging 5 minutes). However, platyfish mated more frequently (in 26.7 percent of observations compared with 13.4 percent in swordtails), and the number of inseminations resulting from copulations was higher in platyfish (86.0 percent) than in swordtails (39.4 percent). In F_1 hybrids the frequency of copulation (29.0 percent) was slightly higher than in platyfish, and the number inseminated after copulation was intermediate (64.3 percent) between parental types. In the F_2 and backcross hybrids these values were much lower. In general, copulatory behavior in F_1 hybrids was either intermediate or more like that of the swordtails. Some features of male sexual behavior in these fishes apparently are under genetic control, but there is no simple mode of inheritance to account for the data obtained; polygenic inheritance should be considered.

Studies on interspecies fish groups reveal almost complete reproductive isolation between platyfish and swordtails when a choice of mates is offered, even though some heterospecific courtship activities may be observed. When no choice of mates is available, heterogamic copulations occur with a relatively low percentage of inseminations (18.2 percent).

Effective reproductive isolation between the swordtail and the platyfish appears to depend on an array of partial isolating mechanisms. None alone is sufficient to ensure isolation, but acting in concert, these factors so reduce the probability of hybridization that under natural conditions reproductive isolation is apparently complete. It is unfortunate that genetic analyses of the sophistication attainable with closely related populations of *Drosophila* (Section 5-3) are not possible here (but see Franck, 1970). For example, within the *D. paul-*

istorum species complex and with *D. pseudoobscura* × *D. persimilis* (Tan, 1946), reproductive isolation (among a whole series of isolating mechanisms, Section 13-2) was shown to be under polygenic control. The same may be so for this pair of fish species. Their known array of isolating factors is as follows:

- Ecological and geographical isolation: Partial segregation of swordtails in swift headwaters and platyfish in the more slowly flowing lowland streams.
- Physiological isolation: Differences in sensory apparatus, threshold for sexual responses, sexual behavior, and responses to stimuli.
- Gametic isolation: Sperm less viable in oviduct of heterospecific females; competition between homospecific and heterospecific sperm.
- Genetic isolation: Hybrid inferiority; partial hybrid sterility.

However, the guppy, *Poecilia reticulata,* is known to display at least three Y-linked male colors that differ from the "wild-type" stock in this species (Farr, 1977). These mutants, known as *maculatus, armatus,* and *pauper,* court females 7 to 13 times per 5-minute interval. Within a half hour, the average female responds positively; she will respond again later to a second male if he differs in coloration, and in genotype, from her first. Live-bearing female guppies apparently prefer rare or novel males (as tested in 9 : 1 proportions), so rare males are significantly more successful in mating (see Section 8-4). Such a sexually selective female preference is believed to be at least partly responsible for the maintenance of this male coloration polymorphism in nature.

Incidentally, *Gold Flamingo,* a common variant occurring in both sexes, available from many retail outlets and perhaps well known to our readers, has a gold body color controlled by a single autosomal recessive gene. Some *Gold* exhibit orange secondary coloration, but here the genetics is not known. Intrasexual competition involving this mutant has also been studied by Farr (1976) and in convict cichlids, *Cichlasoma nigrofasciatum,* by Barlow (1973) and the Webers (1975). In this latter species it seems that one particular color morph, autosomal recessive *white,* is predisposed toward imprinting as reflected in the males chosen as mates by white females. They choose randomly if reared in mixed schools, but select white males if reared in white schools. Holzberg and Schroder (1975) and Schroder (1978) have shown how experimental irradiation can compensate for a lack of such marker genes in chromosome mapping.

In Trinidad, West Indies, guppies taken from springs and isolated headstreams tend to be large and the males brightly colored. The females may outnumber males by as much as 4 : 1. The fish tend to be dispersed over the stream bed and show relatively poorly developed avoidance responses to a variety of disturbances (Liley and Seghers, 1975). Associated with these features are clear, fast-running water, relatively low temperature, and a virtual absence of aquatic predators other than the small cyprinodontid, *Rivulus hartii.* In the lower sections of streams and rivers, guppies tend to be small and the males less brightly colored and as numerous as females. They show vigorous avoidance responses and may form small schools along the edge of a stream or river. In this case, these features are associated with conditions of slow-moving (often turbid) water, high temperatures, and an absence of shade. There are numerous other fish present, including several relatively large species known to be predators.

Hence the two groups of guppies differ in a number of morphological and behavioral characteristics. Experiments in which guppies of the two different populations are raised at different temperatures indicate that the interpopulation differences in size of adult guppies, especially males, are partly determined genetically and partly as a response to temperature conditions. It is suggested that size-selective predation may be one factor involved in the evolution of interpopulation differences in the size of mature fish. Experimental evidence has been obtained in confirmation. Thus, complex behavioral and morphological phenotypes have developed as results of natural selection in the two habitats from which these populations were derived. It is fascinating that this should include variation in schooling, a form of social behavior which protects individuals from predation.

In conclusion, as for the crickets discussed in the previous section, the study of fish species in the laboratory and in nature is illuminating general problems on the evolution of behavior. In other words, some of the *Drosophila* conclusions reached in previous chapters are becoming generalized to different taxa.

10-10 FROGS AND TOADS

Rana pipiens is a superspecies, a species complex North American in distribution, consisting of many variations of the leopard frog theme (Moore, 1975 for review). Some 11 species and/or subspecies have been identified with only their unique male calls fully differentiating them. Differences in morphology (pigmentation patterns, bodily folds, internal reproductive system ducts), in physiology (allozymes, temperature and climate preferences and limitations), or in capacities for hybridization (routine diploid interpopulational hybrids as well as haploid ones involving the experimental removal and later replacement of egg nuclei) are simply not as powerful indices of genetic relationships as are the calls of courting males. (See Littlejohn and Oldham, 1968; and Littlejohn, 1969, where other genera are also considered.) The precise function(s) of such vocalizations remain unproven.

A number of congeneric (members of the same genus) sympatric species groups have now been analyzed and compared both in America and Australia. In many cases the signals are strikingly distinct, differing in several components so that only intensive programs of call synthesis and discrimination tests will permit determination of the critical information-bearing components (Littlejohn, 1969). Alternately, other closely related species differ primarily in just a single component—the pulse repetition rate, which may vary by factors of two or more. The similarity of this situation to that of the crickets discussed above should be clear.

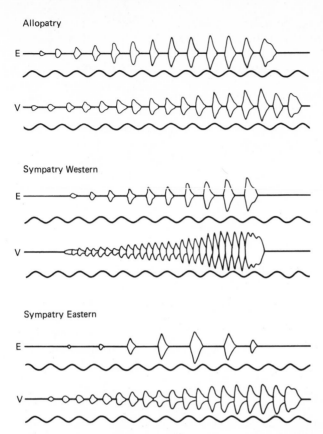

Figure 10-16 Oscillograms of *Hyla ewingi* (E) and *H. verreauxi* (V) mating calls, allopatric and sympatric populations. Only one note of each call is shown. The lower trace in each oscillogram indicates a 50-cycles-per-second time base. (*From Littlejohn, 1965.*)

Littlejohn (1965) studied the Australian tree frogs *Hyla ewingi* and *H. verreauxi,* which have overlapping geographic ranges in coastal southeastern Australia. It was found that remote allopatric populations of the two species have very similar mating calls, while sympatric populations are quite distinct, differing especially in pulse repetition frequency and extent of amplitude modulation. Just a comparison of the oscillograms in Figure 10-16 illustrates this point dramatically. Apparently the marked differences in the sympatric population have resulted from the direct action of selection for increased reproductive efficiency, so that the small differences observed in allopatric populations have been reinforced in sympatric ones (compare Section 5-3 where similar comparisons are made among the *D. paulistorum* subspecies).

Blair (1974) summarized the evidence on mating call as a premating isolating mechanism in various frogs and toads and concluded that:

1 Male vocalization as a species-specific mate attractant has been shown in taxa representing most anuran families.

2   In the few cases where experiments have been performed, the competitive disadvantage of hybrid males in attracting mates has been shown.

3   Mating call is more differentiated in sympatric compared with allopatric populations for the few species pairs that have been intensively studied.

4   In a species group, comparisons between sympatric and allopatric species show greatest differences between sympatric ones.

Finally, there is evidence that over a period of 30 years two American species of toads, *Bufo americanus* and *B. woodhousei*, evolved from a hybridizing situation presumably caused by human disturbance of the environment to form quite discrete breeding populations; they thereby provide convincing evidence for the reality of reinforcement of isolating mechanisms in anuran amphibians leading to speciation.

10-11   QUAIL

An autosomal recessive gene *sg*, when homozygous, causes star gazing in the Japanese quail *Coturnix coturnix japonica* (Figure 10-17). This abnormal behavior involves primarily the withdrawal of the head caudally. The motion occurs when the birds are suddenly confined or if an opaque object is placed above them. It becomes more pronounced as the birds age, rarely being detected until at least 3 weeks of age. By then, the exaggerated motion is obvious, sometimes resulting in a circular movement or squatting on the hocks with the head resting

Figure 10-17   Star gazing Japanese quail at 1 month of age. (*From Savage and Collins, 1972.*)

on the floor. This does not appear to affect a bird's ability to eat, and affected birds grow normally. Males and females homozygous for the *sg* gene are fertile. Other than this, there are now a few additional quail mutants known, and diallel crosses have been practiced on five purebred *Coturnix* lines for behavior related to the solution of a detour learning task. More than 12 generations of bidirectional selection have been tried for behaviors and physiology related to mating (Kiker, Siegel, and Hinkelmann, 1976; Cunningham and Siegel, 1978). The latter produced a variety of magnitudes of responses, mostly small, while the diallel cross indicated profound differences among purebreds in most components of the learning task, for example, in total correct responses and in consecutive correct responses.

In recent years as Kovach (1974) points out, the Japanese quail has been distinguished by a phenomenal growth in popularity as a laboratory subject, including studies on the genetics of behavior. For example, Kovach has successfully carried out directional selection for visual approach preferences. More recently quail chicks have been studied for visual approach preference using a maze designed similarly to the phototaxis choice mazes used with *Drosophila* (Section 8-3). Quick responses to bidirectional selection for preferences between red and blue colors have been obtained using this procedure (Kovach, 1977, 1978).

Since Japanese quail reach sexual maturity at 4 to 8 weeks and the cost of maintaining them is low, it appears that this species will assume rapidly increasing importance, perhaps becoming a "type" organism for behavior genetic studies on birds. There are a reasonable number of genetic markers known at this stage, and there is the possibility of studying evolutionary questions using an array of wild races.

10-12 CHICKENS

For economic reasons there has been much work on broodiness (attention given to already-deposited eggs) in the domestic fowl. This is simply because hens that do not incubate their eggs after laying them tend to give their owners a higher financial return. It is not surprising that there has been artificial selection by farmers against broodiness. For example, Fuller and Thompson (1960) report on an 18-year selection experiment at the Massachusetts Agricultural Experiment Station where the average number of broody episodes per broody individual fell from 3.5 to 1.1 and the percentage of broody fowl was reduced from 86 to 5. A correlation between the behavior of mothers and daughters was found as would be expected for a heritable trait. Certain breeds, for example White Leghorns, are known to be nonbroody. Crosses between various breeds show that sex-linked genes have an important effect on broodiness, but this may not apply to all breeds. (Note that in chickens, males are homogametic and the females, heterogametic.)

Male mating behavior was recorded at a mature age (31 to 34 weeks) by placing one lone male with a flock of eight young hens for eight 10-minute observation periods. The cumulative number of completed matings was recorded (Siegel, 1972). The initial observations were made on an unselected Athens-Canadian random bred population. Selection for high and low numbers of matings carried out over 11 generations gave considerable divergence (Figure 10-18). At this eleventh generation, related behaviors such as treading (where the male covers the female), mounting, courting, and relative aggressiveness shown in paired encounters were measured. The results for these behaviors are nearly identical in ordering to the number of completed matings (Table 10-3), which is as expected since their genetic correlations with the number of completed matings were found to be high. Additive autosomal gene effects influence high levels of these traits, while low levels are controlled by a system involving dominance (Cook, Siegel, and Hinkelmann, 1972).

Courting behavior was assessed by number of courts (Table 10-3). This behavior has a dual role in chickens in that it is both agonistic and sexual. The genetic correlation between it and the number of completed matings was considerably lower than for the number of treads and mounts. The inheritance of courting, while related to the number of completed matings, is quite different. This is reasonable since the data suggest that the aggressive component of courting may be primary to its sexual aspect (Siegel, 1972). The data in Table 10-3 should be considered in the light of these comments.

Relative aggressiveness expressed as a percentage of wins in initial encounters between paired males shows little relationship to male mating behavior (Table 10-3); the biometric correlation between aggressiveness and male mating behavior was close to zero. There is therefore little if any association between sex drive and aggressiveness in males. Body weights at 8 weeks, and two reproductive traits: age at first egg and hen-day egg production, were also assessed for the mating combinations in Table 10-3. All mating combinations

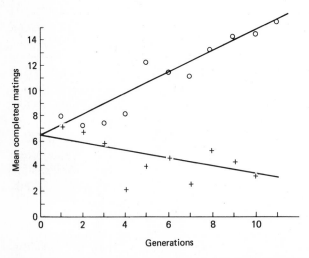

Figure 10-18 Bidirectional selection for cumulative number of completed matings. Linear regressions were fitted to generation means in both directions. (*From Siegel, 1972.*)

Table 10-3 Means for Number of Completed Matings and Related Behaviors, Male Aggressiveness, Body Weights, Age at First Egg, and Egg Production for the Various Mating Combinations H (High-mating), C (Control-mating), and L (Low-mating)

| Mating combination* | Completed matings | Behavior | | | | | 8-week body weight | | Reproductive performance | |
		Treads	Mounts	Courts	Male aggressiveness		Male	Female	Age at first egg (days)	Hen-day egg production
HH	14.3	14.9	16.6	88	42		826	655	180	62
HC	11.7	12.1	13.5	79	57		844	684	181	58
CH	10.2	10.6	11.2	93	55		854	716	166	62
HL	7.5	7.8	9.0	76	61		873	667	178	63
CC	5.9	6.3	7.0	81	49		875	713	188	53
LC	5.8	6.0	6.2	59	37		858	731	174	64
LH	4.8	5.2	5.6	70	52		906	753	174	62
LL	3.3	3.5	4.0	66	43		844	702	175	61
CL	0.8	1.4	2.6	60	40		823	688	172	62

* The first letter denotes the sire line and the second the dam line.
Source: Adapted from Cook, Siegel, and Hinkelmann (1972).

gave rather similar results, so that it is reasonable that these traits should have little association with male mating behavior.

Numerous breeds of fowl are classified according to place of origin, such as Asiatic, Mediterranean, English, and American breeds (Guhl, 1962). The origins of the domestic fowl are lost in antiquity, although four "species" have been recognized in southeast Asia and India. Crosses between wild jungle fowl cocks and domestic hens regularly occur at least in India (Fisher, 1930). The familiar *Gallus domesticus* could well have been produced by hybridization from several wild "species," then becoming distinct from them. This would no doubt have been assisted by fanciers in the early part of this century, who bred fowl for exhibitions based upon colors, plumage, comb characteristics, and range of body weights (see Guhl, 1962).

Among domestic animals, chickens are subjected to the most specialized of husbandry procedures; they are maintained indoors in compact groups and/or isolated in laying cages. In addition, with trends toward mass production and efficient management, the social behavior of these birds has commanded considerable attention. Turn ahead to Table 10-5 near the end of this chapter where a list of behavioral characteristics adapting species to domestication is given.

It is to be expected that detailed studies have been carried out on behaviors such as those preceding egg laying (Wood-Gush, 1972). Two strains, White Leghorn and a Brown strain of Rhode Island Red and Light Sussex origin, were studied in battery cages. The Brown strain sat significantly longer than did the White strain, which showed an excessive amount of pacing during the prelaying period. These differences were not affected by enclosing the battery cages and darkening the room. Both strains responded to frustration in a feeding situation by excessive pacing, from which it is apparent that White females were generally upset during the prelaying period. However, some White strain females did sit, suggesting the likelihood that selection for sitting in the prelaying period may be successful.

Also, it is not surprising that levels of aggressiveness have been studied in different breeds since the domestication of chickens is historically related to cockfighting, which suggests early selection for aggressiveness and related traits. For example, fighting cocks have been observed to be shiftier, faster, and less clumsy than domestic cocks, and among varieties of fighting cocks there are differences in their methods of attack. It is also reasonable, then, that selection for high and low levels of aggressiveness has been successful in White Leghorns (Guhl, Craig, and Mueller, 1960). A consequence of variable levels of aggressiveness in a flock is the peck order, which is the ranking of individuals according to the number of individuals each flockmate dominates by unidirectional pecking or threatening. The intensity of these contests is greater among cocks than hens. To what extent there are genetic differences among birds in a peck order is at present relatively unexplored, but from considerations elsewhere in this book, such differences are likely. It has been reported however, that males with pea combs and walnut combs assume lower social positions when intermingled with single comb flockmates (see Siegel, 1979). Isolating birds in laying cages eliminates peck orders, although possibly dominance-subordination relations could occur between birds in adjoining cages. The adaptation of birds to these various procedures is part of domestication, a continuing process that varies in goals in different places and times.

While we have just selected a few topics for this discussion, the literature on the behavior genetics of chickens is now quite substantial. Indeed Siegel (1979) has most competently reviewed the field from several vantage points; the process of domestication, behavioral effects on selection, and in the context of genotype-environment interactions (for this latter see also McBride, 1958). He lists in addition a number of mutants with behavioral effects. An indication of the diversity of studies comes from heritability estimates of 0.09 for imprinting, 0.58 for tonic immobility, 0.28 for eating, 0.82 for drinking, 0.54 for standing, and 0.51 for resting. He stresses the point that poultry breeders have consciously and subconsciously changed behaviors through selection, and that an interfacing of all of the chicken studies is necessary for the understanding and facilitation of the process of adaptation to the environments in which domestication is placing them.

We conclude this section by quoting part of the "Nuns' Priest's Tale" from Chaucer's *Canterbury Tales,* because this may be the earliest allusion to non-random mating in chickens:

> And in the yard a cock called Chanticleer,
> In all the land, for crowing, he'd no peer,
> His voice was merrier than the organ gay
> On Mass days, which in church begins to play;
> More regular was his crowing in his lodge
> Than is a clock or abbey horologe.
> By instinct he's marked each ascension down
> Of equinoctial value in that town;
> For when fifteen degrees had been ascended,
> Then crew he so it might not be amended.
> His comb was redder than a fine coral,
> And battlemented like a castle wall.
> His bill was black and just like jet it shone;
> Like azure were his legs and toes, each one;
> His spurs were whiter than the lily flower;
> And plumage of the burnished gold his dower.
> This noble cock had in his governance
> Seven hens to give him pride and all pleasance,
> Which were his sisters and his paramours
> And wondrously like him as to colours.

Indeed nonrandom mating in chickens is probably widespread. For one example, the dimorphic plumage color pattern of the Brown Leghorn cock is an important component in the identification by females of that breed, leading to homogamy at the interstrain level. Even so, homogamy is modifiable by certain types of juvenile and adult social environments, but still one may correctly conclude that homogamy is common in chickens and in other poultry too (Lill, 1966, 1968).

10-13 GEESE

In the Canadian Arctic, the lesser snow goose occurs in two color phases, blue and white. This plumage dimorphism (Figure 10-19) is determined by a single autosomal gene with the blue form dominant over the white one (though traces of white on a blue animal indicate the heterozygote). Although the blue goose and the lesser snow goose are still sometimes incorrectly classified as different species, they are both *Anser cerulescens*. The rare instances where blue goslings hatch from nests belonging to two white parents are due to dumping, the deposition of eggs in a nest by a female not incubating on that particular nest (Cooke and Mirsky, 1972).

Positive assortative mating (more matings between like phenotypes than expected under random mating; see Chapter 2) occurs in these birds so that of 3,480 lesser snow goose families surveyed between 1968 and 1970 in Manitoba, Canada, the following was found:

Parents	Observed number of families	Expected number of families assuming random mating
White female × white male	3,099	3,036.5
White female × blue male	195	257.5
Blue female × white male	109	171.5
Blue female × blue male	77	14.5

Figure 10-19 Mixed pair of snow geese at nest; blue phase ♂ and white phase ♀. (*From Cooke, 1978.*)

Note the excess number of matings between likes and the corresponding paucity of matings between unlikes. This pattern can be explained if birds select
their mates according to the color of their parents and/or their siblings. To
investigate this experimentally, Cooke and McNally (1975) tested three captive
flocks for color preferences considering (1) bird-to-bird approach responses, (2)
association preferences, and (3) mate selection. Their findings can be summarized as follows:

- *Approach response:* Young birds placed in a choice situation had a significant preference for birds of the color of their parents. Sibling color when
different from parental color appeared to modify choices. If parents were removed during adolescence, early color preferences could be altered. The most
recent association determined these preferences and no differences were detectable in the responses of goslings to maternal as opposed to paternal colors.
- *Association preferences:* In an open-field situation, birds usually associated with their peer group (siblings and nonsiblings) at both 1 and 2 years of
age. The degree of this association faded from 1 to 2 years of age. When birds
associated with non-peer-group birds, they showed a distinct tendency to asso-
ciate with birds the same color as their peer group.
- *Mate selection:* In a flock raised as a single large group with virtually no
parental contact, pair formation did not depart from randomness in terms of
color, suggesting that nonrandom mate selection in lesser snow geese is additionally a function of early experiences. In other than captive flocks where
parents and offspring were the same color, mate selection reflected preferences
for familial color; where parents and offspring were of opposite colors, either
parental or offspring color was chosen. Similar results were found in marked
wild birds under field conditions.

If the parent is removed (as happens in the wild), color preference may be
altered but is more likely to be maintained through association with birds of the
familial plumage color. Directly or indirectly, parental color influences mate
selection. However, the relative strengths of color discrimination in the two
sexes is currently unknown, as is the reason for the occasional renegade bird
who prefers a mate of a color inappropriate to its pedigree (Cooke, 1978).

Here then is a serviceable example of the role of genetically fostered early
"learning" in self, mate, and species recognition. This was investigated with
captive animals maintained under conditions mimicking wild ones as closely as
possible, at the expenditure of great effort and resources. Recently, Cooke (1978)
has extended the results directly to field conditions. Note the cogent combination here of data harvested from wild and captive animals. This represents one
of the rare experimentally verified mechanisms explaining assortative mating in
a wild bird species under field conditions. But we still know neither the origins
of this mating system nor its evolutionary sequelae. Cooke, Finney, and
Rockwell (1976) ask about the relative Darwinian fitness of the various types of
pair bonds; for example, are white × white and blue × blue parents more reproductively successful than white × blue ones? (In this context it should be
noted that in the large seabird, the Arctic skua, there is a color polymorphism

[O'Donald, 1976, 1977] apparently maintained by a combination of breeding age, mating preferences and sexual selection differences associated with territory size variations among the color phenotypes—see Section 13-4 for discussion.)

10-14 TURKEY-PHEASANT HYBRIDS

Early posthatching mortality is high for ringneck pheasant female × bronze turkey male F_1 offspring brought into existence by means of artificial insemination (Asmundson and Lorenz, 1955). But mortality is low for hybrids from the reciprocal cross, with some survival differences being due to husbandry. Hybrids were raised in mixed flocks with several species and hybrids. The smaller pheasant female × turkey male hybrids from pheasant eggs fared poorly at first, while the larger hybrids out of turkey eggs did better. Later mortality was due to severely pendulous crops, remediable by surgery. Hybrids never engaged in any sort of mating behavior, however preliminary. There was some secondary sexual modification of the skin on the head of males, but this was all. Clearly, normal behavior characteristic of turkeys and of pheasants was totally disrupted by hybridization.

The total sterility of these F_1 generations precludes further genetic analysis (by means of backcrosses, etc.). However, if hybridization is resourcefully attempted and reattempted, this particular, sometimes fruitful, avenue of behavior genetics analysis can be exploited, as shown in Chapter 5 for some sibling species and races of *Drosophila* and below for duck hybrids. A prime example in which the hybridization technique was particularly rewarding was the study of nest-building behavior by lovebirds, discussed in Chapter 5.

10-15 DUCKS

Species-specific behavior has been used for taxonomic purposes; interspecific hybrids, if they can be brought into being in sufficiently viable numbers, allow the employment of this behavior for genetic purposes. Ten male pintail duck, *Anas acuta,* were confined with the same number of mallard, *A. platyrhynchos,* females (Sharpe and Johnsgard, 1966). Most of the eggs deposited never hatched, but a total of three females and four males were reared. These F_1 interspecific hybrids were bred with one another; 18 F_2 female plus 23 F_2 male birds were produced. The males of this F_2 generation were analyzed for pintail- versus mallardlike plumage and behavior. This material is presented for two reasons: (1) because it represents a rare behavior-genetics analysis of birds and of a bird other than the best-known domestic chicken, and (2) because Sharpe and Johnsgard (1966) utilize an interesting hybrid index (see Table 10-4) for the semiquantitative evaluation of these precious interspecific hybrids as regards not only their anatomical but also their behavioral phenotypes scored for values in the range 0 to 4 for plumage and various behaviors. Table 10-4 provides some examples. Extending the number of such characteristics, a com-

Table 10-4 Plumage and Behavior Indices of Hybrid Mallard (A. platyrhynchos) × Pintail (A. acuta) Ducks

Scale	Plumage	Behavior				
		Bill pointing	Turning	Nod swimming	Down-up drinking	Burp call
0	Wholly mallard phenotype	Mallardlike with tail raised	Toward female, mallardlike, partial turn	Mallardlike fully expressed	Present, mallardlike	Absent, mallardlike
1	Appearance approaching mallard phenotype			Intermediate		
2	Appearance intermediate between two species			Rudimentary		
3	Appearance approaching pintail phenotype	Pintaillike with tail lowered	Toward female, pintaillike, full turn	Absent, pintaillike	Absent, pintaillike	Present, pintaillike
4	Wholly pintail phenotype					

Source: Sharpe and Johnsgard (1966).

plete mallardlike phenotype has a value 0 and a complete pintaillike phenotype a value of 15 for morphology and behavior. Figure 10-20 graphs the relations between the two hybrid indices; no F_2 hybrid displays were atypical of one or the other parental species. (Note that the cumulative behavior indices for 11 F_2 males range from 3 to 15.) We may safely conclude that both sets of features— plumage phenotype and behavior—clearly are under genetic control. (Grunt-whistling, a male vocal courtship display aimed at pair formation, is not included in either Table 10-4 or Figure 10-20 because it is very similar in both parental species and the hybrids grunt-whistled like their ancestors.) The $r = +0.756$ line in Figure 10-20 represents a significant positive correlation coefficient between the inheritance of these behavioral and plumage characteristics in the hybrid F_2 generation. It would be interesting to consider how such a close relationship might be exploited by the behavior geneticist.

The Aylesbury duck is a domesticated bird descended from the mallard (*A. platyrhynchos*); it therefore offers the opportunity to make direct comparisons between domesticated and wild species. Desforges and Wood-Gush (1975a) compared the habituation rates of domestic ducks and wild mallards when a stuffed hen was placed in their home pens, and when given novel food in a normal container or in a novel container. The domesticated ducks habituated

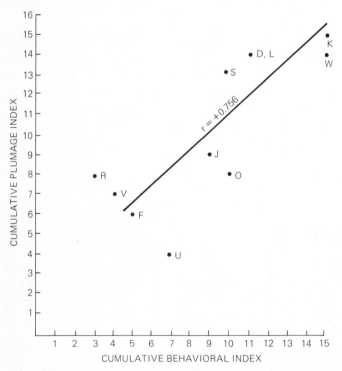

Figure 10-20 Plumage and behavioral hybrid indices correlation for 11 mallard (*A. platyrhynchos*) × pintail (*A. acuta*) F_2 duck hybrid males (letters). (*From Sharpe and Johnsgard, 1966.*)

faster than the mallards. Furthermore, in tests involving escape from a human handler, mallards ran away significantly faster than did domestic ducks. It has been suggested many times (see Hale, 1962 and Table 10-5) that fitness in a captive environment is enhanced by low reactivity and, because of this, the propensity to approach and investigate novel objects would differ between domestic and wild animals, simply because there would be a higher probability for domesticated animals to encounter artificial situations in their peopled world. Low reactivity would seem to be important in suiting animals to a captive environment.

Desforges and Wood-Gush (1975b) compared the individual distances between members of flocks of domestic and mallard ducks. Even though the Aylesbury duck is three to five times heavier than the mallard, individual distances (cm) are smaller in the former than in the latter whether considering like or unlike sexed pairs while feeding or when resting:

	Feeding pairs			Resting
	♂♂	♀♀	♂♀	
Domestic	30.5	0	0	77.5
Mallard	45.7	30.5	17.8	129.5

This indicates that during domestication there has been selection against aggression and for reduced individual distance. Since the domestic chicken has been domesticated far longer than has the Aylesbury duck, and has even been selected for adaptation to an intensively agricultural environment (Section 10-12), it would be of great interest to compare domestic and wild jungle fowl in this way.

A final set of comparisons concern sexual behavior (Desforges and Wood-Gush, 1976). Changes in intensity of the social displays of the Aylesbury duck tend to reduce the attention-catching features of these displays, and the down-up display is performed less frequently than it is by mallards. During domestication there is no premium upon promoting social displays that function in part as interspecific ethological isolating mechanisms, since in the environment created by people, the possibilities of hybridization disappear. Biologically the situation is similar to geographical isolation in the wild (see Section 10-10 on frogs). In addition, female Aylesbury ducks incite several drakes, while mallards incite only one, so that only mallard ducks form pairs. Indeed while copulatory behavior tends to be promiscuous in Aylesbury ducks, it is not completely so, because males and females direct most of their sexual activity to only two or three individuals. As Hale (1962) and others have pointed out, it is of obvious importance to human beings that domestic animals should at least tend to be as promiscuous as are these ducks.

10-16 CATS

We have already recommended the employment of cats as subjects for surveys of the behavioral effects of sex chromosome aberrations, for example, XXY (Tobach and Rosoff, 1978). And again we earnestly note how well *Felis catus* (sometimes called *Felis domestica*) with abundant races may serve as models of human conditions of special interest to the behavior geneticist. Beadle (1977) offers an authoritative and delightful compendium. For but a few interesting examples, Todd (1962, 1978; Morrill and Todd, 1978) has reported the existence of an autosomal dominant gene controlling the expression of a cat's response to catnip, *Nepeta cataria,* though such innate responses are easily modified by a given animal's age, environment (newness dilutes responses), and emotional state (sudden disturbances may eliminate responses altogether). Susceptible cats are sent willingly into repetitive frenzies by catnip and can detect the presence of its volatile oil, *trans-cis*-nepetalactone, in concentrations as dilute as one part per billion.

In cats, too, one may see the rough equivalent of Waardenburg's syndrome wherein congenitally deaf people (some 5 percent of all congenitally deaf people) have eyes discordant for color, abnormalities of the eighth cranial nerve (the auditory nerve), and white-streaked hair, often prematurely gray (Bergsma and Brown, 1971; see Sections 11-5 and 11-7 of this book).

The genotype of Siamese cats includes albino genes. Like other animal and human albinos, Siamese cats possess an abnormally large number of retinal nerves that cross over via the optic chiasma to brain hemispheres opposite their origins. Once there they link up incorrectly with recipient lateral geniculate nucleus cells. As Beadle (1977) pondered:

> Because the information sent on by these differently located receptors is different, the paired cells would transmit conflicting messages to the visual cortex. The resulting confusion would be reflected in the cat's behavior: it would be unable to locate objects in space. The conditional "would" was used just above because Siamese cats obviously can locate objects in space. They can catch mice, steer a safe course around furniture, and avoid falling into holes. How do they overcome the neural disability that afflicts them?

Guillery (1974) finds that such a cat relies upon one of two possible ploys to salvage itself: the cat's brain either suppresses discordant geniculate nucleus cell messages or it reroutes nerves from this junction to its brain cortex so that messages from each hemisphere, from each geniculate nucleus, arrive at different times. Look at a Siamese cat. If it employs the latter strategy and reroutes discordant cell messages, it will be cross-eyed. Why cross-eyed? Siamese kittens learn to squint and cross their bewildering eyes within their initial 2 months or less. Any resulting "visual sanity," however, falls short of normal visual acuity and Siamese cats see less binocularly than do other cats. (They also mature more slowly and have characteristically and frequently-used hoarse voices.)

Recently, Blake and Camisa (1978) have considered how inputs from our two eyes are normally unified by our brains to produce binocular vision. If partner eyes disagree in what they each see monocularly, we employ what is known as *binocular rivalry*—alternating periods of single eye dominance and suppression of the other eye wherein the sensitivity of the suppressed eye is temporarily decreased and we see mostly what our dominant eye supplies. This suppressed eye input is not completely lost however, because we also apparently practice a sort of fusion, though where "potential ambiguity exists between our two eyes, binocular vision is monocular."

10-17 HORSES

In horses, the trotting gait is autosomally dominant to the pacing gait. In order of increasing speed, normal gaits are walk, singlefoot or rack or amble, canter, pace, trot, and gallop. Where Lf means left foreleg; Rf, right foreleg; Lh, left hind leg; Rh, right hind leg; and 0, no feet on the ground, we can write

Pacing = Rh, Rh Rf, 0, Lh, Lh Lf, 0, Rh
Trotting = Rf, Rf Lh, Lh, 0, Lf, Lf Rh, Rh, 0, Rf

Horses must, of course, be trained to race in harness, but there is an inherent tendency for them to move in one or the other distinct fashion. This seems to be simply genetically determined (Snyder and David, 1957). Since this is not new data, we recommend Cunningham (1975) for a survey of the general genetics of horses including the use of blood groups to ascertain the origin of current strains and breeds. He also assays the genetic contribution of outstanding sires and dams to present animals.

Although the racing of thoroughbred horses may be considered an industry in more than one country, little is known about the heritability of racing performance or how stud, dam, or both aspects of selection "protocols" may contribute. Heritabilities calculated in assorted ways (for one example, offspring on sire) range from 5 to 55%, with 35% reported by O'Ferrall and Cunningham (1974) using track records published in the year 1970. These authors suggest that the judicious selection of stallions is more closely reflected in performances than is the selection of mares, but such assumptions are wholly preliminary pending the harvesting of much more data. Little it seems, is known about the behavior genetics of any sort of horse though we also recommend Eldridge and Suzuki (1976); Hafez, Williams, and Wierzbowski (1969); and Schoen, Banks, and Curtis (1976) on Shetland and Welsh ponies for initiation into the literature.

10-18 CATTLE

With regard to nonsexual behavior in cattle, Payne and Hancock (1957) investigated the effect of tropical climate on the performance of European type (Jersey and shorthorn) cattle utilizing six sets of identical twin heifer calves. The twins (one per set) observed in Fiji did not uniformly respond to stresses imposed upon them by a tropical climate: two cows performed (in milk production, butterfat production, nonfat solids production, feed intake, water intake, rectal temperature, and in respiratory rates) as did their cotwins in nontropical environments, while four did not. The authors believe that "this suggests that there are profound differences in the reaction of individual temperate-type dairy cattle of the same breed to a tropical climate, and that these differences must be based at least in part on genetic differences between individuals.''

Hancock (1954) has also intensively investigated grazing behavior in six sets of uniformly treated lactating monozygotic cattle twins. The animals were observed in a 1-acre paddock for 8 days at approximately monthly intervals, each occasion covering a 24-hour period (Figure 10-21). The following data were collected on each day for each cow: (1) time spent grazing, (2) time spent standing or walking and loafing, (3) time spent lying down, (4) distance walked, (5) number of defecations, (6) number of urinations, (7) number of drinks, plus some observations of (8) number of bites of grass, (9) time spent ruminating, (10) number of boluses ruminated, and (11) number of bites to each bolus.

Figure 10-21 Observation methods used by Hancock in studying monozygotic cattle twins. (*Courtesy Ruakura Agricultural Research Centre, Hamilton, New Zealand.*)

Striking statistically significant correlations between cotwins were calculated for time spent grazing, a crucial activity of a dairy cow, just as found for many human behavioral traits where comparisons are made between monozygotic twins (Chapter 7). Hancock comments on the habit of cotwins to graze simultaneously and to stand close together most of the time. "Twins provided a special case of gregariousness," he notes, in that they seek each others' company while grazing and when congregating to lie down. Such inclinations could lead to greater apparent similarities, perhaps imitative ones—a situation faced again, when human cotwins are raised together. (See also pertinent comments by Kilgour, 1975.)

In summary, the mean differences are negligible within sets of identical twins with regard to time spent grazing (4 minutes), loafing (7 minutes), and lying down (8 minutes), while the between-set differences are very great for time spent grazing (138 minutes), loafing (114 minutes), and lying down (60 minutes). These results must surely indicate genetic components in the determination of these traits, even though it is impossible to disentangle environmental effects.

Fortunately, one set of monozygous triplet bulls has been examined for behavioral likenesses, among other traits (Olsen and Peterson, 1951), albeit briefly. All three shorthorn males were reported as being alike in stubbornness and lack of interest in serving cows. They were, however, brought into service after some effort, at 13.5 months of age. At that time they were capable of producing but one ejaculate a week; 4 months later this unimpressively increased to two—low for bulls in general. With reference to these few ejaculates, volume, density, total number of sperm per ejaculate, motility, and presence of abnormal sperm were statistically not consistently different among these brothers. Olsen has studied three sets of monozygous triplet bulls (one milking-shorthorn and two Guernseys), plus two sets of monozygous twin (Guernsey) bulls (Olsen and Peterson, 1952). Twinning occurs more frequently in cattle than in human beings, so there is more material of this sort available for use in behavioral and genetic studies. (Of all births in dairy cattle, 0.11 percent, and of all like-sexed twins, 10.6 percent, are estimated to be monozygotic twins; Hancock, 1954. In human beings, 0.0035 percent is the frequency of monozygotic twinning considering all populations; Levitan and Montagu, 1971.) Only cattle produce an infertile unlike-sexed dizygotic twin called a *freemartin.* This occurs when the female cotwin is masculinized by anastomoses of blood vessels delivering male hormones into her fetal circulatory system.

Finally, few realize that cattle and at least four additional phyla of multicellular animals provide animal models of the hereditary porphyrias (Section 11-4); they should be exploited as such (Mangum, 1978).

10-19 PRIMATES

Two hybrid female offspring, born at one year intervals, were the unexpected product of a gibbon male, *Hylobates moloch,* and a siamang female, *Symphalangus syndactylus,* reared together in captivity. Karyotypic analysis of the hybrids revealed a complement of 47 chromosomes, representing the combined haploid numbers of the parents, 22 for the gibbon and 25 for the siamang (Myers and Shafer, 1979). The second hybrid died at 4 months of age of an infection unrelated to her hybrid nature.

The surviving hybrid has the general appearance of a siamang (Figure 10-22). But closer analysis reveals a mosaic of gibbon and siamang features (Wolkin, 1977). The hybrid has the black pelage of the siamang, yet as an infant she also displayed a hint of a lighter ring of hair around her face, a gibbon characteristic. The hair pattern on her head is similar to that of a siamang whereas the pattern on her arm is clearly the same as the unusual pattern on a gibbon's arm. Elements in her facial structure include the large eyes of the

Figure 10-22 Hybrid offspring from a mating of gibbon and siamang. (*Photo courtesy of Sister Moore, Atlanta.*)

gibbon as well as the long thin nose of a siamang. She also displays the lengthy siamang digital webbing from which the name syndactylus is derived but not the highly visible throat sac of that species. The lack of a throat sac may be one reason that her vocalizations do not have the variety of the more complex vocal repertoire of the siamang.

Anatomic measurements provide us with a similar picture. She has the large torso of the siamang but the slender appendages of a gibbon. She also has the proportionally shorter arms and legs of a siamang. This may account for the resemblance of her locomotor behavior to the more halting brachiation of the siamang.

This birth raises several questions, two of which are: How adaptive is an animal with these characteristics in the natural habitat of gibbons and siamangs? Second, what are the implications of this birth for evolution and speciation?

10-20 CONCLUSIONS

With this serendipitous hybrid we end this literature survey. We are concerned lest we have omitted worthy efforts, but the literature of behavior genetics is disparate and requires cataloging. This is in itself a good sign; it is indicative of a certain, albeit still not impressive, bulk to that literature. Our hope is that we participate in the stimulation of an increase. In particular, the question of the behavior genetics of domesticated species is one that may be of much greater importance to the animal breeder than previously realized. For an early appreciation of this, see Hafez (1968) on interspecific hybridizations, normal and aberrant mating behaviors, and selection for twinning, among other behaviors in domesticated animals. (In 1935, as an even earlier example, Hodgson briefly considered the effects of inbreeding on swine behavior.) We also recommend Hafez's (1969) anthology, *The Behavior of Domesticated Animals*—specifically Chapter 3, "The Genetics of Behavior," by J. Fuller; Chapter 12, "The Behavior of Horses," by E. Hafez and J. Signoret; and especially Chapter 13, "The Behavior of Rabbits," by V. Denenberg, M. Zarrow, and S. Ross. Section III B of this last chapter is a good summary of the behavior genetics of rabbits. These include a variety of maternal behaviors: nest building and nest lining, cannibalism, aggressive protection of the young, changes in housing, retrieving, and suckling. In sheep, there are isolated observations of interest, for example a report of homogamic mating preference in a merino ram (Hayman, 1964). Another example is the effect of audience on the mating behavior of rams when dominant rams viewed by two submissive rams showed no alteration in mating behavior compared with their performance when tested alone, while submissive rams mated and ejaculated less when viewed by dominant rams than when tested alone (Lindsay et al., 1976). The practical significance of this latter observation is that even if rams are sufficiently separated, adequate space must be provided in breeding programs to allow low ranking rams to avoid communication with higher ranked competitors.

In 1975, Hafez published the third edition of his unique anthology, *The Behaviour of Domesticated Animals*. See the expanded section on species characteristics.

Undoubtedly domesticated species, willfully regulated in their reproductivity by people for their own purposes, provide rewarding material for future study. Genetic changes that occur during domestication have been little studied especially at the behavioral level. For example, what happens to behavioral traits during the domestication process in the formation of the laboratory strains of mice, rats, and rabbits that we study? Answers to this question will be important in the study of domesticated animals of economic importance, where domestication is associated with favoring the production of desirable phenotypes at the expense of less desired phenotypes. Although the prime objects of selection are usually traits of morphology or production (for example egg-laying), as pointed out in Section 10-12, correlated responses to selection may include deleterious behavioral components that lead to a fall in fitness until the response may cease.

Table 10-5 gives a list of behavioral characteristics adapting species to domestication as summarized by Hale (1962). The behavior-genetic study of many of these would be rewarding. For example, as Siegel (1979) points out, behaviors that generally are favored in the domestication of chickens include a hierarchical group structure, promiscuity, well-developed young at hatching, nonspecialized dietary habits, limited agility, and an ability to adjust to different environments—all listed in Table 10-5. While studied less, the domesticated duck apparently shows similar behavioral features. Genetic studies of traits of this nature are few. However, the process of domesticating animals, which involves breeding in a goal-oriented sense, must surely be aided by such studies. For example, biometric experiments to assess whether behavioral traits are predominantly subject to directional selection in nature (learning and mating speed are examples repeatedly discussed in this book) or to stabilizing selection must be significant in attempting to assess the likelihood of a goal being attained. In an evolutionary sense, it appears that domestication can most successfully be carried out when genetic programs, as defined in Chapter 1, are relatively open. Animals with long life spans and extended parental care are therefore those in which domestication is likely to be most readily achieved.

Note the phylogenetic series incorporated into the different sections of this chapter, especially when combined with Chapters 8 and 9. Organisms "lower" on this scale are currently offering increasing information on the chemical bases of behaviors. The suggestion was made especially in nematodes that such behaviors are relatable to habitats in the wild. This reasonable working hypothesis gains support from the behavior of the wormlike larval stage of *Drosophila* (Section 8-5). Higher in the phylogenetic series such associations with habitats in the wild are less direct, as will be seen especially in Chapter 13.

The material in this chapter encompasses many of the various methodologies of previous chapters. Since our knowledge of the genomes of many of the organisms is extremely limited, especially those higher in the phylogenetic series, simple quantitative analyses are often reported. In many instances these

Table 10-5   Behavior Fostering Adaptation to Domestication

Favorable characteristics	Unfavorable characteristics
1. Group structure	
a. Large social group (flock, herd, pack), true leadership	a. Family groupings
b. Hierarchical group structure	b. Territorial structure
c. Males affiliated with female group	c. Males form separate groups
2. Sexual behavior	
a. Promiscuous matings	a. Pairbond matings
b. Males dominant over females	b. Male must establish dominance over or appease female
c. Sexual signals provided by movements or posture	c. Sexual signals provided by color markings or morphological structures
3. Parent-young interaction	
a. Critical period in development of species-bond (e.g., imprinting)	a. Species-bond established on basis of species characteristics
b. Female accepts other young soon after parturition or hatching	b. Young accepted on basis of species-specific phenotype (color patterns, pheromones)
c. Precocial development	c. Altricial development (requiring relatively prolonged parental care)
4. Responses to human beings	
a. Short flight distance	a. Extreme wariness and long flight distance
b. Least disturbed by human ubiquity and extraneous activities	b. Easily disturbed by people or sudden changes in environment
5. Other behavioral characteristics	
a. Catholic dietary habits (including scavengers)	a. Specialized dietary habits
b. Adapt to a wide range of environmental conditions	b. Fixed or unique habitat needs
c. Limited agility	c. Extreme agility

Source: Hale (1962).

do not go beyond hybrid analyses, which while being informative cannot provide much detail at the genetic level. However, such analyses are important provided that their limitations are borne in mind. Many of these analyses give suggestive correlations between morphologies and behaviors (including variations in group behavior in fish)—not only in this chapter but also in previous and subsequent ones. We have noted in Section 9-3 in particular that behavioral phenotypes so revealed need to be analyzed where possible with fuller analyses of the types discussed in Chapter 6. Of course in many cases in this chapter we are considering closely related species where such analyses may be impossible due to hybrid sterility or hybrid inviability. The material therefore gives the impression of "opportunism" in approach; this often reflects the ease of experimental subjects for certain studies or in some cases may be an offshoot of a research program on animals of economic significance. The studies do, however, establish the generality of the conclusions derived from the extensive studies on *Drosophila* and rodents and put these results into a phylogenetic context.

In particular, some data are beginning to put learning studies into a phylogenetic context. In nematodes there is evidence for habituation for movement along chemical gradients. When we go to the Diptera, there is good evidence for genetically controlled learning by conditioning in blowflies and *Drosophila,* while in vertebrates such as mice learning has been repeatedly demonstrated in various ways. *Drosophila* also provides evidence of learning in mating choices based on previous experience (Section 8-4). In this regard the observation that cyclohexamide—a protein synthesis inhibitor that disrupts the memory of learned tasks in mice, rats, and goldfish—affects *Drosophila* mating choices altered by experience (Pruzan, Applewhite, and Bucci, 1977) is of great significance. The comparative study of learning among organisms and, if it does occur, its genetics is a topic worthy of more attention across the diversity of organisms discussed here. We predict this as a future trend and as one justification for our approach in this chapter.

SUMMARY

The literature of behavior genetics covers an enormous array of organisms including bacteria, Protozoa, nematodes, insects, fish, amphibians, birds, and domestic mammals. Apart from the specific examples of the previous two chapters, genomes of most organisms are not well known. Yet it is important to put the laboratory-oriented studies of these previous chapters into a phylogenetic context.

The types of behaviors exhibited by organisms are wide and varied. Chemotactic behavior is extensively studied in bacteria, Protozoa, nematodes and even the larval stage of insects. In adult insects, spinning behavior, mating behavior, parasitism levels, and calling song types are examples of traits considered. Indeed in blowflies, good evidence shows that learning from conditioning is under genetic control. In vertebrates, mating behavior studies are extensively reported.

In lower organisms, the relationship of chemotactic behavior may be directly related to habitats in the wild. However, higher in the phylogenetic series, associations with habitats are less direct, and learning becomes more important.

An applied aspect that emerges from this phylogenetic survey is the study of genetic and behavioral changes consequent upon domestication, for example in chickens and ducks. Clearly the "behavioral phenotype" of domesticated populations of a species differs in many ways from that of initiating wild populations. This behavioral phenotype may be constantly changing in noneconomic species such as cats, according to fashion at a given time. This is less likely in organisms domesticated for economic reasons, where behavioral uniformity may be an optimal strategy, at least in the short term.

Homo sapiens: Certain Discontinuous Traits

In this and the next chapter we consider *Homo sapiens.* Consequently, these two chapters include what are assuredly the most complex materials with which a student of behavior is obliged to cope. This complexity stems from our inability to mold and to extend pedigrees to extract maximal genetic information. We simply cannot make the appropriate crosses and backcrosses with human subjects, nor can we obtain information of the type that can be derived from inbred strains and their hybrids and from selection experiments. We lack another advantage awarded by experimental animals—the capacity to control and define the environments in which our experiments take place.

Chapters 11 and 12 encompass the limitations of the methodologies available to the behavior geneticist studying human beings. The stringency of these limitations can be assessed by comparing the methods usable with experimental animals. Probably the major difference is the possibility of designing environments for experimental animals and the impossibility of defining environments for people. If our readers can objectively assess data concerning the possible genetic bases of human behavior, we have achieved one of our primary aims.

Our primer for analyzing behavioral traits in *Homo sapiens* is Chapter 7. There, distinction is drawn between threshold (present-absent) traits and continuously varying traits. Although in some instances this distinction is cloudy, the present chapter is concerned with the former category and Chapter 12 with the latter. Complete coverage is not attempted; only representative examples are cited. Because *Homo sapiens* is anthropocentric, the literature is vast. We are necessarily forced to be extremely selective in our choice of examples. Unfortunately that means that much of the available literature worthy of inclusion is left out. However at the end of this chapter we give a list of General Readings longer than those offered for most other chapters.

If a trait is both hereditary and familial, it still may be cultural inheritance rather than genetic inheritance that is observed. An obvious example of nongenetic inheritance would be great wealth, which has been observed to "run in families." And so we commence with the complex history of kuru to illustrate the difficulty inherent in interpreting heredofamilial traits.

11-1 KURU

Kuru is a degenerative, always fatal neurological disease peculiar to a small area of New Guinea peopled by the Fore tribe and their neighbors. In the Fore language *kuru* means trembling, as from cold or fear, and describes the first symptoms of the disease. This cerebellar malfunction has now been identified as having a viral etiology. The infectious agent is one of the smallest microorganisms found to date—much smaller than most viruses—and is transmitted through cannibalizing the brains of infected persons. It has been termed a "slow virus" because clinical symptoms develop remote in time from its entry into the host. Chimpanzees inoculated with extracts from the brains of kuru victims show equivalent symptoms 10 to 50 months after injection. Human victims live only about a year after clinical symptoms are present (Gibbs and Gajdusek, 1978). (Additional human diseases will probably be linked to slow viruses in the future, as diagnostic techniques become more sophisticated.)

The first physicians to describe kuru reported 14 adult female victims to each adult male victim. Roughly speaking, three-quarters of the victims were women and the remaining quarter embraced children of both sexes evenly. This led to kuru being assigned a wholly genetic basis, with sex differences in heterozygote expression (Stern, 1973):

Ku Ku females and males suffered early onsets of the disease.
Ku ku females had late onsets.
Ku ku males were healthy.
ku ku females and males were healthy.

The genetic interpretation was logical considering the data then available.

The difficulties of studying human populations are compounded in primitive groups whose cultures differ markedly from our own and are obscure to us for that reason. In the case of kuru, the genuine etiology was illuminated only after anthropologists provided information concerning the practice of eating the brains of the dead and sharing them among relatives. Pedigrees suggestive of a genetic basis for kuru were based on family meals. However, the rate of spread was inconsistent with any reasonable genetic hypothesis except that of differential disease resistance, while all observations accord with the spread of a virus by cultural means. It is remarkable how closely the pedigrees interpreted by the genetic hypothesis parallel the cultural-viral reality. Further evidence against a genetic basis for kuru lies in the fact that the disease has been virtually eradicated since laws against cannibalism have been enforced.

Again at the end of the next chapter it will be shown how cultural and biological inheritance frequently display similar patterns. Adoption studies are most effective in separating some components of cultural and biological inheritance—and they are utilized for this purpose later in this chapter. This, however, is not a method generally applicable to primitive groups.

The reader is referred to a discussion by Harper (1977) wherein he shows the ease with which familial environmental factors can be confused with a truly hereditary basis, despite the fact that specific examples of such effects have been well recognized in nonhuman species for many years. In particular, we can cite the mammalian example of scrapie, which is caused by a "slow virus" (Gibbs and Gajdusek, 1978). Like kuru, scrapie causes fatal cerebral degeneration. Seen mainly in sheep, occasionally in goats, scrapie can be experimentally transmitted to a variety of other mammals. Both the clinical and pathological findings are similar to those of kuru, and Gajdusek (1977) has suggested that the two diseases may even be the result of the same agent. In the natural state, scrapie is transmitted predominantly through the mother, and infection appears to be prenatal. There is known to be pronounced variability in susceptibility in sheep that is genetically determined, and the same is true of mice. This lends support to the theoretical existence of genetic factors at play in human "slow virus" dementias such as familial Alzheimer's disease and Creutzfeldt-Jakob disease.

Since familial environmental factors can be so readily confused with traits with a biological basis, it seems plausible that other human familial disorders may be caused by infective agents. How can one avoid confusing these effects with the patterns resulting from true mendelian inheritance? This is not simple, as the history of interpretations of kuru shows. Following Harper (1977) we can however say:

1 Vertical transmission of a disorder through several generations does not necessarily indicate dominant inheritance.
2 A tendency toward familial aggregation without a clear mendelian pattern does not necessarily mean multifactorial inheritance.
3 Predominantly maternal transmission should raise the suspicion of an intrauterine or comparable environmental factor.
4 If mendelian inheritance exists, this does not exclude the operation of important environmental factors influencing the expression of the disease.

At this stage, we can already see the difficulties in interpreting data on our own species where breeding experiments cannot be done, as compared with experimental animals.

11-2 SPEECH DISORDERS

Stuttering, also called stammering, is a particular type of speech disorder beginning during childhood and persisting into adulthood in about 20 percent of cases (Van Riper, 1971). Sex has a marked effect, with males affected almost four times as often as females. Stuttering is also strongly familial; it "runs in families." Although little is known of its cause, environmental (nongenetic) factors play a demonstrable role since monozygotic twins are not always concordant for stuttering. In the presence of such environmental influences, the phenotype shows no mendelian patterns of genetic transmission. The observed familial pattern can often be "explained" by either a multifactorial or a single-major-locus model. This is especially so if only presence or absence of the trait is considered (Kidd, 1977).

The possibility of wholly cultural inheritance must be considered. Cultural transmission is, after all, much more flexible than the biological sort; it is capable of rapid and distinct alteration from generation to generation. Children may mimic an older or one or two colateral relatives with whom they have sufficient contact. There is also great variation in degree and frequency of stuttering, both among individuals and temporally for one individual, probably in response to environmental input, particularly in its emotional aspects. This view of the "inheritance" of stuttering has generally been ascendant in recent decades, but Van Riper (1971) has concluded that it does not adequately explain all aspects of the familial and developmental patterns of stuttering.

Garside and Kay (1964) noted that stuttering females have a higher frequency of stuttering relatives than do stuttering males. Although they could not exclude single-locus inheritance, they preferred a polygenic model with two thresholds—a higher one for females. With a higher threshold, females would be less frequently affected, but those affected would be more "genetically loaded" than affected males and, hence, have more affected relatives. Similar observations and explanations exist for cleft lip and palate (Woolf, 1971) and for other congenital anomalies: males are more commonly affected, the trait clusters familially, but there are no overt mendelian patterns of transmission. (These traits are assuredly not X-linked because affected father plus affected son is a common family situation.)

More recent and more sensitive genetic analyses utilizing the sex effect in searches for differences in dominance variances (Kidd, Kidd, and Records, 1978) have tried to discriminate between the alternatives of multifactorial (or polygenic) and single-major-locus transmission. Although Kidd and coworkers could not exclude either mode of inheritance, the single-major-locus model gave a better fit to family data. Using the same concept of two thresholds but only one locus with two alleles and the requisite environmental factors, their analysis suggests that the gene for stuttering occurs as follows:

- Frequency of the "stuttering" gene = 0.040 ± 0.007.
- Frequency of stuttering among individuals lacking the "stuttering" gene = 0.005 ± 0.003 in males, 0.0002 ± 0.0002 in females.

• Frequency of stuttering among individuals with one copy of the "stuttering" gene = 0.378 ± 0.025 in males, 0.107 ± 0.019 in females.
• Frequency of stuttering among individuals with two copies of the "stuttering" gene = 1.00 in males, 1.00 in females.

This model not only accounts for the greater frequency of affected relatives of female stutterers (a fact used by Garside and Kay, 1964, to support polygenic inheritance) but also explains the observed high incidence of affected sisters of female probands, This latter observation cannot be explained by polygenic inheritance. Additional research, such as twin, adoption, and linkage studies (the last most difficult), is needed to resolve these uncertainties about stuttering. Nevertheless, Kidd's sort of inquiry, though producing no definitive explanation of stuttering yet, has pertinence as a method of genetic analysis for other human traits that show a sex effect and familial concentration, but no mendelian patterns–for example, dyslexia, hyperkinesis, enuresis, and the congenital anomalies mentioned above.

Let us consider dyslexia (Greek *dys* = impaired or bad + *lexia* = speech) a bit more. It is the inability to read with understanding, though this inability is often coupled with above normal intelligence scores. Performance is therefore far below capacity and may involve a preponderance of aural, visuospatial, or written deficits. (Secondary dyslexia is due to brain damage.) Diagnosis has not yet been formalized and is rendered even more difficult because the deficits often cannot be found in adulthood.

Hand, eye, and foot dominance are not ipsilateral (same-sided) in the individuals afflicted, most often boys and men (see Section 5-7), and estimates of population frequencies are higgledy-piggledy. Although rare in Japan, dyslexia is not specific to any one language group. See Herschel (1978) for an excellent review (and references therein); he concludes:

> The hypothesis of an autosomal dominant mode of inheritance with reduced penetrance and variable expressivity cannot be refuted because it fits in nearly any pedigree constellation, as Krüger (1972) showed in a concise paper. Polygenic inheritance seems more plausible, since the disturbed processes are very complex and many environmental factors can modify the course of the disorder.

This interpretation, however, shows the problem in interpreting data of this type, since a mendelian mode of inheritance with reduced penetrance and variable expressivity gives almost the same results as polygenic inheritance, assuming penetrance and expressivity levels to be controlled genetically. In addition, while many environmental factors do not make a trait polygenic, they do however tend to make a discontinuous distribution continuous.

Familial dyslectics share delayed speech development 60 percent of the time. Four separate studies of MZ twins (Herschel, 1978) record all 36 MZ twins as concordant for dyslexia. In minor contrast, Bakwin (1973) scored but 26 concordant out of his 31 MZ pairs. And these same researchers diagnosed 31 out of 97 DZ twins as concordant for reduced reading, spelling, and writing abilities. Here however a caveat—there are to our knowledge no investigations of dyslectic twins reared apart, and twin studies per se are confounded by environmental obfuscations concerning learning, as discussed in Chapter 7.

11-3 ALCOHOLISM

Alcoholism is a "socially important phenotype" (DeFries and Plomin, 1978) which we may profitably survey via adoption studies so that this sort of approach to the genetics of human behavior and this far-from-rare addiction are simultaneously sketched.

Alcoholism may be acute or chronic, but in either instance the excessive, obligatory indulgence it implies is much more often (3 to 4 percent in a "general" population) manifest in men than in women. If one's father is an alcoholic, the risk statistically rises to 26 percent while an addicted mother produces a morbidity of 2 percent. Similarly, a brother results in 21 percent, a sister in 0.9 percent—all of this, regardless of one's own sex though more males are included (Amark, 1951; McClearn and DeFries, 1973). So, yes, alcoholism does "run in families."

As discussed in Chapter 7 briefly and in detail in the next chapter, adoption studies help to "disentangle genetic and environmental factors common to members of natural families by studying genetically unrelated individuals living together (to test environmental influences common to family members) and genetically related individuals reared apart (to test genetic influences)" (DeFries and Plomin, 1978). For the first of two specific examples, 55 Danish males, each with one bioparent formally diagnosed as alcoholic, were all adopted before they reached 6 weeks of age by a nonrelative, which precluded subsequent parental contact. At the time of inspection (Goodwin et al., 1973), 18 percent of these subjects averaging 30 years of age had already become alcoholic (here strictly defined as serious enough to involve legal, social, or marital, and employment difficulties due to alcohol). In the control groups (same sex, age range, adopted, and from similar backgrounds but having no alcoholic bioparents), only 5 percent became alcoholic.

In the second example, Schuckit, Goodwin, and Winokur (1972a,b) canvassed (via interviews) some two hundred alcoholic probands (at an average subject age of 40) plus their half-siblings. This design permits estimations of the relative influences of biological parents compared to rearing parents. The following percentages of alcoholism were found:

Biological parent alcoholic, rearing parent alcoholic	46%
Biological parent nonalcoholic, rearing parent nonalcoholic	8%
Biological parent alcoholic, rearing parent nonalcoholic	50%
Biological parent nonalcoholic, rearing parent alcoholic	14%

Clearly the state of (at least one) biological parents is crucial. Conditions of rearing are apparently less influential in producing or in deflecting from alcoholism than is inheritance.

See Kaij (1957) and Partanen, Bruun, and Markkanen (1966) for twin studies and for scales of needs for alcohol (0 = abstainers to 4 = chronic alcoholics; frequency, density, and control or lack thereof). These plus evidence provided by research with rodents (Sections 9-3 and 9-5) and even insects (Section 8-5) lead inexorably to the conclusion that there is a biological basis for alcohol dependency. Still, in human beings it is not yet possible to make more formal comments on the genetic architecture of alcoholism.

Unfortunately, few aspects of the biochemistry of alcoholism are understood, but attention is being focused upon two enzymes involved in the initial detoxification of alcohol, liver alcohol dehydrogenase (ADH) and aldehyde dehydrogenase (ALDH) (Section 9-5). Little is known of the relevance of any of the metabolites (ethanol and compounds derived from it such as acetaldehyde and acetate) to the mechanisms of intoxication or addiction (Schuckit and Rayes, 1979). Genetical research had been confined to four broad aspects putatively correlated with alcoholism (see Oakeshott and Gibson, 1980, for review):

1 *Associated psychiatric disorders* which are more common among relatives of alcoholic propositi than expected.

2 *Susceptibility to intoxication* as assessed by a variety of physiological and behavioral measures of intoxication (e.g., vasomotor flushing, heartrate, blood pressure, stomach discomfort, and drowsiness) is significantly more frequent among Chinese, Japanese and Koreans, Eskimos, and Amerindians, than among Caucasians. Since such differences occur even among newborns, genetic differences are likely. In addition, twin studies reveal a significant genetic component for susceptibility to intoxication (Propping, 1977) as measured from electroencephalographic patterns (Section 12-8) after ethanol ingestion.

3 *Detoxification of ethanol*—no general conclusions have followed from many recent studies with various racial groups although caucasoid twin studies indicate a significant genetic component.

4 *Alcohol dehydrogenase*—variants have been suggested as being associated with assorted susceptibilities to intoxication, but the evidence cannot be regarded as any more than suggestive.

In conclusion, studies of half-siblings and adoptees provide strong evidence that alcoholism is in part inherited, but more quantitative analyses are essential. Further investigation of factors such as those listed above is necessary; so far it can merely be said that susceptibility to intoxication is partly determined by rates of ethanol metabolism.

11-4 ENZYMES AND BEHAVIOR

What are the relationships between the genetics of behavior and of enzymes— the latter being ever so much better known? Eiduson et al. (1964) considered this question in their excellent chapter on biochemical genetics and behavior. At that time they compiled a list of 20 hereditary metabolic disorders culminating in neurological and/or behavioral disturbances. One decade later the National Foundation–March of Dimes published a list of metabolic disorders involving known specific enzyme defects. (See Table 11.1; Bergsma, 1979, should be consulted for descriptions of each syndrome.) Of the entries in Table 11-1, some

Table 11-1   Hereditary Metabolic Disorders Involving Known Specific Enzyme Defects

Acatalasemia
Acid maltase deficiency*
Acid phosphatase deficiency
Adenine phosphoribosyl transferase deficiency
Adenosine deaminase deficiency
Adenosine triphosphatase deficiency
Albinism-oculocutaneous, tyrosinase negative
Alkaptonuria
Argininemia*
Argininosuccinic aciduria*
Brancher deficiency*
Carnosinemia*
Citrullinemia*
Cystathioninuria*
Disaccharide intolerance
Fabry disease*
Fructose-1-phosphate aldolase deficiency*
Fructose-1,6-diphosphatase deficiency*
Fructosuria (marker)
Galactokinase deficiency
Galactosemia*
Gaucher disease
Globoid cell leukodystrophy*
Glucose-6-phosphate dehydrogenase deficiency
Glutathione peroxidase deficiency
Glutathione reductase deficiency
Glycogenosis, type I
Glycogenosis, type III
Glycogenosis, type VI
G_{M1}-gangliosidosis, type 1*
G_{M2}-gangliosidosis with hexosaminidase A and B deficiency*
G_{M2}-gangliosidosis with hexosaminidase A deficiency*
Goitrous cretinism
G-phosphogluconate dehydrogenase deficiency
Hexokinase deficiency
Histidinemia*
Homocystinuria
Hydroxyprolinemia (marker)
Hyperammonemia*
Hyperoxaluria
Hyperprolinemia (marker)
Hypervalinemia*
Hypophosphatasia
Isovalericacidemia*
Juvenile G_{M1}-gangliosidosis, type II
Juvenile G_{M2}-gangliosidosis, type III
Lactose malabsorption
Lesch-Nyhan syndrome*
Lysinemia*
Maple syrup urine disease*

**Table 11-1 Hereditary Metabolic Disorders
Involving Known Specific Enzyme Defects
(*Continued*)**

Metachromatic leukodystrophies
Methemoglobinemia
Methylcrotonylglycinuria
Methylmalonic acidemia*
Mucopolysaccharidosis I-H*
Mucopolysaccharidosis I-S
Mucopolysaccharidosis II*
Mucopolysaccharidosis III*
Myophosphorylase deficiency*
Niemann-Pick disease*
Nucleoside phosphorylase deficiency
Oroticaciduria
Pentosuria (marker)*
Phenylketonuria*
Phosphofructokinase deficiency
Phosphoglycerate kinase deficiency
Phosphohexose isomerase deficiency
Phytanic acid storage disease*
Porphyria*
Propionic acidemia*
Pyroglutamic acidemia*
Pyroglutamic aciduria
Pyruvate decarboxylase deficiency
Pyruvate kinase deficiency
Saccharopinuria
Steroid 11 β-hydroxylase deficiency
Steroid 17α-hydroxylase deficiency
Steroid 17,20-desmolase deficiency
Steroid 18-hydroxylase deficiency
Steroid 18-hydroxysteroid dehydrogenase deficiency
Steroid 20–22 desmolase deficiency
Steroid 21-hydroxylase deficiency
Steroid 3 β-hydroxysteroid dehydrogenase deficiency
Sulfite oxidase deficiency*
Thiolase deficiency
Transglucuronylase, severe deficiency
Triosephosphate isomerase deficiency
2,3-diphosphoglycerate mutase deficiency
Tyrosinemia
Wolman disease*
Xanthinuria
Xanthurenic aciduria

* Disorders with known behavioral effects ranging from taste aversions through seizures to mental retardation.
Source: Bergsma, 1979.

thirty are diagnosable prenatally (Littlefield, Milunsky, and Jacoby, 1973). Moreover, recent technological advances are resulting in enzymes being used as medications per se and in the production of drugs as well (Arehart-Treichel, 1978).

Some of the conditions listed are discussed elsewhere, for example, phenylketonuria in Section 2-5 and lactose malabsorption in Section 3-4. Gaucher disease and the porphyrias are discussed here.

There are at least two forms of Gaucher disease, both due to unrelated autosomal recessive genes (Philippart, 1979). In the acute or infantile form, where survival rarely exceeds a couple of years, afflicted children display a sort of palsy coupled with strabismus, laryngeal spasms causing problems swallowing, seizures, cerebellar difficulties with impaired balance and bodily positioning, and a gradually increasing dementia if they live beyond infancy (this may then be due to a third, juvenile form of Gaucher disease). There is a tissue deficiency of the enzyme β-glucosidase, coupled with stored glycolipids produced by the normal breakdown of aged red and white blood cells. Apparently the catabolism of these routinely degraded and replaced cells cannot be carried to completion as it constantly is in healthy children and adults. Ninety percent of all people with this hereditary disease have its adult or chronic form. The initial symptoms then appear at approximately age two but do not significantly shorten life expectancy. Fortunately all types of Gaucher disease can be accurately diagnosed before birth by amniocentesis (involving extraction, culturing in vitro, and analyses of total cells).

Porphyria (Greek for purple, referring to the color of urine containing excreted porphyrin, a derivative of the respiratory pigment hemoglobin) is interesting in the current context. Enzymatic defects are involved in all forms of porphyria, and there is some information about the behavioral abnormalities associated with acute toxic intermittent porphyria and variegated chronic porphyria. A third type, porphyria erythropoietica, is not associated with behavioral disorders. Porphyria may also be phenocopied by infection.

Abdominal pain is often the inital complaint in autosomally dominant acute porphyria; less frequently, the first symptom is partial paralysis. In this case, the neurological component is obvious from the clinical beginning. Repeated attacks leave residual damage including dementia (Stevenson, Davidson, and Oakes, 1970). Acute overt attacks may be foreshadowed by years of nervousness, hysteria, and psychoneurosis. During attacks, neurosis gives way to psychotic episodes and manic-depressive behavior (Section 11-9); delirium with hallucinations occurs. Korsakoff's psychosis is now seen: disorientation, polyneuritis, muttering, insomnia, plus pain over areas more extensive than the abdomen, namely, into the extremities.

Abdominal discomfort plus neurological involvement also characterize variegated chronic porphyria, but photosensitivity is prominent in this form. The symptoms associated with both these porphyrias (also autosomal and dominant) may be aggravated by the ingestion of certain drugs such as barbituates or sulfonamides.

Porphyrins are fundamental to various kinds of cellular metabolism in that these compounds participate as intermediates in the synthesis of hemoglobin, myoglobin, cytochromes, catalase, peroxidases, and even plant chlorophyll (Eiduson et al., 1964; Levine and Kappas, 1973). There is no cure for porphyria except treatments of the sort that tranquilize nervousness. But "in the patient with known disease who has been warned about the precipitating factors, the prognosis is now much better than this . . . mortality rate of 24 percent over a five-year observation period" (Tschudy, 1979).

In South Africa in 1688 Ariaantje Jacobs married Gerrit Jansy. She had been sent to this Dutch farmer from a Rotterdam orphanage. Eight children resulted: four of them ancestors of the 8,000 porphyria gene-bearers manifest today in South Africa. Figure 11-1 is a pedigree of three interrelated European royal houses. George III was the British monarch at the time of the American Revolution. His mental disorder, then called sporadic madness, was actually hereditary porphyria variegata; it altered British, and what was to become American, history and led to the formal establishment of psychiatry. It was during a supposed spell of mental incompetence that George III permitted enactment of the notorious Stamp Act; this was later offered as proof of his insanity. He eventually had to be confined to Windsor Castle, though with intervals of clarity, and was replaced by a regent (himself more mildly afflicted); George III lived to be eighty-one. "With a good diet, avoidance of medication with drugs and generally rational treatment his attacks of delirium might have been curtailed." (Macalpine and Hunter, 1969; see also Lerner and Libby, 1976.) Think of the historic sequelae that might have been brought about!

11-5 TASTING ABILITIES AND OTHER SENSORY PERCEPTIONS

We agree with Kalmus (1967, and references therein) that the behavioral consequences of inherited differences in sensory perception is a subject so vast that one can only endeavor to provide suitable references and to survey partly appropriate materials. In Section 11-6 vision is considered briefly. The reader is referred to McKusick's (1978) excellent itemization of the known genetic bases of sensory and other defects, especially the section on hereditary deafness.

Phenylthiocarbamide (PTC) tasting is the best-known human polymorphism for tasting ability, with three possible PTC genotypes (*TT, Tt,* and *tt*) and two possible PTC phenotypes (T, taster; or tt, nontaster). Taste differences involved in food choices, for instance, are likely to involve little or no PTC discrimination, but nontasters are reputedly less discriminating. PTC tasting depends ultimately upon $=\!N\!-\!C$ recognition. (Snyder and Davidson, 1937,

$$\overset{\|}{S}$$

and Barrows, 1945, have investigated other genetic variables concerning diphenylguanidine and brucine tasting deficiencies, respectively.)

Cavalli-Sforza and Bodmer (1971) have commented on the difficulties of ascertaining taste thresholds in subhuman animals. Fisher, Ford, and Huxley (1939) apparently attempted to ascertain taste thresholds with primates in the London Zoo. There, a chimpanzee spat in Fisher's face after ingesting a bit of PTC, which is bitter to tasters but neutral to nontasters. So apparently this

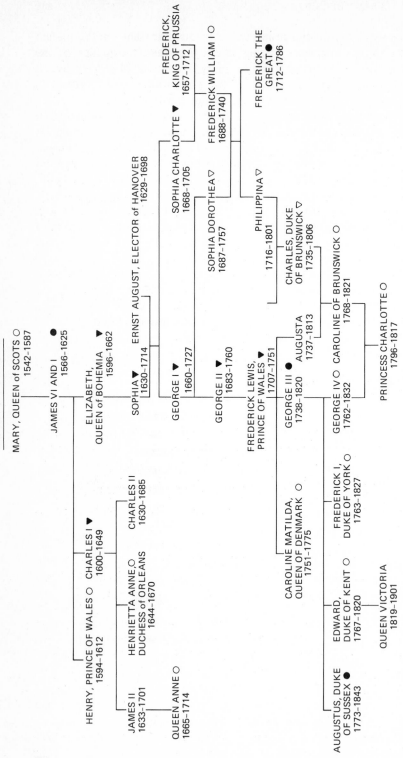

THREE ROYAL HOUSES

MARY, QUEEN of SCOTS ○
1542–1587

JAMES VI AND I ●
1566–1625

ELIZABETH,
QUEEN of BOHEMIA ▼
1596–1662

HENRY, PRINCE of WALES ○
1594–1612

CHARLES I ▼
1600–1649

HENRIETTA ANNE, ○
DUCHESS of ORLEANS
1644–1670

CHARLES II
1630–1685

JAMES II
1633–1701

QUEEN ANNE ○
1665–1714

ERNST AUGUST, ELECTOR of HANOVER
1629–1698

SOPHIA ▼
1630–1714

SOPHIA CHARLOTTE ▼
1668–1705

FREDERICK,
KING of PRUSSIA
1657–1712

GEORGE I ▼
1660–1727

SOPHIA DOROTHEA ▽
1687–1757

FREDERICK WILLIAM I ○
1688–1740

GEORGE II ▼
1683–1760

PHILIPPINA ▽
1716–1801

FREDERICK THE
GREAT ●
1712–1786

FREDERICK LEWIS,
PRINCE of WALES ▼
1707–1751

AUGUSTA
1737–1813

CHARLES, DUKE
OF BRUNSWICK ▽
1735–1806

CAROLINE MATILDA,
QUEEN OF DENMARK ○
1751–1775

FREDERICK I,
DUKE of YORK ○
1763–1827

GEORGE III ●
1738–1820

GEORGE IV ○
1762–1832

CAROLINE OF BRUNSWICK ○
1768–1821

EDWARD,
DUKE of KENT ○
1767–1820

AUGUSTUS, DUKE
OF SUSSEX ●
1773–1843

QUEEN VICTORIA
1819–1901

PRINCESS CHARLOTTE ○
1796–1817

Figure 11-1 Porphyria: This abbreviated pedigree shows the reconstructed route of porphyria through the royal houses of Europe. Rings mark those who showed some signs of this hereditary disorder; solid circles, those whose urine was known to be abnormal. Solid triangles mark transmitters who apparently did not suffer from the disorder; open triangles, those who may have been transmitters. Mary, Queen of Scots, was the first monarch in whom symptoms of porphyria appeared. Mary's son, James had "colics" that he claimed were inherited from his mother. His diaries record his urine was "the color of my favorite wine." George III's sister, Queen Caroline Matilda of Denmark and Norway "died at 23 of a mysterious illness that was featured by rapidly progressive paralysis." George IV, appointed regent when George III was deemed too ill to reign, had "gout" and his daughter Charlotte died of an acute attack of porphyria during childbirth. Her uncle Augustus, Duke of Sussex, had a florid case including the typically discolored urine. Queen Victoria's father suffered repeated severe colic attacks and died during one of them. George I's sister introduced this gene into the royal Brandenburg-Prussia house and there it claimed Frederick the Great of Prussia. The disease still exists in today's descendants of George III, as examined by Macalpine and Hunter (1969) from whose work this pedigree is derived.

particular tasting polymorphism exists in other primates as well as human beings (Figure 11-2). (See Section 9-7 for variations in taste testing in rats and mice.)

Approximately 25 percent or fewer human beings have the *tt* (nontaster) geno- and phenotype. This is not to imply that the *t* allele is completely recessive; nontasting is almost always a recessive condition, but recent studies are based on the response to serial dilutions of PTC, whereas older ones employed a single dilution, usually in the form of PTC crystals or PTC impregnated paper. Utilizing these older testing conditions, Rife (1938) reported an almost 4 percent discordance in the PTC-tasting ability of monozygous twins.

What, then, is correlated with the inability to taste PTC? Adenomatous nodular goiter occurs more often in nontasters than in tasters, and there are some indications that nontasters are more common among people who dislike

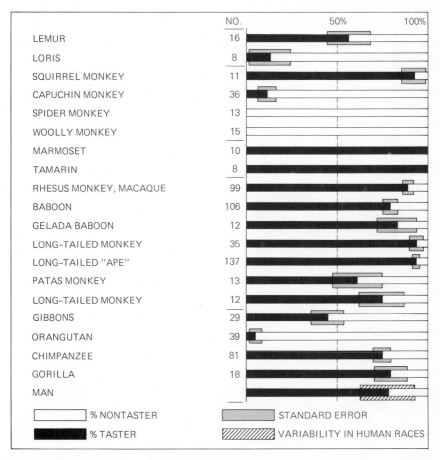

Figure 11-2 PTC tasting in primates. Percentages of tasters and nontasters in different primate genera. Note that no tasters were found among spider monkeys or woolly monkeys. (*From Chiarelli, 1963.*)

the flavor of alcoholic beverages. Other types of goiters (e.g., toxic diffuse goiter) occur more often among tasters. Rimoin and Schimke (1971) have commented on Shepard's (1961) data, again documenting the increased incidence of PTC nontasters in families with athyreosis (a form of thyroid gland dysgenesis resulting in cretinism) and other thyroid conditions, but finding no link via tasting perception among these conditions.

Fischer et al. (1961) and Fischer and Griffin (1960) tested for relations between taste thresholds and food dislikes (with 118 different foods) or for what may be more formally called the genetic aspects of gustation. They discerned the possibility of three loci for taste: one for quinine-like compounds including taste competence for sucrose and sodium chloride, one for 6-n-propylthiouracil and related compounds, and one for hydrochloric acid and other substances. The lower the taste threshold for bitter substances (including quinine and 6-n-propylthiouracil), the more foods disliked. These and other hereditary differences in taste acuteness may reflect "generalized" drug responsiveness. We agree with Spuhler and Lindzey (1967) that the frequencies of the nontaster allele in different human populations (see Garn's 1961 survey) are too high to be maintained by the traditionally implicated mutation-selection balance of forces. This is truly balanced polymorphism for sensory perception, though its intricacies are still obscure. We may then inquire: Why?

Families vary profoundly in their taste thresholds, with reports ranging from fivefold intrafamilial differences to identical taste thresholds in identical twins. Hirsch (1967b) reviewed this and other familial data regarding, for example, auditory acuity and discrimination. Ehrman (1972) has considered phenotypic assortative mating on the basis of sensory perception, especially with regard to auditory normalcy. Such positive assortative mating for hereditary sensory defects (e.g., deafness) may have the most profound genetic consequences. See, for instance, the pedigree in Figure 11-3.

Vandenberg (1967,1972) has offered suggestions for research work, commenting that practically nothing is known about the genetics of smell or of kinesthetic perception, a continuous trait defined in Chapter 12. His chapters and those of Hirsch (1967b) and of Spuhler and Lindzey (1967) are recommended as surveys of what is still to be done. What has been done is enticing, but we note the concentration of efforts upon PTC tasting because of its ostensible genetic simplicity. After all, humankind samples environments via senses—before responding behaviorally. The genetic basis of cues responded to are important in three respects (Ginsburg, 1967): (1) as a clue to the evolutionary history of the species, (2) as a potential for further evolution, especially if conditions of life change, and (3) as a means of understanding individual variability and of dealing effectively with it in a given situation.

11-6 COLOR AND OTHER VISION

Defective color vision of the kind generally known as red-green color blindness was recognized as early as the eighteenth century. (See Kalmus, 1965, and

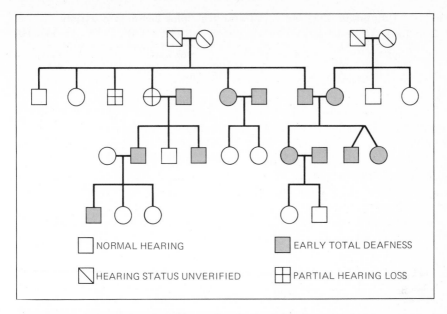

Figure 11-3 Phenotypic assortative mating for deafness demonstrated in a sample human pedigree. (*From Ehrman, 1972, and Sank, 1963.*)

Cornsweet, 1971, on the physiology of normal and restricted color vision.) Normal people can match color by the mixing of colors from three spectral regions: red, green, and blue. Hence, we can refer to normals as trichromats. In the most severe form of the red-green defect, the subjects are able to match color only when two hues are involved. These subjects are therefore known as dichromats. There are two types of dichromats, the protanopes (red-blind individuals) and the deuteranopes (green-blind individuals). The two corresponding anomalous (or weakened) trichromatic conditions are termed protanomaly and deuteranomaly: affected individuals are partially red-blind or green-blind, respectively.

These conditions are controlled by sex-linked recessive alleles at two closely linked loci, one for red-blind alleles and the other for green-blind alleles. Because the alleles are sex-linked, the frequency of the conditions is much higher in males than in females (Section 2-3). A characteristic incidence of color blindness in males is about 8 percent (0.08), so in females the expected incidence of $(0.08)^2 = 0.0064$ or 0.64 percent. In addition to the above conditions, there is a rare dichromatic defect, tritanopia and the corresponding tritanomaly, in which color discrimination in the blue-green region is affected. Autosomal incomplete dominance or autosomal recessive inheritance seems to be involved here.

The frequency of color-blind males varies from population to population and may be greater than 10 percent (Table 11-2). This proportion is far too high to be maintained by mutation alone and so suggests a genetic polymorphism,

Table 11-2   Proportion of Males with Defective Color Vision in a Variety of Populations

Population	Percent*	Population	Percent*
Europe		*Africa*	
English	6.8–9.5	Bechuana	3.4
Scottish	7.5–7.7	Bugandan	1.9
French	6.6–9.0	Bahutu	2.7
Belgian	7.5–8.6	Batutsi	2.5
German	6.6–7.8	Congolese	1.7
Swiss	8.0–9.0	*North America*	
Norwegian	8.0–10.1	U.S. White	7.2–8.4
Czechoslovakian	10.5	U.S. Negro	2.8–3.9
Russian	6.7–9.6	Amerindian	1.1–5.2
Jewish (Russian)	7.6	Eskimo	2.5–6.8
Finnish (Leningrad)	5.7	Canadian White	11.2
Turkish (Istanbul)	5.1	Mexican (urban)	4.7–7.7
Asia		Mexican (tribal)	0–2.3
Tartar	5.0–7.2	*South America*	
Chinese	5.0–6.9	Brazilindian	0–7.0
Japanese	3.5–7.4	"White" Brazilian	6.9–7.5
Indian (caste Hindu)	0–10.0	"Dark" Brazilian	8.8
Indian (tribal)	0–9.0	Brazilian Japanese	12.9
Israeli	2.1–6.2	*Australia*	
Druse (Israel)	10.0	White	7.3
Filipino	4.3	Aborigine	2.0
Fiji Islander	0–0.8	Mixed	3.2
Polynesian (Tonga)	7.5		

* Ranges of percentages represent different samples.
Source: Kalmus (1965).

although as yet there is little understanding of the selective factors involved. These factors surely concern, however, interaction between genetics first, and culture as a selective force, second. Color blindness is more common in societies that long ago abandoned hunting and gathering; its incidence seems to increase in industrialized societies, suggesting that selection for normal color vision must have been relaxed. Note, for instance, the three entries for Australia in Table 11-2. Could there have been, as Neel and Post (1963) suggested, a "transitory positive selection" for color blindness? Color-blind hunters (and soldiers) are said to possess a sharpened sense of form and awareness of edges; they "see through" camouflage. Judd (1943) proposed that normal-visioned observers be provided with filters to confer upon them, when needed, enhanced color-blind capacities to distinguish elements of and in terrain. Would, therefore, a mixed scouting troop be advantageous—one including a small proportion of color-blind hunters (Pollitzer, 1972)?

The performance of color-blind drivers has been investigated by Cole (1970) and their traffic accident frequencies found to be insignificantly different from those of normal-visioned individuals (Gramberg-Danielson, 1962). This must in part be due to the more conspicuous use of blue in green traffic lights.

Operations involving color coding in industry require four classes of judgment (Cole, 1972): comparative judgment of colors, connotative recognition of colors (learning associations such as green as leaves), denotative recognition of colors (the correct naming of a simple color), and aesthetic judgments.

Dunlop (1943) pointed out that mental and even nutritional maladjustments may affect color vision, so that a phenocopy of hereditary color blindness may be produced. *Phenocopies* are phenotypic modifications produced environmentally that mimic genotypic modifications (Section 2-5). Taylor (1971), while investigating the effects of defects in color vision on employment, found to his surprise that of 613 teenage boys with defective color vision only 224 voiced suitable [that is, not requiring color vision] career choices. He "wondered then whether there was something about colorblindness which made the subject gravitate towards the wrong career." Color-defective art students have been studied by Pickford (1972):

> The influences of temperament and personality are important. Perhaps they may be summarised as follows: If a student is bold and ignorant of his defect or insensitive about its presence, he may be able to use colours in a striking way which may seem original, and he might suggest that he could start a new mode of colouring in painting. If he is sensitive about his defect he may become involved in attempts to learn selfconsciously to compensate for it or avoid its effects. This may lead to considerable anxiety about his work and examinations, especially if he feels that his defect will not be understood so that allowance can be made for it. . . . Consequently, some collaboration between schools of art and psychology departments will be called for if a proper handling of the whole subject is to be achieved.

The reader should also be aware of an array of genetic anomalies wherein albinism is combined with abnormal, often cross-eyed, vision. Similar abnormalities have been described in the "white tiger," Siamese cats (Section 10-16), ferrets, hamster, mink, and many other mammals (Guillery and Kass, 1973, and references therein).

11-7 SOME MUTATIONS IN *HOMO SAPIENS*

Exploiting McKusick's comprehensive efforts (1978), Table 11-3 itemizes a few of the mutations, in the broadest sense, recorded in human subjects that are known to alter behavior. These alterations are *not* normally the primary effects of the mutant genes, but for us they assume no less than the greatest importance.

We adhere to McKusick's (1978) method of classification but omit itemization of hereditary deafness (referring our readers to pages *xxiii–xxv* in McKusick, 1978, and to Section 11-5 in this book; also see Jay, 1974, on ophthalmic genetics as correlated with the McKusick catalog). We have arbitrarily selected 10 examples in each genetic category to provide a sample of the diversity proffered by mutation. They demonstrate the awesome variety of mutant material influencing sensory appreciation of our environments.

11-8 THE EPILEPSIES

Although epilepsy is one of the earliest recorded medical problems, it is not yet understood adequately. Actually, epilepsy is not one condition but many, and it can result from a variety of different causes. Furthermore, it is a symptom of an underlying brain disorder, not a disease in the usual sense.

About 10 percent of the general population will have one or more seizures sometime in their lifetimes. These may result from conditions such as infectious diseases of the brain, injury to the head, cerebrovascular diseases, poisoning, high fever, alcohol or drug withdrawal, or an imbalance of body fluids and metabolites, although in many individuals none of the above factors seem to be obviously implicated as a cause of the seizures.

Seizures have many manifestations, but are divided into two major categories. Generalized seizures affect the entire body. Symptoms of generalized seizures (tonic-clonic, also called grand mal) usually lasting 2 to 5 minutes consist of a loss of conciousness and stiffening of the body, with alternating spasm and relaxation of the muscles. Patients with generalized seizures lasting 10 to 30 seconds (petit mal) may exhibit a rapid blinking of the eyes or a blank stare, which may be mistaken for daydreaming. Partial seizures indicate malfunction of a focal part of the brain. Elementary partial seizures (focal seizures) involve a part of the body, while complex partial seizures (temporal lobe or psychomotor) result mainly in disturbances of thinking and behavior.

When repeated seizures occur, not related to specific environmental events, the term epilepsy is used. The cumulative probability of developing

Table 11-3 Some Mutations Affecting Behavior in Human Beings

Autosomal Dominant Phenotypes

10430 Alzheimer disease of brain
       Presenile dementia, sometimes with parkinsonism, like Pick's disease (lobar atrophy)
10850 Ataxia, periodic vestibulocerebellar
       Vertigo, diplopia (double vision) and slowly progressive cerebellar ataxia in some
11340 Brachydactyl-nystagmus-cerebellar ataxia
       Nystagmus, mental deficiency, and strabismus
11530 Carotinemia, familial
       Nightblindness
12620 Disseminated sclerosis (multiple sclerosis)
       Neurological disorder, narcolepsy
12640 Double athetosis (status marmoratus or Little's disease with involuntary movements)
       Infantile cerebral palsies
12770 Dyslexia, specific (congenital word blindness)
       Speech defects associated in many instances
12820 Dystonia, familial paroxysmal
       Paroxysmal dystonia, unilateral dystonic postures without clonic movements or change in consciousness
13040 Electroencephalographic peculiarity
       Occipital slow beta waves (16 to 19 per second) replace alpha waves
13630 Flynn-Aird syndrome
       Neuroectodermal syndrome with visual abnormalities including cataracts, atypical retinitis pigmentosa, and myopia; bilateral nerve deafness, peripheral neuritis, epilepsy, and dementia

Table 11-3 Some Mutations Affecting Behavior in Human Beings (*Continued*)

Autosomal Recessive Phenotypes

20130 Acro-osteolysis, neurogenic
       Abnormality of peripheral sensory nerves, perhaps insensitivity to pain
20420 Amaurotic family idiocy, juvenile type (Batten's disease in England. Vogt-Spielmeyer's disease on European Continent)
       Rapid deterioration of vision and slower but progressive deterioration of intellect Seizures and psychotic behavior
20700 Anosmia for isobutyric acid
       Inability to smell isobutyric acid (sweaty odor)
20790 Argininosuccinicaciduria
       Mental and physical retardation, convulsions, and episodic unconsciousness
20910 Atonic-astatic syndrome of Foerster
       Muscular hypotonia, static ataxia, monotonous speech
21450 Chediak-Higashi syndrome
       Photophobia and nystagmus
21870 Cretinism, athyreotic
       Endocrine disorder (thyroid malfunction with profound mental and physical consequences)
21890 Crome's syndrome
       Congenital cataracts, epileptic fits, mental retardation, and small stature
22180 Dermo-chondro-corneai dystrophy of Francois
       Skeletal deformity of hands and feet, corneal dystrophy, abnormal electroencephalograms with seizures
23070 Gangliosidosis GM (2), type III or juvenile type
       Ataxia between ages of 2 and 6 years followed by deterioration to decerebrate rigidity, sometimes blindness occurs later

X-Linked Phenotypes

30050 Albinism, ocular
       Fundus is depigmented and choroidal vessels stand out strikingly, nystagmus, head nodding, and impaired vision
30160 Angiomatosis, diffuse corticomeningeal, of Divry and Van Bogaert
       Demyelinization
30170 Anosmia
       Inability to smell
30370 Color blindness, blue-mono-cone-mono-chromatic type
30540 Faciogenital dysplasia
       Hypermobility in cervical spine with anomaly of the odontoid resulting in neurological deficits, ocular hypertelorism
30700 Hydrocephalus due to congenital stenosis of aqueduct of Sylvius
       Mental deficiency and spastic paraplegia
30990 Mucopolysaccharidosis type II (Hunter's syndrome)
       Mental retardation and deafness
31170 Periodic paralysis, familial
31300 Spatial visualization, aptitude for
31330 Spinal ataxia
       Incoordination of limb movements

recurrent seizures (epilepsy) is about 1 percent by age 20, rising to 2 percent by age 40, and 3.5 percent by the end of life. Of these cases of epilepsy, only 30 percent can be ascribed to a reasonably clear-cut cause, the remainder being described as idiopathic; an inborn tendency of dysrhythmic cerebral activity coupled with an absence of detectable structural abnormality. Unfortunately the conditions grouped under idiopathic epilepsy are not homogeneous, thus making genetic investigations difficult.

The risks of epilepsy are somewhat higher for the siblings or offspring of a proband with epilepsy. Such risks are in the range of 6 to 8 percent (as compared with 2 to 3 percent for the general population) as indicated in Figure 11-4. In specific cases (when the onset in the proband is at a younger age or when one of the parents is also affected), the risks for siblings may be higher.

Seizure disorders present a number of interesting problems and challenges for genetic analysis. Some of these difficulties are shared with other common medical problems, while others are unique to the epilepsies.

There are well over 100 single locus (mendelian) traits associated with an increased risk of seizures. These include disturbances of amino acid metabolism, storage diseases involving the accumulation of metabolic intermediates in brain cells, changes in the vascular supply to the brain, and tumors of brain tissue. Clearly, the normal functioning of the brain can be altered by a variety of changes in its anatomy and biochemistry. If there are associated congenital malformations, it may be necessary to consider the possibility of a chromosomal anomaly. Most of the detectable changes in chromosomes (involving additions or deletions) have some effect upon the nervous system, including an increased risk of seizures.

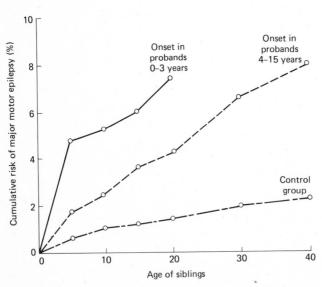

Figure 11-4 Cumulative risk of major motor epilepsy in siblings of probands with major motor epilepsy by age at onset in the probands. (*From Anderson, 1977, after Eisner et al., 1959.*)

Considerable research effort is being directed toward the identification of the heterogeneity in a large remainder of cases of epilepsy. The cases that emerge from the analysis of similarities within sets of affected relatives may be important in this search (Anderson, 1977, and references therein).

It is true that anyone may develop seizures in response to a very severe head injury, but there may be individual variability in seizure threshold. Those who develop seizures in association with head trauma, alcohol withdrawal, or febrile illness may differ genetically from those who do not.

Studies of the febrile seizures associated with high temperatures in young children provide some evidence for this view. In three studies of probands with febrile seizures the risk of febrile seizures among siblings ranged from 8 to 11 percent, as compared with 1 to 2 percent for control groups. The probability of developing epilepsy after a febrile seizure also is higher when there is a family history of seizures.

Some individuals will develop seizures in response to a flickering light of a particular color and flicker frequency. Similar stimuli can induce photosensitive changes in the EEG pattern. In a study of probands who had both seizures and a photosensitive EEG (Doose et al., 1969), 8 percent of the siblings had seizures. Among those siblings who had a photosensitive EEG, the risk of seizures was 20 percent.

Several studies in progress (Hauser and Kurland, 1975) are testing the possibility that the chances of seizures following head trauma may be higher for those with a family history of seizures than for those without affected relatives.

Specific EEG patterns associated with epilepsy appear to increase in frequency to a peak in childhood or adolescence (depending upon the type of pattern) and then decrease. For example, in the studies of probands with epilepsy and spike-wave EEGs, the frequency of spike-wave changes among siblings (with or without epilepsy) is highest in the 5 to 15 year age group; spike waves represent transient, sharp changes in EEG voltage polarity and as such, reflect pathology (Tsuboi and Endo, 1977). EEG patterns are also discussed elsewhere throughout this chapter.

Such conclusions usually have been based upon cross-sectional studies, however, and appropriate longitudinal studies are needed for a more reliable analysis of the actual changes within individuals. This applies just as pointedly to manic-depressives and to schizophrenics discussed in the next two sections. Eventually we may understand the mechanisms that lead to the development of the EEG anomalies as well as the maturational processes that later reduce them.

In the past the role of genetic factors in epilepsy often was exaggerated, as reflected in laws prohibiting marriage for persons with epilepsy. Such laws overlooked the heterogeneity involved and so assumed a risk for children that is higher than the facts would indicate. Fortunately these restrictions have been repealed or modified (Anderson, 1977, compiled for the U.S. government).

As with other questions for genetic counseling, an important first step is careful medical evaluation and diagnosis. In a small proportion of cases, an underlying mendelian trait or chromosomal anomaly will be detected. In these situations the risks for subsequent children can be calculated with precision (Anderson, 1977, and Eisner, Pauli, and Livingstone, 1959).

The second step is to inquire about possible environmental precipitating events or types of brain injury. Information is also needed about the seizure manifestations and EEG patterns. The presence or absence of seizures in relatives (at least parents, siblings, and children) should be recorded.

With these data in hand, it is usually possible to provide an estimate of the risk of seizure or epilepsy for close relatives. Data are more accurate for siblings of probands than for their ancestors, but the results from a few recent good studies show that the risks for children of probands are similar (Metrakos and Metrakos, 1969, for review; Anderson, 1977).

In a majority of situations involving a person with recurrent seizures (epilepsy) it is reasonable to assume a risk of epilepsy for a child or sibling in the range of 6 to 8 percent. This estimate becomes somewhat higher if one additional close relative already is affected. Genetic counseling for women with epilepsy who are in the child-bearing age must consider the additional possibility that anticonvulsant medications during pregnancy may be harmful to some developing fetuses.

Anatomically, it is clear that lesions in the hippocampus can be associated with seizures in human beings. Studies in mice have shown genetic control of synaptic patterns in the hippocampus (a specific ridge in the brain) and of other hippocampal variations associated with susceptibility to audiogenic seizures. At the biochemical level, there are many investigations of neurotransmitter substances and neuronal receptor sites in experimental models of epilepsy. Other studies (generally quite separate) deal with genetic variation in membrane receptors and in enzymes controlling neurotransmitter levels. Animal models may play important roles in all such endeavors (see Section 9-5 where audiogenic seizures in mice are discussed).

11-9 MANIC-DEPRESSIVE MENTAL ILLNESS (AFFECTIVE DISORDERS)

Kraepelin's (1896) description is still valid:

> Manic-depressive insanity includes on the one hand the entire province of so-called periodic and circular insanity, and on the other hand, the simple manias, the largest part of clinical pictures designated as melancholias, and also a not inconsiderable number of cases with amentia. Finally, we count as well certain mild and very mild, partly periodic, partly enduring morbid pictures with similar coloring, which may start out as grave disturbances, but which alternatively may pass over without clearly defined boundaries into the realm of deviant personality organization.

Unipolar affective illness is depression alone, without the alternate manic phase characterized by inordinate exaltation, elation, and excitement; bipolar illness is manic depression. Both represent harmful extremes of emotion due to profound mood fluctuations.

Manic-depressive mental illness occurs in 0.6 percent of the general population as defined by Lynch (1969) and Stern (1973). Depression alone (unipolar) has a minimal frequency of 1 to 5 percent in the general U.S. population, but incidences are somewhat variable between different populations (Rosenthal, 1970). The bipolar form (manic depression) is rare (1.6 per 1,000) in one northern Swedish isolate consisting of a few hundred people, where schizophrenia (discussed later in this chapter) occurs in relatively high frequency, approximately 9 per 1,000 (Böök, 1953; Fuller and Thompson, 1978). *Isolates* are small discrete populations whose members are much more likely to breed with one another than with nonisolate individuals because of religious, ethnic, or other limiting reasons. Manic depression is more prevalent within another isolate, the Hutterites of the western United States and western Canada, in whom schizophrenia is rare (Eaton and Weil, 1955). Manic depression has an incidence of 4.6 per 1,000 Hutterites, or 9.3 in persons aged 15 or older; schizophrenia has an incidence of 1.1 per 1,000 Hutterites, or 2.1 in those aged 15 or older. The Hutterites represent a sect founded in the 1500s by Jacob Hutter. Persecution caused their migration from Europe (Moravia, and then Hungary and Russia) to South Dakota in 1874. The Hutterites favor a kibbutzlike socially-collective life-style distinct from the geographically remote socially-withdrawn northern Swedish isolate.

There is no doubt at all that heredity plays a major role in the etiology of manic depression as shown by the twin studies summarized in Table 7-2, but debate continues on questions such as: (1) Are unipolar depression and bipolar manic depression manifestations of separate genetic entities themselves? (2) Is the mode of genetic transmission polygenic or monogenic, and are these disorders instances of heterogenous genetic control?

With reference to the first question, it has been observed that the genetic prediposition is greater when bipolar episodes are observed. We might also ask if it is greater in instances of earlier onset so that a gradient can be envisioned, in order of increasing psychopathology and genetic predisposition: (1) unipolar and of late onset, (2) unipolar and of early onset, (3) bipolar and of late onset, and (4) bipolar and of early onset. Such a gradient is most easily interpreted in terms of a polygenic model or even a simple major locus with two thresholds for onsets, but what if bipolar and unipolar illnesses are not related genetically? No monozygotic twin pair has ever been diagnosed as one unipolar manic plus one unipolar depressive, although there are many recorded monozygotic twins with one bipolar and one unipolar member (Zerbin-Rudin, 1969). Bipolar index cases often have unipolar relatives, and there is a 23.4 percent repeat risk for the parents of a bipolar patient. Twin studies show an impressively (sometimes approaching 70 percent or more) high bipolar and unipolar concordance rate if the twins are monozygotic, and 26.3 percent if they are dizygotic.

With reference to the second question, Gershon and collaborators (1976, 1977, 1979) cogently summarize:

> Early onset of affective disorder is associated with increased morbid risk of the disorder in relatives, but age at onset is not itself a transmitted factor. Female relatives have higher prevalence of illness, but sex of the ill person does not appear to be a factor in transmission. Genetic models of multifactorial or of single gene

autosomal inheritance are compatible with some but not all of the family history studies reported. The hypothesis of sex-linked transmission of bipolar illness has been proposed, and some pedigrees compatible with X-linkage have been reported, but family studies do not suggest that this is generally present. (Mendlewicz, Fleiss, and Fieve, 1972 and see Eisenberg, 1973 and Winokur, 1973.) Other possible modes of inheritance remain to be tested.

Comings (1979) has discovered a common polymorphism of a human brain protein, "Pc 1 Duarte," which just may be a major gene in depressive disorders. Then, it would have to act in conjunction with an environmentally controlled threshold effect such as that discussed just below for schizophrenia (by Kidd and collaborators, 1973).

Although other models can be tested, we feel that detailed pharmacological, biological, and behavioral studies are essential. This is because the variety of genetic hypotheses may imply a level of heterogeneity in genetic bases, just as is apparent for epilepsy. To this end, Gershon et al. (1976) report on studies of certain enzymes, among them monoamine oxidase, which are responsible for the enzymatic inactivation of catecholamines (see Section 9-5); these neurotransmitters are hypothesized to have altered functional activity in the affective disorders. In addition, abnormalities of cation activity have been reported, and a pharmacological specificity of response to lithium carbonate occurs. Although progress is not likely to be fast, an approach through biochemical or pharmacological genetics may ultimately be productive.

11-10   SCHIZOPHRENIA

Schizophrenia, which appears to be an array of profound behavioral disorders, presents an urgent and exceptionally difficult problem, as difficult as any we consider in this text. The history of the classification and treatment of schizophrenia is itself rather schizoid. The United States' cost of schizophrenia has been estimated at $11.6 to $19.5 billion annually. About two-thirds of this cost is due to lack of productivity by schizophrenic patients and about one-fifth to treatment costs. The estimate might be considerably higher if better figures were available on the cost of maintaining patients in the community (Gunderson and Mosher, 1975). Wienckowski (1972) commented:

> More than 2 million Americans at one time or another have suffered from the tragic mental disorder of schizophrenia. Half of the Nation's beds in mental hospitals are now occupied by schizophrenic patients. It is estimated that 2 percent of the population will have an episode of schizophrenia some time during their lives; in certain social settings--the urban slum, for example--the prediction rises to 6 percent, or more than 1 in 20 persons (as loosely diagnosed, see Dunham, 1965 and McNeil, 1970).

Rosenthal (1970) remarked, "I was asked to write about the promise that genetics holds for the understanding, prevention and treatment of mental illness in general, and schizophrenia in particular. That is better than being asked to write about man in relationship to his universe, but not much better."

What advice can the geneticist offer physicians and therapists? Schizophrenia is characterized by disturbances of thinking (delusions, bizarre or illogical responses), perceptions (auditory and visual hallucinations), and affective responses (loss of interest, volition, and the capacity to experience ordinary pleasures). Kraepelin (1896) described the four classic subtypes, included here for historic reference:

Catatonic. Stuporous, often mute, remaining in one position for hours to weeks.

Hebephrenic. Disturbed thinking, notably shallow affective responses, childlike behaviors and mannerisms.

Paranoic. Characteristic persecutory or grandiose delusions, often auditory hallucinations.

Simple. Gradual loss of interest and personal attachments leading to an indifferent, apathetic person who is almost totally isolated from human society.

These subtypes may be compounded in one individual at various times. The schizophrenic tends to withdraw from reality to the degree that he is unable to distinguish clearly between his inner fantasies and the realities of his physical environment.

The theories that attempt to explain the genetic component(s) underlying the manifestation of schizophrenia constitute three broad schools: monogenic (hypothesizing a single crucial genetic locus), polygenic, and that of the comprehensive, genetic heterogeneity. Numerous studies documenting families with histories of schizophrenia demonstrate a high rate of concordance (occurrence of the disorder in two or more individuals of a family or group) for siblings (10 to 15 percent, Ödegaard, 1963) and higher concordance rates for monozygotic twins (50 percent, Gottesman and Shields, 1966). Evidence such as this indicates that a functional genetic component of some sort or sorts is present in schizophrenia, though such a component need not be omnipresent. These may even be heterogenic, that is, "schizophrenia" may be the end product of environmental interactions with several monogenic conditions, a currently favored and sensible approach to research strategies.

We consider here only one of the monogenic theories of the inherent basis for schizophrenia. Older monogenic proposals have been thoroughly amended so that they incorporate minor modifying genes, thereby contradicting any monogenic hypothesis. Heston, a psychiatrist whose publications (1966 on adopted children of schizophrenic mothers, 1970, 1972) on these matters are highly recommended, has offered data (Figure 11-5) which show that the recorded cases of schizophrenia *plus* schizoidia approximate the expectations for control being a single dominant gene. *Schizoidia,* part of the schizophrenic spectrum, may be defined as a pre- or potentially schizophrenic mental state characterized as "very withdrawn," "always frightened," "feels worthless," "quite regressed—wants to be treated like a baby, and despises oneself for these inclinations" (Landis, personal communication; see also Landis and Tauber, 1972). The schizoid patient is in touch with reality and realizes that there is trouble. For contrast, consider the withdrawal from reality permeating

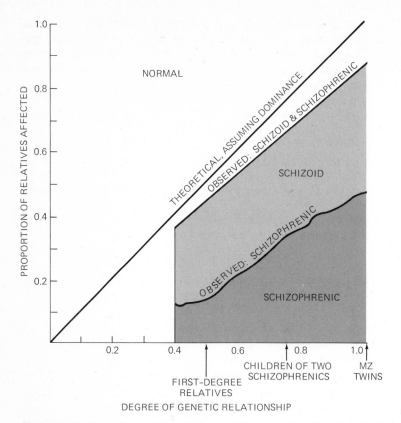

Figure 11-5 Basis for monogenic theory of cause of schizophrenia. Observed cases of schizoidia plus schizophrenia approximate proportions expected if major influence of a single autosomal dominant gene is assumed. (*From Heston, 1970. Copyright 1970 by the American Association for the Advancement of Science.*)

the following case history (Beckett and Bleakley, 1968) of a paranoid schizophrenic. Note the *autism* (condition of being wholly self-centered in thought and behavior, dominated by the subjective) that is displayed and realize the poor prognosis.

The patient was a twenty-three-year-old, single male:

Present Illness: The patient had just completed his M.A. degree in electronic engineering. (This level of educational attainment is not typical.) During his final examinations he had the flu and began to worry excessively about his heart. He complained of general fatigue and of his brain being numb. He became very confused, felt that his food was being poisoned, and refused to wash. When his parents insisted upon his washing, he became impulsive and aggressive. He repeatedly referred to his mother as a murderess and a witch and said that she was hypnotizing or poisoning him. He was hospitalized after unsuccessfully attempting to stab her.

Past History: The patient was the elder of two siblings. He was a healthy infant and was easily toilet-trained. Although he was a brilliant scholar, he was inclined to be shy and sensitive with other children. At seventeen he suffered from pains in the legs and throat and was out of school for a year. During this time he was irritable and wanted to be left alone. Since he had entered college his only known social contacts had been with his family at mealtimes. The remainder of his time had been spent in his room in study or walking about the streets by himself.

Interview: On the ward he had been observed to have sudden fits of rage, while at other times he would sit in a corner convulsed with laughter. In appearance he was haggard and untidy, his movements seemed uncoordinated, and he talked to himself in a rapid and incomprehensible rush of words. He accused the interviewer of interfering with his thoughts by constantly sending coded electronic messages by microwave which he could not decode, and as a result he would not talk during the interview.

The diagonal in Figure 11-5 represents the theoretical expectation for control by a single dominant gene. (Note that one gets the same configuration for a polygenic trait determined entirely by additive genes.) The small unshaded area in the lower left corner is blank for degrees of genetic relations (as defined in Section 6-9) lower than 0.4. Note that the expected coefficient for first-degree relations (between parent and offspring, sibs, and dizygotic twins) is 0.5. Relations between two parents and their offspring have an expected coefficient of $\sqrt{2} \times 0.5 = 0.71$ (Section 6-9). In the case of monozygotic twins the expected coefficient of relationship is 1.0.

The polygenic models of the etiology of schizophrenia generally concur with diathesis-stress theory (Table 11-4) proposed by Rosenthal (1971) and by Gottesman and Shields (1971, 1972, 1973). According to this theory (not in total contradiction to that devoted to the single causative gene) the schizophrenic does not inherit schizophrenia itself but is genetically predisposed to develop the condition. The environment of the individual determines the likelihood of the affliction being manifested. Environmental stress induces schizophrenia in the predisposed individual. Essentially, the diathesis-stress theory is based on a threshold polygenic model. The genetic basis for such a model is the additive effect of a number of relatively minor genes that exhibit little or no dominance or recessiveness; or perhaps several different loci control several different components of behavior, and these, when collectively disturbed, crystalize to what we know as schizophrenia.

Erlenmeyer-Kimling (1968) says it best. She describes genetic heterogeneity for schizophrenia as:

A heterogenous [different genetic bases for phenotypically similar conditions] collection of entities stemming from a number of different, independently acting genic errors. . . . According to this model, a number of different primary enzymatic defects could feed into a final common pathway or intermediary mechanism. . . . The extent to which the final pathway is disrupted might be different, however, depending upon the route taken in reaching it, so that variations in predisposition could exist between the different genotypes. Multiple allele series at a given loci are also possible—with different alleles producing different degrees of effect. The activity of the various genes will, furthermore, be modulated by the total genotypic background against which the alleles are placed. . . . Finally, both the degree of predisposition and the influences of environmental factors will cooperate to deter-

Table 11-4 Comparisons of Two Major Theories About the Genetic Basis for Schizophrenia

Aspects of the illness	Monogenic biochemical theory	Diathesis-stress theory
Biological unity	Homogeneity: one gene, dominant, recessive, or intermediate. Trait is qualitative, discontinuous.	Homogeneity or heterogeneity. Trait may be qualitative or quantitative.
"What" is inherited?	A specific but as yet unknown error of metabolism due to a mutant gene.	(1) A single gene. (2) Several major genes. (3) Polygenes in any case, a "constitutional predisposition."
Manifestation	Very high: almost everyone (67–86%) with the genotype, but some have constitutional resistance to expression	May be considerably lower than monogenic biochemical. Depends on whether predisposed schizophrenic encounters sufficient stress and how predisposed he is.
Role of environment	No discretely stressful environments needed to precipitate illness. (May be cumulatively stressful?) Proponents like to cite a constant rate of schizophrenia in all cultures.	Necessary to precipitate the illness. The stressors are seldom defined: head trauma, disease, alcohol, parturition, exhaustion, etc., but usually psychological.
Clinical subtypes	Of secondary interest, usually thought to reflect other inherited or constitutional factors that influence the form in which the illness is expressed.	Usually holds that they represent different predispositions interacting with different kinds of stressors. (An unnecessary complication?)
Severity of illness	Reflects the degree of metabolic disturbance.	Reflects the amount of inherited predisposition and the intensity of the stressor.
Remission	For some reason, the effect of biochemical disturbance clears, but personality defect remains.	Either the physiological aspects of the disease process are reduced or the stressors are reduced.
Premorbid personality	Varies in usual ways. When aberrant, the deviations are thought to be early signs of the metabolic disturbance.	Can provide clues about the nature of the predisposition inherited, as introversive personality or high anxiety.
Research strategy	(1) Search for the biochemical aberrancy and its amelioration or correction, if any. (2) Estimate gene frequency in population, mutation rate, mode of inheritance, etc.	Learn about the nature of the predispositions, the stressors, and the nature of their interactions.
Example of problems posed by previous findings	Why does the distribution of illness in kindreds vary so markedly, showing dominant, recessive, or intermediate patterns?	Why does the illness continue when the ostensible stressor has been removed?

Source: Rosenthal (1970) and see Matthysee and Kidd (1976).

mine whether schizophrenia, psychological disturbances of lesser sorts, or perhaps no symptomatology becomes manifest at the behavioral level.

Table 11-4 summarizes the two major theories about the genetic basis of schizophrenia. Note the overlap, and hence confusion, manifest in at least the first two aspects listed.

We, in turn, can now summarize in Table 11-5. Some calculated degrees of kinship and inbreeding are offered for explanation and for more general use. Coefficients of relationship (Section 6-9) are included for comparison. The morbidity risk for schizophrenia does not uniformly reflect the coefficient of relationship; half-sibs, and nephews, nieces, grandchildren, and even aunts and uncles share a close coefficient of relationship, but not similar morbidity. Could this reflect the varying power of the environment? Half-sibs are more likely to share the same or at least partly the same environment than are members of a different generation (e.g., grandchildren), yet those latter have risks similar to those of half-sibs! Differences in environment may account for a large part of the range in concordance evident among individuals with the same coefficient of relationship.

Fischer (1973) has continued the work of Harvald and Hauge (1965) with the same Danish twins (see Section 7-1). She has found a 56 percent concordance rate for MZ and 26 percent for DZ cotwins. For MZ twins discordant for schizophrenia, Fischer finds no significant differences in the proportions of schizophrenic offspring they each produce. The affected proband and the unaffected MZ cotwin produce roughly the same proportion of schizophrenics in the next generation.

We conclude this most difficult of sections by summarizing a list of tenents quoted in part from a review appropriately entitled, "Schizophrenia and Genet-

Table 11-5 Risk of Schizophrenia among Relatives of Schizophrenic Individuals

Relation to propositus	Coefficient of relationship	Observed morbidity* (%) for schizophrenia
Unrelated	0.00	0.85
Stepsib†	0.00	1.80
Half-sib	0.25	3.20
Sib	0.50	7.0–15.0
Parent	0.50	5.0–10.3
Child‡	0.50	7.0–16.4
Grandchild	0.25	3.0–4.3
Nephew or niece	0.25	1.8–3.9
Cousin	0.125	1.8–2.0

* Ranges of risks represent summaries of many studies, where morbibity data sources are cited, with the coefficient of relationship added.

† Stepsibs are genetically unrelated. They are the offspring of two spouses from their marriages to other partners.

‡ Of one schizophrenic parent. Coefficient is 0.71 for a child of two schizophrenic parents.
Source: Modified from Stern (1973).

ics: Where Are We? Are You Sure?'' by Gottesman, 1978. (We also refer our readers to the whole issue of the Schizophrenia Bulletin, vol.2, no. 3, 1976.)

1   No specific genotype has yet been identified with schizophrenia; attempts—even ours—to fit assorted models of genetic transmission are at best, ambiguous.

2   Risks of schizophrenia to the relatives of index cases increase with degrees of genetic relatedness (25%, 50%, 100%) even without shared environments.

3   Risks to relatives of schizophrenics vary with severity of the proband's schizophrenia, the number of other relatives already affected, and, in the case of offspring, with the status of the other parent, for example, from 1.8 percent (simple schizophrenic × normal) to 46 percent (schizophrenic × schizophrenic).

4   Gender/sex is not relevant in schizophrenia except for age at onset (earlier in males): paternal half-siblings of index schizophrenic adoptees are as often schizophrenic as maternal half-siblings; offspring of male schizophrenics are as often schizophrenic as offspring of female schizophrenics; the sex ratio of schizophrenia is even by the end of the risk period; female MZ twin pairs are not significantly more concordant than MZ males and opposite-sexed fraternals are as concordant as same-sexed fraternals.

5   Identical twin concordance rates for schizophrenia are about three times those of fraternal twins and at least 30 times the general population rate.

6   More than one-half the MZ pairs in recent studies are discordant for schizophrenia despite sharing all their genes; MZ and DZ twins as such are not at a higher risk for schizophrenia than are singletons.

7   Identical twins reared apart from childhood are as concordant for schizophrenia as those reared together.

8   Children of normal parents fostered into homes where a foster parent became schizophrenic do not show an increased rate of schizophrenia.

9   Children of schizophrenics placed for adoption when very young still develop schizophrenia at rates considerably higher than general populational rates, sometimes as high as for those children reared by their affected parents.

10   Adoptive relatives of schizophrenic adoptees do not have elevated rates of schizophrenia, but biologic relatives of the adoptees do have high rates.

11   Schizophrenia occurs in both industrialized and undeveloped societies; in the former the lifetime risk (with conservative diagnostic standards) is usually about 1 percent by age 55.

12   In large urban communities there is a social class gradient in the prevalence of schizophrenia, most of which can be attributed to a downward social drift of predisposed persons.

13   No environmental causes have been found that will even with moderate probability produce schizophrenia in persons who are unrelated to a schizophrenic.

We conclude by recommending the survey of genetic approaches to the study of schizophrenia compiled by Erlenmeyer-Kimling (1978a), in which existing genetic evidences are presented as a potential tool in research efforts.

It must finally be noted that the evolutionary consequences of the monogenic and polygenic models are rather different. The monogenic model would only be sustainable if heterozygote advantage or some other mechanism occurred to maintain the population incidence at the high levels found, since the hypothetical gene would occur too frequently to be explained by mutation alone. This means that a schizophrenic gene in a heterozygous form may give its bearer some advantage (Caspari, 1961; Huxley et al., 1964).

According to Erlenmeyer-Kimling (1978b, in translation):

> There is no evidence that the mothers of schizophrenics have a greater number of offspring than reproducing women in the general population and there appears to be no evidence supporting the idea that siblings of schizophrenic patients have a greater than average reproductive rate. In fact, the reverse may be, or at least may have been, true, with the siblings producing fewer children than expected for their time and place. The problem of selective forces that are responsible for maintaining schizophrenia in the population is thus unsetttled.

The biochemistry of schizophrenia has been the subject of extensive research in recent years. The resulting literature details the intricate methodology applied to the investigation of the various theories currently proposed (see examples in Kety, 1967; Omenn and Motulsky, 1972; McGaugh, 1972; Keith et al., 1976). Two classes of biochemical variables with particular relevance to coping with stress are the catecholamines and the adrenal steroids. Pollin (1971) has suggested that a high level of catecholamines is related to the schizophrenic genotype, but the elevated steroid values are related to the schizophrenic phenotype. However, Fuller and Thompson (1978) suggest that these data should be regarded with caution. Even so, the biochemical and pharmacological approach may slowly unravel the necessary brain chemistry needed to understand schizophrenia, and as a byproduct, the controversies as to modes of genetic control should then be resolved. As for the other mental illnesses discussed, biometrical methods are of great importance in ascertaining a genetic component initially, but as soon as actual loci can be identified and followed, such methods become less important and can be replaced by pedigree studies.

11-11 SEX

As a relief from the complexities of schizophrenia, we offer a stellar lone example of the intimate relation between that which is passed from generation to generation via gametes and that which we know as behavior—the disorder called testicular feminization or androgen insensitivity. It is male (affecting only that sex) pseudo (false) hermaphroditism (the characteristics of both sexes are combined in one individual). Afflicted individuals perceive themselves, and are accepted by their families from birth on, as females. They have a normal life span. When they are wed, it is to men, but their unions are wholly sterile. They possess blind-ended, small vaginas accompanied by the full feminine development of the mammary glands. In addition, however, testes are present; their herniation is often the initial symptom invoking the diagnostic genital-pelvic examination; amenorrhea is the next symptom. Bergsma (1973) cautioned that "the diagnosis should be suspected in all girls presenting with 'inguinal hernia(s)'; before exploration and herniorhaphy 'surgery' a buccal smear 'to count the number of X chromosomes' should be performed." (Also see Rimoin and Schimke, 1971.)

The pedigrees constituting Figure 11-6 are consistent with two types of inheritance, and it is still impossible to distinguish clearly between them:

An autosomal dominant such that a *Trtr* female (*tr* from the word transformed) is normal, though development of secondary sexual characteristics may be tardy and the individual often has sparse axillary and pubic hair. If she conceives XY offspring, half of them are intersexual because a *Trtr* male is feminized.

An X + X^{tr} female appears as above and is essentially normal and fertile. An $X^{tr}Y$ male is feminized.

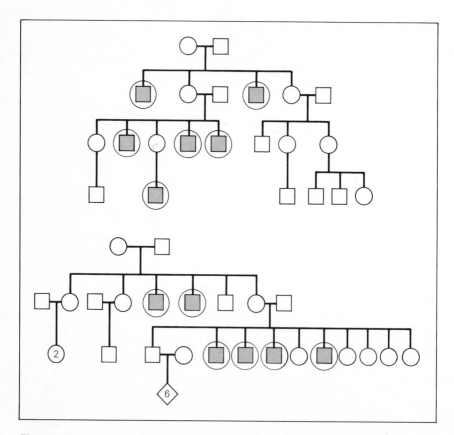

Figure 11-6 Two types of inheritance of testicular feminization. Afflicted individuals, depicted as squares (male) inside circles (female), are genetically male but phenotypically female. The gene appears to be dominant and autosomal, or recessive and sex-linked. If autosomal, it is sex-limited (to genetic males) in expression. (*From* Principles of Human Genetics, *3d ed., by Curt Stern, W.H. Freeman and Company. Copyright © 1973.*)

No individual homozygous for this gene is known nor have linkage studies proved fruitful (Hölmberg, 1972), so under what circumstances would it be possible to choose between these two alternatives?

Since Morris first described this syndrome in 1953, its cause has been postulated to be the insensitivity of peripheral target tissues to androgenizing secretions of the testes and adrenal glands; there is no response to circulating testosterone, even to injections of this hormone. There is a sharp deficiency of germ cell maturation in the small nodular testes, a lack of male secondary sexual characteristics including those of personality, total sterility, and the absence of facial and body hair. A more recently described milder autosomal dominant form is rectified at puberty when male virilization occurs (Imperato-McGinley et al., 1974).

And what of the psychotherapeutic aspects of this gross syndrome? What sex does or should a feminized male relate to? Gayral et al. (1960) commented on 11 cases in five families (note the incidence of repeats here). In all instances the family physicians elected not to inform their patients of the true, currently incurable, nature of their affliction. Instead, the "women" were told of ovarian abnormalities resulting in lifelong sterility. Responses ranged from ostensible indifference to overcompensation to neurotic depression. In every case, the patient was feminine in appearance, bearing, and temperament despite the absence of any aspect of menstruation. The sexual drive and capacity for con-jugality were normal and the authors concluded that too much information, however accurate, would only result in at the least confusion and at the most severe depression. The parents of feminized males might therefore be coun-seled to the extent of citing the possibility of the production of additional similarly sterile offspring without being given more information. We wonder how our readers judge such profound decisions. Should all psychotherapeutic efforts be directed toward reassuring the individual afflicted with testicular feminization that she is a social, psychological, and emotional human female?

11-12 CONCLUDING REMARKS

Considering the diversity of traits discussed, analysis is relatively simple when there is an apparent correlation between enzymes and behavior, and for vari-ous other rare defects that can be followed in pedigrees. This applies also to certain sensory defects, for example, PTC insensitivity and color blindness. However, we must note the complexity of the whole phenotype "taste" and the difficulty of its genetic analysis. The same comment applies to vision, hearing, and smell.

For the above traits, complications due to family environments are not generally great; however, when we turn to speech defects and alcoholism, such complications assume importance. At this stage we see a need to consider "cultural" as well as biological factors and the utility of adoption studies in disentangling the two issues to be considered further in the next chapter. The oddest example of cultural inheritance is kuru, where problems of disentangling biological and cultural factors were made difficult by the problems of working with a primitive group, at least until the viral nature of the disorder was recognized.

The heredity-environment issue becomes even more complex when we turn to mental illness, since in many cases defining the phenotype itself becomes an additional problem (e.g., schizophrenia). If more biochemical bases for such conditions are eventually made known, they will greatly aid our comprehension. For now though, this may be too much to hope for, but it does provide a noble goal for researchers. In any case progress is not likely to come quickly, in spite of enormous resources invested, especially because of the difficulties of collecting and interpreting data in a species within which breeding experiments cannot be undertaken.

These sorts of complications in themselves have led to a most profuse documentation, especially on the mental illnesses. In many cases about all we can say is that there is a heritable component (although frequently just about every possible genetic hypothesis has been promulgated at one time or another). In spite of this voluminous literature, we have here only provided comments appropriate for a general text on behavior genetics. And since this means a very selective and incomplete coverage, we present below a selected assortment of references.

SUMMARY

Detailed work on traits such as speech disorders and alcoholism, indicate influences of genotype and environment, but it is often difficult to unravel the relative importance of these two components. This is because the methodologies of human studies are extremely difficult, because appropriate crosses cannot be carried out as in experimental animals. For many disorders, e.g., the mental disorders, epilepsy, manic-depressive illness, and schizophrenia, hypotheses are many and varied as to genetic bases; they range through an array of single and polygenic interpretations for the conditions considered. Biometrical methods are of great importance in establishing a genetical component. However, as soon as actual loci can be identified and followed, pedigree studies replace such methods. There are many mutants having a wide array of behavioral and physiological effects in humans which can be followed in pedigree studies. These include simple traits such as color blindness and the ability to taste phenylthiocarbamide, for which populations are polymorphic.

GENERAL READINGS

DeFries, J. C., and R. Plomin. 1978. Behavioral genetics. *Ann. Rev. Psychol.* **29**:473–515. This paper is an amply documented review of it all, placing topics touched upon in our eleventh chapter within the context of behavioral genetics.
Gottesman, I., and J. Shields. 1972. *Schizophrenia and Genetics. A Twin Study Vantage Point.* New York: Academic. This is the most authoritative compilation of what the study of twins offers to efforts to clarify the etiology of schizophrenia.

Penrose, L. S. 1963. *The Biology of Mental Defect,* 3d ed. London: Sidgwick & Jackson. We list this for its value as an historical review.

Rosenthal, D. 1970. *Genetic Theory and Abnormal Behavior.* New York: McGraw-Hill. This is concise and the most useful of its sort; see our eleventh chapter for references to this well-tabulated survey.

Wynne, L. C. (ed.). 1978. *The Nature of Schizophrenia.* New York: Wiley. This is a compendium encompassing the psychology and psychiatry of the illness.

Homo sapiens: Continuous Traits

12-1 INTELLIGENCE: GENOTYPE AND ENVIRONMENT

The attribute most important for achievement in school is what both psychologists and lay people call *intelligence*. Motivation, personality, and interpersonal relations at home and in the school are also significant factors, affecting both achievement in school and performance on tests designed to measure "intelligence." Empirical correlations have established that the IQ test as a measure of intelligence reflects ability to learn in school in many societies. It is clear too, from their academic performance, that schoolchildren vary in their capability for learning, especially abstract learning. The IQ score is an attempt at quantifying intelligent behavior (Section 7-3). Stern (1973) wrote: "Intelligent behavior is regarded as behavior which, on the basis of inherited capacity, makes good use of the social inheritance, such as language and numbers or scientific and moral concepts." Psychologists have divided mental abilities into distinct, so-called primary abilities, such as ability to visualize objects in space, to memorize, or to reason inductively. There may also be a general ability underlying intelligence, in addition to these primary abilities. Since primary abilities may vary somewhat independently of each other, individuals with the same overall IQ may differ in their mixtures of primary abilities. Although most work on intelligence is in terms of single scores obtained from intelligence tests, primary abilities are emphasized in some newer studies.

First considered are generalized intelligence tests. The designers of intelligence tests have attempted to make them independent of environmental influences within a given society. The interpretation of results of a test given to individuals in different societies is highly complex, since a different society implies a differing environment at least, and possibly a different average genotype. Refer to Section 7-4 for a discussion of large genotype-environment interactions obtained in rats, which could be quantified, since both genotype and environment can be defined for experimental animals. This is not possible for human beings. Even within a fairly homogeneous group it seems impossible to achieve complete freedom from nongenetic influences. Human intelligence always operates in a cultural organization, and so it may not be possible to obtain tests that are entirely culture free. One aim of investigators has been to devise "culture-fair" tests in which the effects of cultural differences for groups within a given society are minimized. Even so, factors such as the greater eagerness of middle-class parents than of lower-socioeconomic-class parents for the intellectual success of their children seem difficult to eliminate. Indeed, in the search for a culture-free test, we may eliminate not only environmental sources of variation but also genetic sources.

Despite these difficulties, some conclusions have been reached. Erlenmeyer-Kimling and Jarvik (1963) carried out a literature review of IQ and certain other general intelligence tests and calculated correlation coefficients between various groups of individuals reared together—unrelated individuals, foster parent and child, sibs, monozygotic (MZ) and dizygotic (DZ) twins. Another set of correlation coefficients was calculated between various groups of individuals reared apart—unrelated individuals, sibs, and MZ twins. The data were based on 52 separate studies, and the approximate median correlation coefficients obtained for each relation group are given in Table 12-1.

Recall from Section 6-9 that the coefficient of relationship is the coefficient of the additive genetic variance (V_A) in the covariance between relatives and

Table 12-1 Approximate Median Correlation Coefficients Between Related and Unrelated Persons Reared Together and Apart for Intelligence Test Scores, and Coefficients of Relationship for Each Category

Persons	Reared apart	Number of studies	Reared together	Number of studies	Coefficient of relationship
Unrelated	−0.01	4	0.23	5	0
Foster parent and child			0.20	3	0
Parent and child			0.50	12	0.50
Sibs	0.40	2	0.49	35	0.50
DZ twins					
Like sex			0.53	9	0.50
Unlike sex			0.53	11	0.50
MZ twins	0.75	4	0.87	14	1.00

Source: Erlenmeyer-Kimling and Jarvik (1963).

reflects the share of genes due to common ancestry. An observed correlation coefficient close to the coefficient of relationship reflects a heritability close to unity if environmental complications can be eliminated and if the dominance variance is small (where relevant). Table 12-1 shows that the closer the degree of relationship, the higher the correlation coefficient, in both the reared-apart and reared-together categories; this implies a likely genetic component. However, comparing those reared apart with those reared together reveals an environmental component, since for the three cases where comparisons are made, the correlation coefficients for those reared together are greater than for those reared apart. The difference is particularly large for unrelated persons, for whom a correlation coefficient of zero would be expected; for unrelated persons reared together the coefficient is +0.23, a figure that shows the effect of environment quite clearly. Thus we can conclude that intelligence is controlled by heredity and environment but that heredity may be more important. This is the conclusion we arrived at in Sections 7-3 and 7-4, where some twin studies are considered in greater detail. The genotype-environment correlation of +0.25 obtained using Cattell's multiple abstract variance analysis (MAVA) method (Section 7-4) shows the relevance of environment; it is also close to correlations obtained between intelligence and social status.

Jinks and Fulker (1970) carried out a full biometrical analysis of a number of sets of IQ data and found dominance to be of importance for high IQ. This implies that there has been considerable directional selection for increased intelligence during human evolutionary history. In particular, an analysis of the IQ scores of 3,558 individuals in a pedigree study carried out by Reed and Reed (1965) on mental retardation agrees with a dominance hypothesis (Eaves, 1973). In fact, assuming a realistic correlation between mates (positive assortative mating) of 0.3 and complete dominance, Eaves computed $V_A = 0.43$, $V_D = 0.215$, and $V_E = 0.18$. The importance of assortative mating is discussed in Section 7-5; it is necessary merely to reiterate here that recent biometrical-genetic analyses of intelligence invoke assortative mating as a significant factor.

Thus far we can conclude that IQ is under genetic control to quite a large extent. Not only is additive gene action relevant but so is directional dominance for high IQ. Furthermore, the value of the additive genetic variance is inflated unless a component for assortative mating is estimated separately. The distribution of IQ in populations is continuous, so it is likely that many genes are involved. Under various simplifying assumptions, the probable number of genes can be estimated. The estimates vary from 22 genes, with an average of 100 genes (Jinks and Fulker, 1970). There is therefore no doubt that IQ is under polygenic control and should show all the features of a polygenic trait, although environmental effects are likely to be larger than for morphological traits. Even so, certain rare single genes such as the one responsible for phenylketonuria (Section 2-1) can reduce IQ dramatically. The same is true of certain chromosomal abnormalities such as Down's syndrome (Section 4-3).

Penrose (1963) showed the effect of these more discrete abnormalities on IQ by using a regression approach (Section 6-10). One group of fathers with an average IQ of 117.1 had children with an average IQ of 109.1. Another group of fathers with an average IQ of 86.8 had children with an average IQ of 92.0. The IQ of the children, therefore, on average regressed nearly one half the way toward the mean of the population. If IQ were determined entirely by additive genes, a regression of one half the way to the population mean would be expected. For sibs, ignoring the complication of dominance, a similar result is expected. For half-sibs, nephews, and nieces, who are more remote relatives, a regression three-quarters of the way toward the mean is expected. For persons having IQs of 50 and above, observed data are in good agreement (Table 12-2).

However, propositi with IQs less than 50 have relatives with IQs considerably higher than expected on an additive gene hypothesis. The interpretation is that these retarded individuals have low IQs because they are homozygous for deleterious recessive genes or are karyotypically abnormal. In other cases, new mutations (mainly dominant) or environmental accidents such as birth trauma may be responsible. In all cases, these rather discrete but rare events lead to a breakdown in additivity. Therefore, for very low IQs, genes of major effect, chromosomal aberrations, and environmental accidents may play a part, in contrast with IQs closer to the expected mean of 100, where polygenic inheritance is normal. This conclusion has been confirmed in other studies such as that of Roberts (1952) based on the sibs of retarded individuals.

This survey of intelligence as assessed mainly by the IQ test shows clearly that genetic and environmental influences play a part. Since our conclusion so far is in favor of a considerable degree of genetic determination of IQ, it seems appropriate to conclude by seeking out various situations in which specific environmental influences might predominate. One important category is adopted children. The review of Erlenmeyer-Kimling and Jarvik (1963) gave a median correlation between foster parent and child of 0.20. This can be argued as resulting from (1) the similar environment of the foster parent and child and possibly (2) selection on the part of adoption agencies for similarity of the foster parents to the biological parents (for example, the chances are that the presumed better-endowed children would be placed in similarly better-endowed adoptive homes). The obvious genetic experiment would be studying adopted children with no open or hidden selective placement.

Table 12-2   Mean IQs of Retarded Patients and Their Relatives

				Relatives' mean IQ	
	Type of relation to patient	Number of pairs	Patients' mean IQ	Observed	Expected on additive gene hypothesis
Patients with IQ ≥ 50	Sib	101	65.8	84.9	82.9
	Half-sib, nephew, niece	143	63.2	89.5	91.8
Patients with IQ < 50	Sib	120	24.2	87.4	61.1
	Half-sib, nephew, niece	90	33.3	95.1	83.3

Source: Penrose (1963).

Despite these difficulties, it is important to consider studies on adopted children. An obvious hypothesis is that if heredity has something to do with IQ, adopted children should be less similar to their adoptive parents than the children of a control group are to their biological parents. For a group of adopted children in Minnesota homes (Table 12-3) there was a continuous decline in mean IQ with the decline in the occupational status of the father from the professional to relatively unskilled occupations. This decrease in IQ was from 113 to 108, which is a rather narrow range, even if we attribute it merely to an environmental effect on test performance. In contrast is the control group of children reared by their biological parents: here there was also a steady decline in IQ according to the occupational status of the father, but the range was three times as great, from 119 to 102. This latter range shows a much more pronounced correlation with the occupation of the father than is the case for the adopted children. In conclusion, it seems reasonable to assume that the greater differences between the scores of biological children compared with those of adopted children are due to the fact that the biological children resemble their parents more than do the adopted children, because they have inherited parental genes.

A particularly instructive analysis carried out by Skodak and Skeels (1949) comes from the study of parent-child resemblance in intelligence during the development of the child. The performances of a group of children adopted during the first months of life were measured and were correlated with educational level of biological mothers and fathers and of adoptive mothers and fathers (Figure 12-1). For comparison, data are available from another study on the correlation of the IQs of children and their biological parents, in whose homes they had been reared (Honzik, 1957). Up to about 2 years of age, there was little correlation between the performance of the child and the education of either the biological or the adoptive parent, irrespective of whether the child was reared by his own or by adoptive parents. However, as age increased, a strong rise in the correlation between the IQ of the child and his biological parents occurred, irrespective of whether the child was reared by his own or by adoptive parents. This correlation reached 0.3 by about 4 years and increased little thereafter. In complete contrast, however, was the low and declining

Table 12-3   Mean IQ of Adopted and Biological Children According to Fathers' Occupation

Occupation of father	Adopted children		Biological children	
	Number	IQ	Number	IQ
Professional	43	112.6	40	118.6
Business, management	38	111.6	42	117.6
Skilled trades and clerical	44	110.6	43	106.9
Semiskilled	45	109.4	46	101.1
Relatively unskilled	24	107.8	23	102.1

Source: Stern (1973).

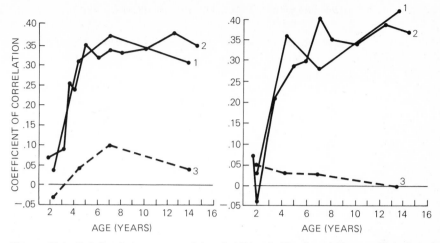

Figure 12-1   Relation between parents' and children's IQs. Correlation coefficients between education of biological and adoptive parents and IQs of children, according to ages of the children. *Left:* Correlation between child's IQ and mother's education. *Right:* Correlation between child's IQ and father's education. The three graphs in each chart are for (*1*) child reared by own parents and biological parent; (*2*) adopted child and nonrearing biological parent; (*3*) adopted child and adoptive parent. (*Graph 1 data from Honzik, 1957; Graph 2 and 3 data from Skodak and Skeels, 1949.*)

correlation between children and their unrelated adoptive parents. It is difficult to escape the conclusion that the intelligence of a child does indeed depend on the genes received from his biological parents. Studies of adopted children, while perhaps showing a small environmental component in the determination of intelligence, clearly show the much greater importance of genetic influences.

12-2   IQ, FAMILY SIZE, AND SOCIAL CLASS

We now consider the relation between IQ and family size. Several family studies have shown fairly consistent negative correlations between intelligence and family size. Correlation coefficients in the region $r = -0.20$ and -0.30 have been found. Similarly, a negative correlation may occur between birth order and IQ based on data from Dutch conscripts (Belmont and Marolla, 1973). Table 12-4 presents data based on a study of IQ and family size carried out at the Minnesota State School and the Dight Institute of Human Genetics (see Maxwell, 1969, for a discussion of Scottish data showing similar results). If studies are confined to particular social classes, smaller but still negative correlations again occur. From these results, predictions have been made that the average IQ of the population should drop from two to four points per generation. However, this has not happened; indeed, there has been a slight tendency for the reverse to occur. The explanation lies in the omission in much of the literature of families with no children. Thus many studies are biased, since both infertility and the probability of nonmating are ignored. Higgins, Reed, and

Table 12-4   Relation Between Family Size and IQ of Children

Family size (children per family)	IQ of children	Total number of children studied
1	106.37 ± 1.39	141
2	109.56 ± 0.53	583
3	106.75 ± 0.58	606
4	108.95 ± 0.73	320
5	105.72 ± 1.15	191
6	99.16 ± 2.17	82
7	93.00 ± 3.34	39
8	83.80 ± 4.13	25
9	89.89 ± 2.94	37
10	62.00 ± 7.55	15

Source: Data from Higgins, Reed, and Reed (1962).

Reed (1962) showed that more than 30 percent of persons whose IQs are 70 or less have no children, compared with 10 percent of those with IQs in the range 101 to 110 and only 3 to 4 percent of those whose IQs exceed 131 (Figure 12-2). This contrasts sharply with the data in Table 12-4. Presented another way (Table 12-5), the reproductive rate of all siblings in families including unmarried sibs is low for IQ values less than 55 and steadily increases to nearly three children for IQ values of 131 and above. Based on a consideration of figures of this nature, Higgins, Reed, and Reed argue that the IQ level of the whole population should remain relatively static from one generation to the next, and certainly should not drop to any degree. The negative correlations disappear

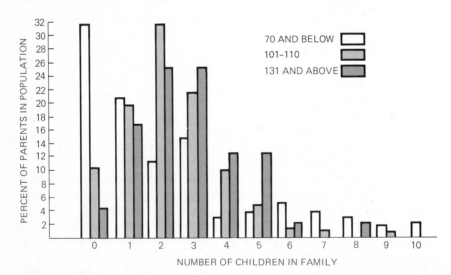

Figure 12-2   Family size and IQ. Distribution of size of family according to IQ of parents for three IQ categories: 70 and below, 101 to 110, and 131 and above, measured in percentages. (*After Higgins, Reed, and Reed, 1962.*)

Table 12-5 Relation Between IQ and Reproductive Rate of All Siblings Including Unmarried Brothers and Sisters

IQ range	Average number of children	Number of children
0–55	1.38 ± 0.54	29
56–70	2.46 ± 0.31	74
71–85	2.39 ± 0.13	208
86–100	2.16 ± 0.06	583
101–115	2.26 ± 0.05	778
116–130	2.45 ± 0.09	269
131 and above	2.96 ± 0.34	25

Source: Data from Higgins, Reed, and Reed (1962).

when a more complete sample is considered. This result points out a continuing difficulty in the handling of human data—the problem of the exact constitution of the sample used.

Indeed, there are many factors affecting IQ whose precise mechanisms of influence are not understood. One related to family size is the well-documented decrease in IQs of twins by about five points and triplets by about nine points, compared with single births. We know that twins have a lower birth weight and higher mortality in the perinatal interval. Hence the IQ effect could be due to prenatal damage. But twins surviving the perinatal death of a cotwin have a normal IQ that contradicts this and that suggests that the cause of the IQ drop lies in postnatal factors. These may be related to the inevitably diminished parental care given to each twin or to more complex psychological factors arising from the rather unusual lives of twins, especially if they are identical. Whatever the explanation, our understanding of genotype and environment in the control of IQ will be assisted by detailed studies of this type.

A few papers (Waller, 1971; Gibson and Mascie-Taylor, 1973) have looked at social mobility in relation to the discrepancies between the general intelligence of sons and the social class into which they were born. Waller's (1971) study involved 173 males and their 131 fathers who were representative of the nonfarm white population of Minnesota. He subdivided the population into five social classes; the average IQs ranged from >120 for the highest social class to 81 for the lowest. Waller's findings support the hypothesis that social mobility is correlated with the discrepancies between the general intelligence of sons as measured by IQ and the social class into which they were born. Figure 12-3 shows the relation between the percentage of sons moving up or down from the father's social class and the differences in father-son IQ. As the difference in IQ score (written as son minus father) increases, so does social mobility. Indeed for IQ score differences of 30 or more the mobility is in the 80 percent region. Therefore, differences in ability, which from the evidence cited must have a considerable genetic component, provide a situation leading to considerable mobility among classes. In this way social classes in an open society are prevented from congealing into castes.

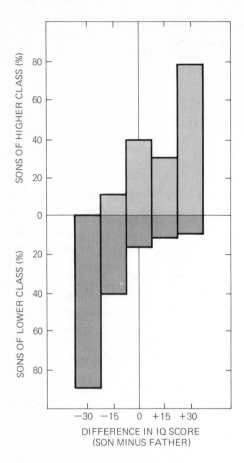

Figure 12-3 Relation between IQ and social mobility. Percentage of sons moving up or down from father's social class, by differences in IQ score. Note that percentages increase as IQ differences increase. (*From Waller, 1971.*)

Gibson and Mascie-Taylor (1973) considered university scientists and their fathers. Again, the differences in IQ between fathers and sons are correlated with the son's mobility on the socioeconomic scale relative to the father's occupation. It is argued that if IQ is correlated with social mobility and has significant heritability, then social mobility will lead to nonrandom transfer of genes from class to class. Hence, it is expected in theory that social classes will come to differ genetically to some extent (Thoday and Gibson, 1970). This, of course, does not exclude the addition of differences due to environmental, including cultural, reasons. The conclusion depends on the assumption that the relation between IQ phenotypes and social mobility implies some significant relation between IQ genotype and social mobility.

12-3 IQ DIFFERENCES BETWEEN BLACKS AND WHITES

One of the features of the range of variation in IQs just discussed is that they are maintained partly by mobility among occupational classes based to some extent on the selection for higher IQ in the higher occupational classes. In the United

States blacks and whites enjoy no comparable mobility; skin color is an effective bar to mobility among races. Let us then look at the IQs of blacks and whites. In North American samples the average IQ of blacks is about 85, while the average IQ of whites is about 100. Many studies have shown results of this nature. Data obtained in one such study, based on IQ tests given to 1,800 black elementary schoolchildren (Kennedy, Van De Riet, and White, 1963) in the southern United States, are shown in Figure 12-4. The distribution is compared with a 1937 sample of the U.S. white population. In this case the mean IQ difference is 21.1, which is fairly extreme, most differences being in the range 10 to 20. Even though there is a considerable degree of overlap between the two distributions, about 95.5 percent of blacks have IQs lower than the mean white IQ of 101.8, and 18 percent of blacks, compared with 2 percent of whites, have IQs less than 70. The IQ difference is usually less for the northern states than for the southern states, and clearly many intangible factors are involved. However, the qualitative point that blacks have lower mean IQs than whites is generally a reproducible result.

Figure 12-4 shows clearly that individual differences in IQ *within* any one race greatly exceed differences between races. The white distribution is, however, more spread out, since the standard deviation for blacks is 25 percent less (12.4 compared to 16.4 for whites). This is an observation characteristic of a number of IQ studies. Even so, there is no denying that the mean IQs of blacks and whites differ; but can this be interpreted genetically? Some writers have

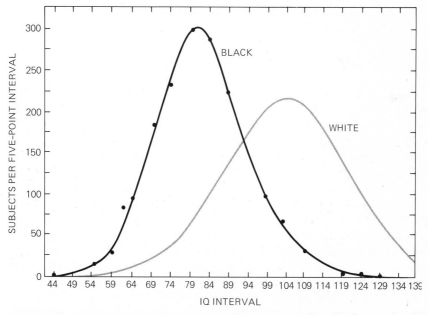

Figure 12-4 IQ in U.S. blacks and whites. Values for blacks were obtained from 1,800 southern schoolchildren. Values for whites reflect "normative" sample of white population. (*From Kennedy, Van De Riet, and White, 1963.*)

asserted that the answer is yes. Jensen (1972, p. 163) and certain others argue that the various lines of evidence produced "make it a not unreasonable hypothesis that genetic factors are strongly implicated in the average Negro-white intelligence difference." Quantitatively, Jensen (1973) believes that between one-half and three-quarters of the average IQ difference between U.S. blacks and whites is attributable to genetic factors. The remainder is due to environmental factors and their interaction with the genetic factors. Jensen has some support for his views, but it comes as no surprise that others argue the differences are largely environmental (Pettigrew, 1971; also Bodmer and Cavalli-Sforza, 1970; 1976).

As we have seen, stratification of occupations in whites is maintained with social mobility across classes. There are, however, no direct comparisons for the mean IQ differences between blacks and whites because of the effectiveness of skin color as a bar to mobility across classes. What then of the environments of blacks and whites? The predominantly black schools of the United States are generally less effective than the white schools, so that an equal number of years of schooling does not mean equal educational attainment. A number of students of child development have noted the developmental precocity of black infants, particularly for motor behavior, which as Jensen (1972) argues could hardly be environmental, since it is present in 9-hour-old infants. This behavioral precocity is paralleled by certain physiological indices of development such as bone development and brain-wave development. Yet after a few years the blacks drop behind (see Coleman et al., 1966). Just as average school environment may differ for blacks and whites, so too may the home environment, in that blacks frequently live in economically deprived areas. The very early home environment may be of substantial importance for intellectual development. Some data clearly show the detrimental effects of severe sensory deprivation very early in life (see Pettigrew, 1971, and below).

Coleman et al. (1966), in a survey of the entire United States, assessed several environmental variables and socioeconomic indices generally thought to be major sources of environmental influence in determining individual and group differences in scholastic performance. Included were factors such as reading material in the home, cultural amenities in the home, structural integrity of the home, foreign language in the home, preschool attendance, parents' education, parents' educational desires for the child, parents' interests in child's schoolwork, time spent on homework, and child's self-concept (self-esteem). Coleman et al. found all these factors to be correlated in the expected direction with scholastic performance within each of the racial groups studied. But comparisons among groups show that the most environmentally disadvantaged group is the American Indian, scoring lower on every environmental index than the blacks. The whites are, as expected, highest. However, the American Indian achievement and ability scores exceeded those for blacks on nonverbal intelligence, verbal intelligence, reading comprehension, and mathematics achievement. It is difficult to interpret this result either genetically or environmentally. The only valid way of interpreting results of this nature is to test different races under identical environments. Such data are unavailable. Although superficially identical environments might be created by having black children adopted into white homes and vice versa, even in this case (because of possible prejudice), the environments could not be regarded as identical.

In spite of the difficulty of interpreting such data, there have been recent studies of IQ test performance of black children adopted by white families. Scarr and Weinberg (1976) studied 130 black/interracial children adopted by advantaged white families, characterized as highly educated and above average in occupational status and income. The adoptees were socially classified by education of natural parents and their precise racial categories. These socially classified adoptees, whose natural parents were educationally average, scored above the IQ and the school achievement mean of the white population, with an average IQ of 106, while children adopted during the first year of life scored an average IQ of 111, which is even higher. The value of 106 represents an increase of one standard deviation above the average IQ of 90 of black children reared in their own homes in the north-central region of the United States where this study was carried out. Even so, the IQ scores of the biological children of these adoptive parents were even higher, which is reasonable since the IQ scores of the parents approached 120. But the high IQ scores of the socially classified black adoptees reveals considerable malleability for IQ according to rearing conditions such that in the group adopted during the first year of life, the IQ change may be as much as 20 points.

In considering relative genetic and environmental contributions to IQ, it can therefore be said that the IQ differences among the black/interracial adoptees are due to placement variables, adoptive family characteristics, and genetic background. Because social and biological variables were confounded, it is very difficult to make a clear comparison, and Scarr and Weinberg (1976) suggest that a confounding of genetic and social variables is usual in families. They go on to suggest that genotype-environment correlations are the rule and that they account for a sizable portion of the IQ variance in the general population (as indeed is suggested by the use of Cattell's MAVA method discussed in Section 7-5, Wilson's twin data in Section 7-4, and earlier discussions in this chapter).

Pettigrew (1971), discussing the role of the black, has commented that a black is not expected to be bright and therefore expects to fail; this leads to a lack of self-confidence, a lack of interest in learning, and a lack of progress. Furthermore, Pettigrew found that blacks give more correct answers when tested by blacks than when tested by whites for certain tests incorporating intelligence. Another likely factor is nutrition. Pettigrew quotes a study by Harrell, Woodyard, and Gates (1956) in which dietary supplementation during the last half of pregnancy had directly beneficial effects on the IQ scores of the children later. Indeed McKay et al. (1978) in a study of chronically under-nourished children from Colombian families of low socioeconomic status have shown that combined nutritional, health, and educational treatment between 3½ and 7 years of age can prevent large losses of cognitive ability, with significantly greater effects the earlier the treatments begin. When a group moves from a restrictive environment to a stimulating environment, average IQ is expected to improve (as predicted from Cooper and Zubek's 1958 experiments with rats—Section 7-4). Perhaps the most dramatic evidence comes from the Osage Indians. This group occupied land on which oil was discovered, a circumstance that afforded them a living standard vastly superior to that of other Indians. On performance and language tests they were found to be superior to the level of comparable whites in the area. Similar increases in IQ have been recorded among whites in mountain areas of eastern Tennessee between 1930 and 1940. This was a period during which broad economic, social, and educational improvements occurred, and the average IQ increased from 83 to 93. These broad general IQ increases are clearly consistent with the adoption studies of Scarr and Weinberg (1976) discussed above.

For all these reasons, we find it difficult to agree with Jensen's conclusions; we do not regard it possible to prove his hypothesis that much of the difference between the IQs of blacks and whites is due to genetic causes. Jensen's argument assumes high heritability of IQ in white populations but heritability cannot be generalized across populations or even among environments. (See Chapter 6 for more on heritability.) On the other hand, we do not regard it possible to disprove his hypothesis. The experimental situation needed for proof is not available—a problem of experimental design inherent in work on our own species. If a controversy of this type occurred in laboratory rodents for example, it would have been resolved long ago because genotypes can be replicated and environments controlled. It does, however, seem worthwhile to increase our knowledge of specific genetic and environmental factors that may control IQ.

Finally, how genetically different are races? A quantitative definition of a race is given in Section 5-3, stressing that a race is characterized by gene frequencies differing quantitatively from another race. Based on 25 loci for blood groups and other such genetic markers, Bodmer and Cavalli-Sforza (1976) showed that the genetic difference between racial groups is indeed small in comparison to that within groups (see also Lewontin, 1972), just as IQ differences between groups are small compared with the within group differences. Using discrete genetic markers, it appears that the difference between Africans and Orientals is a little larger than that between Africans and Caucasians, or between Caucasians and Orientals; Caucasians are intermediate between the two groups. (See Bodmer and Cavalli-Sforza, 1976, for an excellent discussion of racial differentiation.) Even so, irrespective of race, throughout our evolutionary history, natural selection would have operated to increase intelligence. Evidence for directional selection in the direction of high IQ has already been presented in this chapter. If the limits to selection are being approached in all human groups, as is likely given the number of generations over which selection has occurred, then similar limits to selection in different races are likely for a trait such as IQ that is apparently under the control of 22 or more genes (Jinks and Fulker, 1970). It could on the other hand, be argued that there may be some differences in sensory, perceptual, and motor processes among races, relatable to the habitats of these races. Studies in this area are, as will be pointed out in Section 12-6, so few that definite conclusions are not possible, even though some racial differences for such traits are known. This discussion is best concluded by a comment by George Gaylord Simpson (1969):

> There are biological reasons why significant racial differences in intelligence, which have not been found, would not be expected. In a polytypic species, races adapt to differing local conditions but the species as a whole evolves adaptations advantageous to all its races, and spreading among them all under the influence of natural selection and by means of interbreeding. When human races were evolving it is certain that increase in mental ability was advantageous to *all* of them. It would, then, have tended over the generations to have spread among all of them in approximately equal degrees. For any one race to lag indefinitely behind another in overall genetic adaptation, the two would have to be genetically isolated over a very large number of generations. They would, in fact, have to become distinct species; but human races are all interlocking parts of just one species.

12-4 PRIMARY MENTAL ABILITIES

A more comprehensive approach to the issue of mental abilities as measured by IQ comes from the construction of tests designed to measure a number of separate abilities (Vandenberg, 1967). One such battery is the Chicago Primary Mental Abilities Tests constructed by Thurstone and Thurstone (1941), which has been used in several surveys. Table 12-6 presents a compilation of the results of four such surveys for verbal, space, number, reasoning, word fluency, and memory scores. The four studies agree on hereditary components as assessed by *H* statistics for verbal score and word fluency score. There is also good agreement for a hereditary component for spatial score (an ability to deal mentally with two- and three-dimensional patterns). A rather lower significance for hereditary factors is apparent for memory score. With regard to the remaining two scores, the British study (Blewett, 1954) disagrees with the American studies. The number score is based on very simple arithmetic tests. No evidence for a hereditary factor was found for the British study, while the other three (American) studies suggested hereditary factors. For the reasoning score the reverse occurred, as Blewett's subjects gave evidence for a hereditary component while none of the three American studies did. Vandenberg (1967) commented on the need for caution in the interpretation of these results because of variations between groups in socioeconomic experience or educational practices and because of simpler interpretations such as variations in sample sizes or different instructions. Even so, on the basis of all the data together, significant hereditary variation is suggested at least for number, verbal, space, and word fluency scores. Vandenberg went on to show that these four components are at least somewhat independent of each other from the genetic point of view. Possibly, educational practices and/or socioeconomic

Table 12-6 *H* Statistics Calculated from Scores of DZ and MZ Twins on Six Items of Chicago Primary Mental Abilities Test

Test Item	Study			
	Blewett (1954)	Thurstone et al. (1955)	Vandenberg (1962)	Vandenberg (1964)
Verbal	0.68*	0.64*	0.62*	0.43*
Space	0.51†	0.76*	0.44†	0.72*
Number	0.07	0.34	0.61*	0.56*
Reasoning	0.64*	0.26	0.29	0.09
Word fluency	0.64*	0.60*	0.61*	0.55*
Memory		0.38†	0.21	

* $P < 0.01$ as measured by an F test consisting of V_{DZ}/V_{MZ}. This test of significance is frequently used by Vandenberg. It is related to the H statistic by $V_{DZ}/V_{MZ} = 1/1\text{-}H$ (see Chapter 7).
† $P < 0.05$.
Source: Vandenberg (1967), where the original sources are specified.

experiences are more important for reasoning and memory than for these four components, leading to a tendency for more ambiguous results.

The finding of four genetic components corresponding to four of the scores is, if substantiated in the future, a result of some considerable significance. It indicates that intelligence is made up of several separate contributions and that the IQ test evaluates an aggregate of these plus other contributions. The trend toward such refined analyses of complex traits should lead to a better understanding of the composition of the evolutionary units underlying human intelligence.

There are many other tests for mental abilities to which subjects have been exposed. Not much can be said about the degree to which the traits examined are under genetic control. Considerable research is being undertaken in this area, especially on Americans of European and Japanese ancestry in Hawaii (DeFries, Vandenberg, and McClearn, 1976), which is important in view of the ongoing controversy about the interpretation of IQ differences between blacks and whites.

12-5 PERSONALITY

Despite a predominant interest in intelligence, there has recently been a growing emphasis on personality. Multifactorial techniques have led to a tendency to concentrate on specific aspects of personality, rather than on personality as a whole. Griffiths (1970) defined personality as "the more or less stable organization of a person's emotional, cognitive, intellectual and conceptual, and physiological behavior which determines to a large extent his adjustments to environmental situations." Defined in this way, intelligence is just one aspect of personality. Advances in the assessment and production of reasonably reliable and valid tests of personality have made the assessment of genetic differences more reliable than previously.

The multifactorial tests include scales purporting to measure specific personality traits. Two examples of such tests are the Minnesota Multiphasic Personality Inventory (MMPI) and the California Personality Inventory (CPI). Gottesman (1965) used the MMPI in a study of 34 MZ and 34 DZ adolescent twin pairs in Minnesota (where a high proportion of subjects were of Scandinavian origin) and in another study of 82 MZ and 86 DZ pairs in Boston. The MMPI consists of 550 questions yielding scores on 10 aspects of personality (Table 12-7). There is reasonable agreement in the rank ordering of the H statistics with the exception of paranoia. Reasonably high H statistics were recorded for social introversion and psychopathy, and also, as might be expected, for the two psychotic scales (depression and schizophrenia). Even so, results varied according to age and sex, and the lack of complete agreement in the ordering of the H statistics argues for an effect due to geographical area or to the origins of the populations studied.

Jinks and Fulker (1970) analyzed neuroticism in the data of Shields (1962) as assessed by a questionnaire designed to give a measure of both neuroticism and extroversion. The subjects were MZ twins brought up together and apart and DZ twins. Estimates of H and E statistics are given in Table 7-4. The general conclusion of Jinks and Fulker was that the data for neuroticism could be explained by a model of additive gene action, with no dominance. This means that intermediate expression for neuroticism is favored; that is, stabilizing selection for an intermediate optimum is likely, extremes being at a reproductive disadvantage. Gottesman (1965) speculated similarly for a number of such traits, since he considered that extremes would be disadvantageous, but previously little evidence was available. Jinks and Fulker found an indication of positive assortative mating in the data, which was not, however, significant. They also considered that cultural and class differences had little or no effect on this major personality dimension.

Table 12-7   *H* Statistics Derived from Scores of MZ and DZ Twins on Minnesota Multiphasic Personality Inventory

Personality trait	Minneapolis study		Boston study	
	H	Rank	*H*	Rank
Hypochondriasis	0.16	7	0.01	10
Depression	0.45	3	0.45	1
Hysteria	0.00	10	0.30	7
Psychopathy	0.50	2	0.39	2
Masculinity/femininity	0.15	8	0.29	8
Paranoia	0.05	9	0.38	3
Psychasthenia	0.37	5	0.31	6
Schizophrenia	0.42	4	0.33	4
Hypomania	0.24	6	0.13	9
Social introversion	0.71	1	0.33	4

Source: Gottesman (1965).

Extroversion was analyzed by the same questionnaire by which neuroticism was assessed. This trait, together with neuroticism, completes a broad two-dimensional view of major personality tendencies as described by Eysenck (1967). The environment was found to be more relevant for extroversion than for neuroticism, in that the introvert phenotype was found to be more modifiable than the extrovert phenotype by the within-family environment. Even so, the degree of genetic determination was high. An interesting point comes from Shields (1962), who discussed how one of a pair of monozygotic twins brought up together assumes the dominant role and thereafter is the leader.

The first study to use factor scores was that of Eysenck and Prell (1951). It marked the beginning of the trend away from single measures to combined measures. Data presented in Table 12-8 for neuroticism, extroversion, autonomic activity, and intelligence show high H statistics especially for neuroticism. Many other studies have been devoted to nature-nurture analyses of personality, especially twin studies (see Mittler, 1971). Eysenck's more recent work is of interest in that some attempt is being made to relate personality to constitutional variables. It is an area of extreme complexity and challenge for both psychologists and geneticists. Not unexpectedly, a recent analysis of twins' personality inventory scores from a test designed to measure psychoticism, neuroticism, extroversion, and the tendency to lie gave genetic variation generally consistent with an additive gene hypothesis (Eaves and Eysenck, 1977). This would reflect stabilizing selection and signify that behavioral extremes are expected to be reproductively less fit than are intermediates.

12-6 SENSORY, PERCEPTUAL, AND MOTOR TASKS

Electroencephalographic (EEG) records are much more similar in MZ twins than in DZ twins. Most early work depended on the visual inspection of EEG records rather than the more precise analyses made possible by the use of computers (Juel-Nielsen and Harvald, 1958). Computer analysis has opened up new possibilities for the study of genetic aspects of the central nervous system. Generally, MZ twins are much more similar than DZ twins, as expected. Mittler (1971) commented that genetic factors could play a more important role in the development of visual-spatial abilities than in the various traits discussed so far, which mainly involve components of intelligence and personality. Precise methods of EEG quantification can be expected from work on evoked cortical

Table 12-8 Intraclass Correlations and H Statistics for Factor Scores of Various Personality Traits

Trait	r_{MZ}	r_{DZ}	H	Source
Neuroticism	0.85	0.22	0.81	Eysenck and Prell (1951)
Extroversion	0.50	−0.33	0.62	Eysenck (1956)
Autonomic activity	0.93	0.72	0.75	Eysenck (1956)
Intelligence	0.82	0.38	0.71	Eysenck (1956)

potentials. Specific signals, such as flashes of light or pure tones, are used, and the exact constituents of the cortical response to these signals are analyzed. For example, Dustman and Beck (1965) reported on a comparison of visually evoked potential to 100 light flashes in 12 pairs of MZ twins, 11 pairs of DZ twins, and a control group of 12 pairs of unrelated twins matched for age. They analyzed the wave components for the first 250 milliseconds and the first 400 milliseconds and compared central with occipital readings. Generally, MZ twins showed higher intraclass correlations than did DZ twins; an H statistic of 0.57 was obtained for MZ twins for the occipital reading for 250 milliseconds. When Lykken et al. (1974) used no less than six EEG parameters on twins, they concluded that "most of the variance in the frequency characteristics of the EEG appears genetically determined."

Variants of sensory and perceptual functions for which a relatively simple genetic mechanism can be found are considered in Chapter 11. A number of detailed tasks involving visual perception have been carried out. As Fuller and Thompson (1960) commented, much of this work was motivated by the idea that afterimages, flicker fusion, and susceptibility to illusions are valid indices of personality. A summary is given in Table 12-9 for comparisons of MZ and DZ twins. A deficiency of much of the data is the failure to determine possible effects of prior experience upon simple perception.

An *afterimage* is assessed by fixating on, for example, a square on a neutral background for a fixed period of time, then judging the size of the afterimage by "projecting" the afterimage on screens at greater or lesser distances than the fixation distance. The data in Table 12-9 are for afterimages projected on screens at 50 and 200 cm after fixation at 100 cm. The H statistics obtained are high to very high.

Table 12-9 Intraclass Correlations and H Statistics Derived from Scores of MZ and DZ Twins on Perceptual Tasks

Task		r_{MZ}	r_{DZ}	H
Size of afterimage	(1)	0.71	0.08	0.68
	(2)	0.68	0.00	0.68
	(3)	0.98	0.22	0.97
	(4)	0.75	0.23	0.67
Eidetic imagery	(1)	0.50	0.10	0.44
	(2)	0.66	0.15	0.60
	(3)	0.67	0.05	0.65
Critical flicker fusion		0.71	0.21	0.63
Muller-Lyer illusion	(1)	0.53	0.39	0.22
	(2)	0.55	0.05	0.52
	(3)	0.51	0.37	0.22
	(4)	0.57	0.28	0.40
Autokinetic phenomenon		0.72	0.21	0.64

Source: Mittler (1971).

Eidetic imagery indices are obtained by using a variety of complex visual stimuli, such as pictures with large colored areas, and recording the degree of image persistence reported by the subject. From the sum of several tests a composite eidetic score can be obtained. Each stimulus is presented with and without a slight flicker. *H* statistics are high, but not on the whole so high as for the size of afterimage. *Critical flicker fusion* also gives a high *H* statistic.

The *Muller-Lyer illusion* is produced by a pair of arrows whose shafts are of equal length, but whose arrowheads point outward or inward (Figure 12-5). The shaft of the arrow with the outgoing heads looks longer, although both shafts are the same length. (See Gregory, 1966, for a discussion of the Muller-Lyer and other illusions.) The subject is asked to judge which shaft seems longer. The *H* statistics for this task are lower than for the others.

In the test for the *autokinetic phenomenon,* the subject is instructed to fixate on a stationary light and describe what he sees. If movement is reported, the subject traces the path and is scored by the length of the line drawn. For this test the *H* statistic is high.

Generally, all these visual perceptual tasks show some genetic basis. The *H* statistics are similar in magnitude to those obtained for mental abilities and personality traits. It is surprising, however, how little work has been done until very recently on the possible genetic basis of such traits. Comprehensive studies such as have been done for mental abilities and personality on a variety of relatives have been little reported for visual perception. We can, however, finally recommend with pleasure: Guttman, 1974; Goodenough et al., 1977; Loehlin et al., 1978; and Rose et al., 1979 for up-to-date reviews on perceptual speed.

For motor skills, twin studies have provided evidence suggesting genetic control—for example, for pursuit motor, spool-packing, and card-sorting tests. It is, therefore, difficult to avoid the conclusion that there must be a substantial genetic component in sensory, perceptual, and motor behavior; whether the

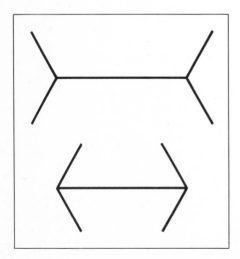

Figure 12-5   The Muller-Lyer illusion. The shafts of the two arrows are the same length.

genetic component is more important than for mental abilities and personality traits is difficult to assess from the evidence reviewed.

Spuhler and Lindzey (1967) discussed variations in sensory, perceptual, and motor processes among races. Although the earliest psychological comparisons of races dealt with simple sensory processes and modes of response, there has, until recently, been little systematic work in this area. Late in the last century it was found that for reaction time to visual, auditory, and tactile stimuli, American Indian subjects had the lowest average latency. They were followed by an African-Caucasian hybrid group, with a Caucasian group the slowest of all to react. Early in this century, the visual acuity of Torres Strait Islanders was found to be superior to that of European groups. Other variations among races for sensory motor processes discussed by Spuhler and Lindzey include weight discrimination, pain threshold, and olfactory acuity. These early studies suggest the possible existence of appreciable racial differences in behavior; little can be said about the degree to which such differences may be genetic. After this series of early studies, the investigation of the more complex processes discussed in previous sections of this chapter predominated in psychology. Only recently has there been any resurgence of interest in sensory-perceptual-motor studies. For example, recent work shows the Muller-Lyer illusion to be almost four times as common among American subjects as among Bushmen subjects. In the face of limited evidence, however, Spuhler and Lindzey (1967) wrote: "If we except PTC tasting, color vision, and certain perceptual illusions, there appears to be little compelling evidence at present either for racial differences or racial equality in simple sensory or motor processes." If the effort spent on analyzing personality and mental abilities is applied to sensory and motor processes, significant advances would surely occur.

12-7 BEHAVIOR AND MORPHOLOGICAL VARIATION

Sheldon and coworkers (1940, 1942) looked into possible relations between human structure (i.e., somatotype) and behavior. He found quite high correlations, but in spite of this, few further studies have been carried out. Lindzey (1967) stressed the reluctance of psychologists generally to give serious consideration to the study of morphology and behavior. Sheldon presented a classification of morphology based on three extreme physical types, with ratings for each dimension derived from a standardized set of photographs. Three extreme phenotypes (somatotypes) are envisaged:

• Endomorphy. The individual high for this component is characterized by softness and spherical appearance associated with an underdevelopment of bone and muscle, a relatively low surface to mass ratio, and highly developed digestive viscera. Since the functional elements of these structures derive primarily from the endodermal layer, the term endomorphy is used.

- Mesomorphy. The individual high for this component is hard and rectangular, with a predominance of bone and muscle, and so is equipped for strenuous and exacting physical demands. The term derives from the dominance of the mesodermal layer in this somatotype.
- Ectomorphy. The individual extreme for this component is linear, skinny, and characterized by a flatness of the chest and delicacy of the body. Therefore, the ectomorph is made up more than the other somatotypes of ectodermal tissue, whence the term ectomorphy is derived. Ectomorphy is a physique poorly equipped for persistent physical action.

Sheldon's classification is not the only one based on body types that has been devised; others are discussed in Hall and Lindzey (1957), Lindzey (1967), and Stern (1973).

Some fairly clear associations exist between behavior and morphology. For example, the frail ectomorph cannot employ physical or aggressive responses with the same effect as the robust mesomorph. Height, weight, and strength put limits upon the adaptive responses an individual can make in a given environment. Lindzey (1967) has quoted evidence for clear and consistent behavioral differences among individuals who vary in morphological development. In general, an individual who is physically extreme in some sense, such as being excessively fat or thin, is exposed to a somewhat different set of learning experiences than someone who is more average physically. Note that "average" varies among ethnic groups.

A striking set of examples comes from the somatotypes of athletes (Carter, 1970). Almost all groups of championship athletes are highly mesomorphic. The most mesomorphic are the weight lifters followed closely by Olympic track and field throwers, football players, and wrestlers. The least mesomorphic men are the distance runners. Women athletes range from the track and field jumpers and runners, who are the least mesomorphic, to the gymnasts, who are the most. It is not surprising that champion performers at various levels of a particular sport exhibit similar patterns of body size and somatotype, but the patterns tend to become more extreme as the level of performance increases. Extremes at the behavioral level correspond to extremes at the morphological level. Conversely, certain somatotypes found in nonathletes are not found at all in groups of championship athletes. Extreme somatotypes in athletes can be made more extreme by training, but training is unlikely to convert a nonathletic somatotype to an extreme athletic mesomorph.

Sheldon (1942) claimed striking associations between morphology and temperament. He selected three components of temperament:

- Visceratonia. Individuals high in this component are characterized by general love of comfort, sociability, gluttony for food, enjoyment of people, and affection. Such a person is relaxed in posture, reacts slowly, and generally is an easy person with whom to interact.
- Somatotonia. A high score is ordinarily accompanied by a love of physical adventure, risk taking, and a strong need for muscular and vigorous physi-

cal adventure. Such individuals are aggressive, with tendencies toward action, power, and domination.

• Cerebrotonia. A high score implies restraint, inhibition, and the desire for concealment. Such an individual is secretive, self-conscious, and afraid of people.

Therefore, we have Sheldon's classification for both physique (structure) and temperament (function). A priori, we can pair them as follows: endomorphy-visceratonia, mesomorphy-somatotonia, ectomorphy-cerebrotonia. Based on 200 subjects in each of the temperament classes, and using scoring systems for physique and temperament, Sheldon obtained the correlations presented in Table 12-10. The results clearly show large positive correlation coefficients for the above pairs and negative correlations elsewhere: there is a marked association between the structure or physical traits of the individual and the expected functional or behavioral qualities. The positive correlations are, however, extremely high and have been criticized by psychologists on the grounds that Sheldon himself executed both sets of ratings. Later studies (for a review see Lindzey, 1967) also show positive associations between morphology and temperament, but at rather a lower level (Child, 1950; Parnell, 1958; Walker, 1962).

For individuals showing criminal behavior, Sheldon found an excess of mesomorphs, in particular endomorphic mesomorphs, among delinquent youths. A number of other surveys (Eysenck, 1964; Lindzey, 1967) confirm this, including one involving female delinquents (Epps and Parnell, 1952). Several investigators have found an association of somatotype with schizophrenia (Heston, 1970; Parnell, 1958); mesomorphs are underrepresented among schizophrenics and ectomorphs are overrepresented. Paranoids on the other hand are often mesomorphic (Parnell, 1958). Finally, for sexual development and behavior, a high ectomorphic rating is associated with later sexual development; in particular, ectomorphs tend to experience first coitus later than either mesomorphs or endomorphs (Martin and Eysenck, 1976).

12-8 CRIMINALITY

In Dostoevski's *The Brothers Karamazov,* the Russian monk advises:

> Remember particularly that you cannot be a judge of anyone. For no one can judge a criminal until he recognizes that he is just such a criminal as the man standing

Table 12-10 Correlation Coefficients Between Physique Components and Temperament Components

	Visceratonia	Somatotonia	Cerebrotonia
Endomorphy	0.79	−0.29	−0.32
Mesomorphy	−0.23	0.82	−0.58
Ectomorphy	−0.40	−0.53	0.83

Source: Sheldon (1942).

before him and that he perhaps is more than all men to blame for that crime. When he understands that, he will be able to judge. . . . If I had been righteous myself, perhaps there would have been no criminal standing before me.

Unsatisfactory home life, poor upbringing, poverty, ignorance, mental deficiency, absence of parents, cultural conflicts, and other environmental inputs have been repeatedly implicated in the cause of criminality and antisocial behavior. What of the genetics?

Stern (1973) asked, in an end-of-chapter exercise:

Among 278 sibs of criminals, Stumpfl found 103 who had a criminal record. This corresponds to 1 criminal out of 2.7 sibs of criminals. Among 62 nonidentical twin partners of criminals, Stumpfl and Kranz found 30 offenders. This corresponds to 1 criminal out of 2.1 nonidentical twin partners of criminal twins. It has been suggested that the last-named higher frequency of criminals (1 in 2.1) as compared to the first-named frequency (1 in 2.7) is due to the greater environmental similarity for twins than for ordinary sibs. (a) What is the statistical significance of the data? (b) What bearing has the answer to the preceding question on the suggested explanation for the different frequencies?

For adult crime, a concordance rate of 71 percent ($N = 107$) has been recently reported for MZ and 34 percent ($N = 118$) for DZ twins. Equivalent figures for juvenile delinquency, however, are 85 percent ($N = 42$) and 75 percent ($N = 25$). These data must not be judged conclusive since it is not possible to disentangle heredity and environment (Eysenck, 1964). The reasons should, at this point in our text, be clear.

The factors discussed below, having some hereditary bases, may be important determinants of the commission or noncommission of crime (Rosenthal, 1971):

• A large number of criminals have low IQs.
• Juvenile delinquents and criminals have a higher incidence of abnormal EEGs than the general population. Rosenthal (1971) cited some prison samples with a 75 percent rate of EEG abnormality. The relation between heredity and both normal and abnormal EEG has been summarized by Vogel (1970; also see Section 12-6). Table 12-11 gives the genetic basis of a number of EEG patterns. On the basis of limited EEG observations, individuals producing either a monotonous tall alpha- or a beta-wave pattern seemed to marry equivalent producers assortatively. For example, 17 out of 56 beta-wave producers married beta-wave producers, while only 5 out of 54 non-beta-wave producers married beta-wave producers. Those who form monotonously tall alpha waves (unusually regular alpha waves of high amplitude) similarly showed an unexpectedly high proportion of marriages with people of the same EEG type. Occipital slow rhythms may be associated with psychopathy; some persons with such an EEG type apparently exhibit an accumulation of psychological peculiarities. If this relation were established, it could be the first physiologically characterized normal variant in human beings showing a qualitative influence on personality without impairing intelligence. Mittler (1971) and Vogel et al.

Table 12-11 Genetic Basis of Variants of Human Electroencephalograms

Rhythm	Genetic basis	Population frequency (%)
Normal alpha (8–13 cps)*	Polygenic	[common]
Low-voltage alpha	Autosomal dominant	7.0
Quick alpha (16–19 cps)	Autosomal dominant	0.5
Occipital slow (4–5 cps)	?	0.1
Monotonous tall alpha	Autosomal dominant	4.0
Beta waves	Multifactorial	5.0–10.0
Frontal beta groups (25–30 cps)	Autosomal dominant	0.4
Frontoprecentral beta (20–25 cps)	Autosomal dominant	1.4

*cps-cycles per second.
Source: Omenn and Motulsky (1972).

(1979) may be consulted on the use of twins to elucidate the genetic components of electrocortical brain activity. Monozygotic twins normally have high concordance for EEG types as would be predicted. (For twin EEG responses to alcohol ingestion, see Propping, 1977, and Section 11-3.)

• Glueck and Glueck (1956) stated that some 60 percent of delinquents are mesomorphs of athletic build. Being more athletic, are such individuals more prone physically to express their dissatisfactions and/or to attempt to eliminate them?

• Some male individuals have an extra Y chromosome and so are XYY (see Section 4-3 on the behavior of males with more than one Y chromosome). Apparently, even an oversized Y chromosome may have deviant behavioral implications. Nielson and Henriksen (1972), studying imprisoned Danish youths, found long Y chromosomes four times more frequently than in control males. Criminal records were also more frequent among the fathers and brothers of these imprisoned subjects than among the fathers and brothers of the controls.

We envision future efforts to elucidate the roles and interplay of matters genetic and of matters environmental in the production of a criminal, much as has been attempted for IQ. For one comprehensive study, see Cloninger et al., 1978, on alcoholism, antisocial personality, and criminality. In many instances it appears that certain physiques and perhaps EEG patterns are associated with criminality. So far, we know more about possible physiques and specific nervous system changes that are associated with criminality than we know about such physical parameters that are associated with intelligence. In other words, we may be closer to the actual genes related to criminality than to the genes related to intelligence.

12-9 GENETIC AND CULTURAL TRANSMISSION OF BEHAVIORAL TRAITS

When discussing heredity and environment in human beings (Section 7-5), mention was made of cultural differences among families and social groups that are maintained by sociocultural inheritance. These differences led to correlations among relatives that are very difficult to distinguish from those due to genetic determination. Several examples of positive correlations between genotype and environment in the 0.2 to 0.3 range have been discussed in this and Chapter 7 for twins, more remote relatives, and comparisons across races in adoption studies. It is not surprising then that, based on an analysis of parent-offspring cultural transmission, Cavalli-Sforza and Feldman (1974) concluded that "cultural" inheritance is almost completely confounded with biological inheritance.

In biological evolution we study rates of evolution, while in cultural evolution the matter for study consists of customs, ideas and beliefs. The rules of biological transmission are understood; indeed this text is based upon such an understanding. The roles of cultural transmission are by contrast poorly understood. Cavalli-Sforza and Feldman began with a blending "inheritance" type model whereby both parents contribute an equal amount to their children but also took into account that many other people contribute to our cultural development. One important conclusion is that the variation between individuals of the same group for a culturally transmitted trait will be lower than if the trait is transmitted biologically. Language is of course the extreme example since considerable homogeneity of the language spoken by people in a given population is necessary for communication. Many human social customs follow a similar pattern of transmission. Another important determinant of behavior is the overwhelming effect of certain individuals—such as social and political leaders and teachers. Indeed Cavalli-Sforza and Feldman consider possible models for the transmission of educability in some detail.

Although the bases of biological and cultural inheritance are completely different, the distinction between the two modes of transmission is not simple. Indeed there is no way of making such a distinction unless one can study adoptions and test the correlations among individuals with both biological and adoptive relatives. More specifically, one should study the correlations between the adoptee and his biological relatives (parents, sibs) on the one hand and those between him and his foster relations on the other. The former of course give the biological and the latter other types of inheritance including cultural. In practice, however, we may not know the biological relatives; an extramarital maternity may be the cause for adoption and the father unknown. Furthermore, since sterility of foster parents is a common cause for adoption, there are likely to be no foster sibs. In addition, adoptees often come from poorer families forming a biassed sample, or adoption agencies use poorly defined criteria for matching the adopting and biological families. Finally, the time after birth at which adoptions occur may influence the adoptee, as we have already seen (Scarr and Weinberg, 1976). These complications, together with the fact that adoptions are relatively rare events, means that such data must be treated with considerable caution as we have already indicated in this book. In spite of these difficulties, adoptions do make essential contributions to testing for biological versus other types of inheritance. They have, however, been used especially for investigating behavioral traits where sociocultural transmission is suspected (Cavalli-Sforza, 1975).

Eaves (1976) has looked at the effect of cultural transmission on continuous variation using a model based on the influence of parents on their offspring. He concludes that whatever the ultimate origin of culturally inherited differences, they are expected to lead to environmental differences between families. This is perfectly reasonable, since parents influence children in a variety of ways through language, social customs, and education. If cultural differences are due in part to genetic differences, he shows that significant covariation of genetic and cultural differences is to be expected, as found in several sets of data discussed already. He argues too that adoption studies are the most powerful way of examining these issues.

It is important to ascertain what traits are likely to show cultural transmission. For example "radicalism" scores on a social attitudes questionnaire displayed a pattern of transmission suggestive of a component of cultural inheritance at least. By contrast, using twins, a personality inventory purporting to measure psychoticism, neuroticism, extroversion, and a tendency to lie provided little support for any major role due to the influence of parents (Eaves and Eysenck, 1977).

One fascinating example of a cultural inheritance model is for disadvantageous traits such as kuru (Section 11-1). This is a disease now thought to derive from a slow virus transmitted by the custom of eating the brain of dead relatives. Here we have the spread of kuru due to pressure on members of the group to conform with this highly disadvantageous cultural procedure. During the peak recorded incidence of the disease in the 1950s, 1 percent of the Fore population of New Guinea died each year, with the prevalence of active kuru reaching 5 to 10 percent of the inhabitants. It is hard to see how a group could long sustain such a disadvantageous custom. (Kuru is thought to have appeared around 1910.) Another example of cultural evolution with much longer term effects is the incidence of lactase deficiency on a worldwide basis (Section 3-4).

As May (1977) comments, "formidable mathematical difficulties stand in the way of a more full understanding of the interplay between cultural and biological parameters." The general equations relating gene frequencies in successive generations are not only nonlinear, but involve frequencies from earlier generations (Feldman and Cavalli-Sforza, 1976). We would agree with May's suggestion that the incorporation of cultural inheritance into quantitative theory is likely to lead to considerable progress, but we stress the point made by Eaves that it is important to find out what traits are likely to show cultural transmission. At that stage, the likely significance of cultural transmission in the field of behavior genetics will finally be clear. Whatever the outcome, the genetic program underlying cultural inheritance must be extremely open. Many of the difficulties in studying human behavior as exemplified in this chapter reside in the input that occurs during the life span of the subject leading to acquired behavior determined by an open program.

SUMMARY

Intelligence within populations is controlled by heredity and environment, but heredity may be more important. This conclusion is derived from intelligence tests on groups of individuals of differing relationships reared together and apart. Similar conclusions occur from considerations of adopted children in comparison with natural children.

Interpreting the well-known IQ differences between blacks and whites is almost impossible, since the appropriate experimental situation of studying both races in identical environments cannot be done. In experimental animals the issue would have been resolved long ago, because genotypes can be replicated and environments controlled.

It is unfortunate that the anthropocentric interest in intelligence has led to a limited literature on the less complex sensory, perceptual, and motor tasks. Significant advances would undoubtedly occur if the effort spent on analyzing personality and mental abilities is applied to these tasks.

Including criminality and antisocial behavior, all continuously varying behavioral traits (except for laterality) are controlled by genotype, environment, and interactions between genotype and environment. An important component of the environment is culture, whereby differences among families and social groups are maintained by sociocultural inheritance. However, the role of cultural transmission in human behavior genetics is as yet far from clear.

GENERAL READINGS

Bodmer, W. F., and L. L. Cavalli-Sforza. 1976. *Genetics, Evolution and Man*. San Francisco: Freeman. Perhaps the best modern account, and presented in a nonmathematical way. A chapter on behavior genetics is included.

Jensen, A. R. 1973. *Educability and Group Differences*. New York: Harper & Row. A presentation of the author's approach.

Loehlin, J. C., G. Lindzey, and J. N. Spuhler. 1975. *Race Differences in Intelligence*. San Francisco: Freeman. A useful overview of this complex area.

Mittler, P. 1971. *The Study of Twins*. Gloucester, Mass.: Peter Smith. A very readable account of twins in behavior-genetic research covering many of the traits considered in this chapter.

Penrose, L. S. 1963. *The Biology of Mental Defect*, 3d ed. London: Sidgwick & Jackson. A classic treatment of mental defect in broadest terms.

Behavior and Evolution

13-1 EVOLUTION

Evolution is the development of organisms through time, by the process of the differential survival in each generation of the progeny of individuals with certain special characteristics. In a recent text Dobzhansky et al. (1977) provide a definition:

> Organic evolution is a series of partial or complete and irreversible transformations of the genetic composition of populations, based principally upon altered interactions with their environment. It consists chiefly of adaptive radiations into new environments, adjustments to environmental changes that take place in a particular habitat, and the origin of new ways for exploiting existing habitats. These adaptive changes occasionally give rise to greater complexity of developmental pattern, of physiological reactions, and of interactions between populations and their environment.

Indeed the theory of evolution is the unifying principle in biology. Until the development of evolutionary theory, the diversity of organisms, their distribution patterns and behavior, adaptations to environments, and interactions with other organisms appeared as an array of uncoordinated observations. The history behind the development of the modern unified theory of evolution is discussed in many places and with many emphases (see the General Readings at the end of this chapter). However, in the 1930s, many of the dissenting theories explaining the causal basis of evolution were fused into a broad synthetic theory of evolution. The synthetic theory was not thought up by one scientist, but itself evolved over 150 years through an accumulation of factual evidence and theoretical conclusions, receiving a major impetus in 1859 when Charles Darwin published his book *The Origin of Species.*

This was certainly the dramatic step in the development of the synthetic theory of evolution. The concept that Darwin recognized and documented was *natural selection;* that is, among the diverse individuals in a population, some have a higher probability of survival than do others. However, Darwin knew nothing of the nature and causes of hereditary variation, and indeed his opinions on the subject were inconsistent. It is one of the curiosities of science and human endeavor that the answers to this dilemma existed in Darwin's time, since Mendel reported on various researches into the inheritance of peas in 1866, which led to the principles of genetics enunciated in modern form in Chapter 2. However Mendel's paper was ignored until its rediscovery more than three decades later, combined with confirmations and breeding experiments carried out in the first decade of this century.

The synthetic theory of evolution is basically the combination of Darwinism and Mendelism. It took 30 years for this synthesis to be realized in the 1930s. Part of the reason for the time taken to develop the synthetic theory was because Darwin worked with continuous traits such as height and weight whereas Mendel and other early geneticists worked with discrete traits such as tall versus short peas. Indeed it was not until the 1930s that the mathematical calculations were completed and understood to show that quantitative traits could be interpreted as being controlled by many separate genes acting simultaneously (Chapter 2). The three scientists effecting this reconciliation were the Englishmen R. A. Fisher and J. B. S. Haldane and the eminent American scientist Sewall Wright.

Natural selection favors certain individuals in a population and, as a consequence, natural selection alters populations of genes controlling traits. This means that the genetic constitution of populations slowly changes due to the action of natural selection. Artificial selection (Chapter 5 and 6), whether it be for various exotic breeds of pigeons or dogs, geotaxis and phototaxis in *Drosophila,* or activity in rodents, acts in the same way. It can be seen from preceding chapters that behavior plays a significant and important role in evolutionary change, both induced by artificial and by natural selection (see especially Chapters 8 through 10). The time is now opportune to discuss more specifically the role of behavior in evolution itself.

13-2 COMPONENTS OF FITNESS IN *DROSOPHILA*

If we define the fitness of a genotype as its relative ability to contribute to future generations, what is the role of behavior in fitness? More precisely, fitness can be regarded as the average number of progeny left by the carriers of a given genotype relative to the number of progeny left by other genotypes. To this we must add the complication that the fitness of a given genotype depends on the environment(s) to which it is exposed. Population cages containing various *Drosophila pseudoobscura* chromosomal karyotypes in pairs usually give stable equilibria at 25°C, since often the heterokaryotypes are found to be fitter than the corresponding homokaryotypes (Wright and Dobzhansky, 1946). This is a situation for which a stable equilibrium is expected, as shown in Section 4-2. Also as expected from theoretical considerations, the stable equilibria occur irrespective of the initial frequencies of the karyotypes. However, at 16.5°C little change occurred in the frequencies in the population cages, and at 22°C an intermediate situation arose whereby some but not all populations showed stable equilibria (Van Valen, Levine, and Beardmore, 1962). These findings demonstrate the dependence of the equilibria, and hence the relative fitness of genotypes, on the environment, in this case temperature variations. Furthermore, fitness estimates are applicable only to genotypes in a given population, since genetic backgrounds vary and influence fitnesses, as is shown by the frequent breakdown of heterokaryotype advantage in between-population crosses in *D. pseudoobscura* (Dobzhansky, 1950). The gene complexes within populations are coadapted within and between chromosomes but not between populations. Therefore, we cannot speak of fitness as a property applying to a specific gene or karyotype without qualification. We may conclude that the dependence of fitness on environments and on the whole genome makes it impossible to define fitness as an invariant parameter associated with a particular genotype or karyotype.

It is not hard to see that most if not all of the behavioral parameters discussed in this book contribute in some way to the overall fitness of an organism; indeed, no behavioral trait can be regarded as neutral as far as fitness is concerned. Even a trait's lack of obvious connection with fitness does not mean that there is no effect; the lack of an obvious effect may just reflect our hopefully temporary ignorance. Further, to consider behavior properly as a component of fitness, it is necessary to go outside artificial laboratory situations to the real world—a problem that presents peculiar difficulties associated with the species selected. It is also necessary to consider the contribution of behavior to alterations in the gene pool as well as the action of genes controlling or adjusting behavior, which has been the main theme of the book so far. This is because fitness is defined in terms of contributions of genotypes to future generations, which must mean that the effect of behavior on evolutionary processes is an issue of central importance. Finally, it will be seen in this chapter that when the researcher moves out of the laboratory into the wild, he often finds it impossible to dissociate behavioral from ecological factors.

Unfortunately, in any one experiment only a few (or only one) fitness factors are normally measured. A question of evolutionary significance concerns relations between fitness factors. There is evidence in *Drosophila melanogaster* that males that mate quickest also copulate more often and more successfully, and leave more progeny (Fulker, 1966). For polymorphic inversions in *D. pseudoobscura* at 25°C, heterokaryotypes are superior in innate capacity for increase in numbers. This is defined by Andrewartha and Birch (1954) as the maximum rate of increase attained by a population under a specific environment. The heterokaryotypes are also superior to the homokaryotypes as regards population size, productivity, egg-to-adult viability, and mating frequency. For mating behavior, the karyotype of the male is important for mating frequency in *D. pseudoobscura* as in *D. melanogaster* (Spiess, Langer, and Spiess, 1966). The results for these various traits in *D. pseudoobscura* were mainly derived by different experimenters in experiments carried out at different times (for references, see Parsons, 1973). The relative associations among these components in a given population have been inadequately explored, even though they are of considerable importance in the study of the overall fitness of organisms.

Prout (1971a,b) described an experimental system for estimating certain components of fitness simultaneously in *D. melanogaster*. Fourth-chromosome recessive mutants, eyeless (ey^2) and shaven (sv^n), were used. This chromosome is very short (see Figure 2-3), and recombination is not relevant as a source of complication. The estimated components of fitness were larval viability in each sex, and from adults two components were obtained, one representing female fecundity and the other male mating ability (virility). The adult components were the more important, so that ey^2ey^2 and ey^2sv^n females were superior to sv^nsv^n, and the heterozygous males were superior to both homozygotes. In males, the depressed values of the two homozygotes varied with the female genotype to which the males were mated, indicating mating interactions. Larval fitness components were small compared with the adult fitness components especially concerning males. Prout, therefore, stressed the need to define a small number of components of fitness that encompass the entire life cycle and that are accessible for experimental evaluation. He tested his fitness estimates by attempting to predict the performance of experimental populations segregating for these same mutants. The results were in reasonable agreement with predictions. Therefore, the fitness estimates can account for most of the performance of the experimental populations. Further work is needed integrating fitness parameter estimates with population performance using an approach of this nature; in particular, generalizations over a series of environments seem essential.

It is perhaps inevitable that in the enormous literature on allozymes there are now experiments adjudicating the association between allozymes and behavioral traits. Åslund (1977) studied the maintainance of leucine aminopeptidase, a polymorphism in *D. melanogaster*, and concluded that the major mechanism for the polymorphism appeared to be heterozygote superiority in mating behavior measured in various ways for both sexes. His data show that the superiority is higher at 25°C than at 16°C; indeed at the latter temperature there is no mating superiority. Table 13-1 shows this for an overall measure of male mating vigor, the number of females inseminated by single males during 24 hours. The low insemination frequency at 16°C is predictable from McKenzie's

Table 13-1   Number of *Drosophila melanogaster* Females Inseminated by Single Males during 24 Hours

Male genotype	Number of inseminated females	
	25°C	16°C
Lap-$A^F A^F$	10.10 ± 0.301	3.60 ± 0.238
Lap-$A^F A^O$	11.30 ± 0.300	3.53 ± 0.361
Lap-$A^O A^O$	9.97 ± 0.323	3.63 ± 0.247

All means are based on 30 repeats.
Source: Åslund (1977).

(1975) experiments on mating frequencies in relation to temperature in *D. melanogaster*. Again we see the dependence of relative fitnesses on environments. (See Åslund and Rasmuson, 1976, for another example—this time the esterase-6 allozyme polymorphism.)

In most of the above examples, male mating behavior is an important component of fitness. This agrees with the earlier experiments of Merrell (1953), who found gene frequency changes in experimental populations of *D. melanogaster* to be predictable from male mating behavior variations. In the neotropical South American species *D. pavani,* males heterozygous for various gene arrangements were superior in mating activity to the corresponding homokaryotypes from the same population (Brncic and Koref-Santibañez, 1964). On the other hand, in *D. persimilis,* both sexes appeared to be important in the mating frequency of various karyotypes (Spiess and Langer, 1964b). Even so, it may be concluded that, at least in the laboratory, male mating behavior differences between genotypes are of importance in modifying gene pools in subsequent generations.

Even allowing for the occasional difficulty in interpreting experiments involving the two sexes (Section 4-2), it is difficult to escape the conclusion that mating behavior, especially of males, forms an important component of fitness. In many, but not all, cases there is associated evidence for heterozygote superiority. However, for a variety of fitness traits including mating, heterozygote superiority becomes more pronounced under extreme environments, especially temperature (Parsons, 1973). Since temperature is a prime variable involved in the distribution and abundance of *Drosophila* (Parsons, 1978a), we must conclude that the real evolutionary significance of the results discussed here are difficult to assess without extrapolation to nature—a task of great difficulty in an insect of the size of *Drosophila*.

However, together with the discussion in Sections 6-5 and 8-2, we can say that there are good data from a variety of sources to the effect that:

1 Male mating speed is subject to directional selection for rapidity of mating.
2 Within a given species, rapid matings tend to be controlled by the

genotype of the male involved, while the genotype of the female may assume importance in slower matings.

3   Mating speed is associated with fertility and number of progeny.

4   Where studied in relation to other components of fitness encompassing the entire life cycle, mating speed is a most important component in drosophilids.

As a consequence, male reproductive success must vary widely, much more than that of females. This is certainly true of laboratory experiments in *Drosophila*. Trivers (1972) documents field examples in a variety of organisms including dragonflies, baboons, frogs, prairie chickens, sage and black grouse, elephant seals, dung flies, and certain lizards. As Trivers points out, the explanation lies in the relative parental expenditure of the sexes in their young. Where one sex, say the female, invests considerably more than the other, then males will compete among themselves to mate with females as is documented in Table 4-1. Trivers goes on to discuss parental expenditure strategies in general, but these are rather abstruse for consideration in a text of this nature.

13-3   HABITAT SELECTION: MAINLY *DROSOPHILA*

Since this is a text on behavior genetics, our discussion of habitat preference is restricted to comparisons within species and between closely related species. Consider *Drosophila* first. A general account of the behavioral and ecological genetics of this genus appears in Parsons (1973). For example, there are food preference differences among species associated with seasonal and geographical variations in distribution (Dobzhansky and Pavan, 1950). One factor seems to be differential attraction to different yeast species (Dobzhansky et al., 1956). In the extraordinarily diverse Hawaiian *Drosophila* fauna, the distribution of a number of plant species and a number of other ecological factors such as wind intensity, humidity, temperature, and light intensity are clearly important (Carson et al., 1970). Moderate wind currents and light intensities, humidities below 90 percent, and temperatures above 21°C seem to be avoided by many species. Therefore, it is not surprising that during overcast weather, when the humidity approaches 100 percent and especially if misty rain is falling, the flies of these species tend to move upward into the available vegetation and can be found on the undersurfaces of leaves and plant limbs up to about 10 feet from the ground. However, on cloudless sunny days when humidity falls, the flies rapidly disappear, presumably seeking out small poorly lighted areas where humidity is high and light intensity is low. Here then we see intimate behavioral adaptations to the prevailing environment.

Drosophila shows a range from species dependent on a particular plant species (monophagous species) to those using a variety of host plants (polyphagous species). A number of the polyphagous species of *Drosophila* can be cultivated on laboratory media; such cultivation is normally a much more difficult proposition for monophagous species. Presumably, monophagous species are adapted to their own specialized niches whereas polyphagous species have less specialized requirements. The species of *Drosophila* on which most behavior-genetics work has been done are widespread and generally fall into the less specialized class as regards nutritional requirements. Even so, subtle behavioral and associated ecological differences are found among certain closely related species.

Some species that are difficult to culture on laboratory media have quite elaborate behavioral patterns. We refer mainly to the Hawaiian species, many of which are geographically very restricted. In some instances specialized forms of behavior perhaps rare elsewhere in the world occur (Spieth, 1958; Carson et al., 1970). Field and laboratory studies indicate that the males of many species patrol and defend a small but definite *lek* (courting and mating territory). The territories are not randomly determined but are at specific sites in the vegetation; each species has preferences apparently controlled by ecological factors such as light, humidity, temperature, and spatial conditions. The territories are close to, but separate from, their feeding sites. Associated with this is the development of sexual dimorphism. These species still show the basic drosophiloid behavior pattern, but superimposed on this are territoriality, aggression, and advertising in males, associated with the spatial separation of feeding and courtship sites. Males are not defensive at feeding sites, where they may be judged as gregarious, but once at their specific leks they become pugilistic. The existence of such leks will presumably have the effect of enhancing variability among males for reproductive success, as discussed at the end of the previous section. Unfortunately, genetic studies on these species are fewer than those on cosmopolitan species, but future behavior-genetics investigations of the complex behavioral patterns of the Hawaiian species and their associated morphologies should contribute greatly to our understanding of the evolutionary biology of the genus. In fact, the Hawaiian flies form a group of species in which the integration of genetics with behavioral and ecological studies is essential.

The Hawaiian species show such diversity that of the world fauna of 1,500 to 2,000 species, up to 500 identified species occur in this and in closely related genera (drosophiloids) in Hawaii, and there are probably 200 or more species belonging to the closely related genus *Scaptomyza* and other related genera (scaptomyzoids). This burst of diversity in the Hawaiian Islands represents an adaptive radiation analogous to that of Darwin's finches on the Galapagos Islands (Dobzhansky, 1968). Probably the adaptive radiation arose from the chance arrival of one or two karyotypically similar species (Carson et al., 1970). Future work on the fascinating diversity of species now developed will clearly be of importance for the behavior geneticist as well as for the evolutionary biologist—two categories that are not mutually exclusive.

Remarkable as the Hawaiian fauna is, recent studies on the Australian *Drosophila* fauna have revealed two or more picture-winged lek species deep inside rain forests utilizing, in contrast with Hawaiian lek species, the undersides of bracket fungi as courting territories (Parsons, 1977c, 1978b). The undersides of the fungi are white, light grey, or light brown, strongly enhancing displays. Bracket fungi often have several flies which are spread out relatively regularly under the fungus. Assuming lek behavior, an excess of males would be expected for total collections under the fungi as shown in Table 13-2. More limited collections from soft forest fungi, the oviposition sites, and larval resources utilized by these species, if anything, favor females. The separation of feeding and breeding sites for the Australian and Hawaiian species represents parallel evolution in *Drosophila* for lek behavior, being in subgenera *Hirtodrosophila* and *Drosophila* respectively. It may be inferred that this parallel evolution depends upon a basic environmental similarity whereby long periods of temperature/desiccation stresses are minimized, which permits the development of complex behavioral patterns. The necessary environmental conditions do indeed occur in Australia deep inside rain forests where bracket fungi occur in microhabitats with low light intensities, often close to permanent water.

We now consider some closely related species. The species *D. melanogaster* and *D. simulans* are morphologically almost identical, and so are sibling species (see Section 4-2 for definition). Although frequently collected in the same locality, they are completely distinct species. This is evinced in the sterility of their hybrids. It is instructive to review (Parsons, 1975) some of the rather subtle behavioral and ecological differences found within and between the two species, since in the laboratory they survive in similar culture systems, implying that their needs are at least similar. Discussed below are some of the relevant studies.

• *Sexual behavior:* Ethological isolating mechanisms almost entirely prevent cross mating. The sexual behavior of the two types of males can be readily dissected into the same basic elements of courtship—orientation, vibration, licking, and copulation—as described in Section 3-2 where differences between mutants of *D. melanogaster* are discussed. However, males of *D. simulans* take longer to begin courtship and consequently have longer bouts of simple orientation; in other words, the courtship behavior of *D. melanogaster* is more active than that of *D. simulans* (Manning, 1959). There is, therefore, no difference in the basic organization of the sexual behavior of the two types of males, but *D.*

Table 13-2   Number of Flies of *D. mycetophaga* and *D. polypori* Collected from the Undersides of Bracket Fungi and in the Vicinity of Soft Forest Fungi

	D. mycetophaga			D. polypori		
	♂	♀	**N**	♂	♀	**N**
Bracket fungi	131	27	158	97	53	150
Soft forest fungi	10	12	22	7	13	20
Total	141	39	180	104	66	170
χ_1^2 for independence		13.88*			5.35†	

* $P < 0.001$
† $P < 0.05$
Source: Parsons (1978b).

simulans males are slower to rise to sexual excitation than are *D. melanogaster* males. Females of *D. simulans* are more responsive to the visual aspects of the male's courtship and less responsive to those stimuli perceived by their antennae than are *D. melanogaster* females. In fact, *Drosophila* species can be grouped in three classes on the basis of components of mating behavior in relation to light-dependence (Grossfield, 1971): (1) species unaffected by darkness, which include a number of cosmopolitan and broad-niched species such as *D. melanogaster*; (2) species inhibited by darkness, so that facultative dark mating occurs, which include *D. simulans*; (3) species in which mating is completely inhibited by darkness. The group in which darkness inhibits mating includes a number of specialized narrow-niche species, including the Hawaiian species, in which, as we have seen, visual cues are of critical importance. Finally, even though isolation is nearly complete, a few hybrids can be brought into being under laboratory conditions. The degree of isolation can be shown to vary among different strains (Parsons, 1972b), but it is consistently strong. Environmental factors known to affect levels of isolation in the laboratory include age, whether single or mass matings are used, and, in the latter case, the proportion of males (for references, see Parsons, 1975).

• *Dispersal activities:* McDonald and Parsons (1973) found that the dispersal activity of *D. melanogaster* exceeds that of *D. simulans*. Comparison of the two species dispersing toward a light source, and without a light source, showed that *D. simulans* is more dependent on the presence of light than is *D. melanogaster,* as was found for mating behavior. Similarly, for phototactic responses along a gradient of light intensities, *D. simulans* shows greater phototaxis than does *D. melanogaster* (see Parsons, 1975; Kawanishi and Watanabe, 1978). *D. melanogaster* shows a more even distribution over the various light intensities than *D. simulans* (Figure 13-1). Therefore, in both cases the behavior of *D. melanogaster* is less light-dependent than that of *D. simulans,* arguing that *D. melanogaster* may be regarded as the broader-niched species.

• *Oviposition:* In competition experiments, *D. simulans* tends to oviposit in the center of food cups and on food having a surface crust; *D. melanogaster* does not. In other words, desiccation may make the medium less favorable for *D. melanogaster* (Barker, 1971). Generally, oviposition data are highly variable; however Kawanishi and Watanabe (1978) found that *D. simulans* preferred to oviposit in areas of higher light intensity than did *D. melanogaster*. In addition, selection of eggs by position in a light intensity gradient similar in design to Figure 13-1 made it possible to segregate a mixed species population into different species populations, since selection for photopositive flies soon eliminated *D. melanogaster* and for photonegative flies soon eliminated *D. simulans*. This result is of potential ecological significance in that *D. melanogaster* tends to be found in darker places than is *D. simulans*.

• *Larval dispersal:* Larvae of both species are found equally in the upper section of the medium, but lower down the proportion of *D. simulans* larvae has been found to exceed that of *D. melanogaster* (Barker, 1971). In addition, the observations on larval behavior in relation to various chemicals (Section 8-5) may be relevant in habitats selected by these two species; this clearly needs additional work.

• *Ethanol in the medium:* *D. melanogaster* is more tolerant of 9 percent ethanol than is *D. simulans* both as larvae and as adults. *D. simulans* adults show

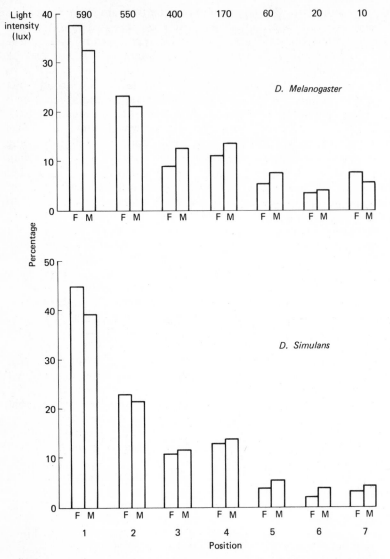

Figure 13-1 Percentages of flies at each of seven light intensities (10 to 590 lux) for the two species *D. melanogaster* and *D. simulans*. Flies were given 4 hours to select a light intensity. (*After Parsons, 1975.*)

an aversion for ovipositing on sites containing 9 percent ethanol; in contrast, *D. melanogaster* shows a slight preference for doing so (McKenzie and Parsons, 1972) and there are similar (but more marked) differences for larval behavior (Section 8-5). This explains the almost exclusive presence of *D. melanogaster* inside a winery near Melbourne, Australia, while immediately outside the winery both species occur, with *D. simulans* usually in greater numbers. Release-recapture experiments during vintage suggest that *D. melanogaster* moves toward the cellar in a regular fashion, while *D. simulans* moves away from it (McKenzie, 1974). Thus the distribution of the two species at vintage may be a function of their dispersal activities. In addition, grape residues outside the winery are initially characterized by active fermentation with the alcohol concentration about 7 percent; at this stage only *D. melanogaster* larvae were present, while larvae of both species occurred at the postfermentation stage (McKenzie and McKechnie, 1979). In agreement with the laboratory experiments, the presence of alcohol in the environment alters the behavioral pattern of the two species in the wild, both as adults and larvae.

• *Temperature and desiccation:* These factors are more ecological than behavioral, but in the avoidance of extremes of high temperature and low humidity, it is clear that behavior must play a part in the selection of nonstressful microhabitats (Parsons, 1978a,b). There are known variations among strains within the two species for tolerance to these two stresses. *D. melanogaster* can tolerate a greater range of temperatures than can *D. simulans* (for references, see Parsons, 1975), indicating that *D. melanogaster* may have a broader niche similar to that for light-dependence in mating behavior, dispersal activity, and phototaxis. Levine (1969) concluded that acclimation to dry heat by *D. melanogaster* depends more on developmental flexibility and physiological acclimation than on genetic differentiation of populations for its adaptation, whereas *D. simulans* shows lower developmental flexibility and depends more on genetic differences. Although behavioral factors are clearly relevant, their relative importance in these instances in the two species is unknown.

• *Some general ecological factors:* Finally, a number of factors are known that have only minor behavioral components but that differentiate the two species. El-Helw and Ali (1970) found that *D. simulans* is more tolerant to natural yeasts in the medium than is *D. melanogaster,* which may be associated with field observations of *D. simulans* in more natural habitats than for *D. melanogaster* (Parsons, 1979a). Minor differences in development rate, survival, fecundity, fertility, hatchability, and adult viability have been found showing a general superiority for *D. melanogaster.* Many of these experiments were carried out at 25°C, a temperature that is often deadly to *D. simulans* in the laboratory (Parsons, 1975). Indeed, in population cages *D. melanogaster* usually displaces *D. simulans* at 25°C, but at 15°C the reverse may occur (Moore, 1952).

No doubt all these effects found within and between these two sibling species are relevant in determining their distributions in the wild. The two species coexist to a fairly high degree in many regions, but the above considerations suggest the existence of subtle behavioral and ecological differences, so that the niches they occupy must differ. The need to study different stages of the life cycle is highlighted. The one possible conclusion is that *D. melanogaster* is more broad-niched than *D. simulans* as regards physical components of the environment including light intensities. This is supported by the observation that genetic variability may be higher in natural populations of *D. melanogaster* than in *D. simulans* (for references, see Parsons, 1975). Arguments and data presented by Beardmore (1970) support the proposition that within a species an association is expected between the ecological heterogeneity to which a population is exposed and its genetic variability. It may be a reasonable argument for comparisons between very closely related species, but certainly not for more distantly related species (Selander and Kaufman, 1973). The comparison between these two sibling species is discussed in some detail to show the subtle interplay of behavioral and ecological factors in determining the habitats and isolation among species.

While the differences between *D. melanogaster* and *D. simulans* (of subgenus *Sophophora*) have been emphasized, their relative magnitudes are put into better perspective by comparison with *D. immigrans,* a widespread species of subgenus *Drosophila*. Atkinson and Shorrocks (1977) investigated resource utilization by studying emergences of domestic *Drosophila* species from 32 fruits and vegetables from an English market; the behavioral trait under indirect study is of course oviposition. The sibling species have extremely similar preferences being fruit specialists, while *D. immigrans* utilizes both vegetables and fruits. For lemons, the proportions of flies emerging were 0.048, 0.010, and 0.102 for *D. melanogaster, D. simulans,* and *D. immigrans* respectively, a result agreeing with observations that *D. immigrans* prefers lemons as a resource in Australian orchards (Prince and Parsons, 1980). Trapping flies in a mixed orchard shows this most clearly. A comparison is given in Table 13-3 of the sibling species with *D. immigrans* (Parsons, 1979a). This shows that in total, by whatever criterion is selected, behavioral or ecological, the sibling species contrast with *D. immigrans*. In addition, the greater diversity of resources utilized by *D. immigrans* may well be consistent with its reasonably frequent occurrence at low population densities in rain forest habitats throughout the world. Comprehensive investigations of species in this way may therefore provide information on evolutionary divergence that may relate to the history of the genus. It is noteworthy too, that using a tree-diagram, Atkinson and Shorrocks found that a major difference for breeding sites occurs between three subgenus *Sophophora* species (*melanogaster, simulans, subobscura*) and three belonging to the closely related subgenera *Drosophila* and *Dorsilopha* (*immigrans, hydei, busckii*), indicating the possibility of evolutionary divergences in resource utilization.

The comparative study of oviposition and hence resource utilization by larvae of different *Drosophila* species remains an open field not only for fruit lured species but also for those exploiting more exotic resources. This is well shown by the Australian endemic *Drosophila* fauna that is made up of species from the four major *Drosophila* subgenera in order of species frequency (in Australia): *Scaptodrosophila, Hirtodrosophila, Sophophora,* and *Drosophila* (Parsons and Bock, 1979). For species where information is available, they can be classified into those (1) attracted to fermented-fruit baits, (2) attracted to rotted commercial mushroom, *Agaricus campestris,* (3) collected in the vicinity of forest fungi, (4) utilizing endemic *Hibiscus* flowers as a resource, and (5) collected by nonspecific sweeping. From a consideration of resources utilized by subgenera and their species groups, a relationship between taxonomic divergence, resource utilization divergence, and species distribution patterns emerges. Behavioral and ecological comparisons of the type in Table 13-3 at

Table 13-3 A Comparison of the Sibling Species *D. melanogaster* and *D. simulans* with *D. immigrans*

	D. melanogaster and *D. simulans* (Subgenus *Sophophora*)	*D. immigrans* (Subgenus *Drosophila*)
Physical environment		
High temperature/desiccation resistance*	High resistance, especially *melanogaster*	Lower
Laboratory temperature preferences	High, especially *melanogaster*	Lower
Cold stress resistance*	Less resistance than *immigrans,* especially *simulans*	Higher
Ethanol and Other Potential Resources (Laboratory)		
Larval responses to ethanol (6%)*	High to moderate preference *melanogaster,* minimal response *simulans*	Avoidance
Threshold for ethanol resource utilization	> 9% *melanogaster* 3–6% *simulans*	$\simeq$ 1.5%
Larval responses to acetic acid, ethylacetate, lactic acid*	High	Moderate
Cholesterol needs	Higher than *D. immigrans*	Low
Resource Utilization (Field Studies)		
Oviposition sites	Fruit specialists	Fruits and vegetables
Lemons	Adults avoid lemons, and larval survival low, especially *simulans*	Adults select lemons, and larval survival high
Ecological Observations		
Occurrence in rain forests	*melanogaster* apparently never, *simulans* rarely	Frequent, rare inhabitant
Parasitism by the wasp *Phaenocarpa persimilis* (in sympatric populations in the Melbourne area)	Highly successful	Unsuccessful

* Intraspecific geographic differences are known for these items for *D. melanogaster* and *D. simulans* (except for ethanol).
Source: Updated from Parsons (1979a, 1980).

various taxonomic levels should provide information of evolutionary significance in this extremely diverse and widespread genus, especially where oviposition and larval studies can be incorporated (for further discussion see Parsons 1978b).

Another pair of sibling species about which we have much information, *D. pseudoobscura* and *D. persimilis,* are widespread in North America and are sympatric in some regions. Isolation is maintained by the following factors:

• The two species have somewhat different habitat preferences. *D. persimilis* occurs in cooler niches and *D. pseudoobscura* in warmer niches.

• The two have different food preferences, including differential attraction to different yeasts.

• Many species of *Drosophila* show high activities in the early morning and evening. As shown in Table 13-4 for flies collected in the Yosemite regions of California, among flies attracted to yeasty baits in the morning, the proportion of *D. pseudoobscura* was lower and that of *D. persimilis* higher than among flies attracted during the evening period of activity (Dobzhansky et al., 1956).

• When the two species are sympatric, the mean photoresponse (affinity for light) was greater for *D. persimilis* than for *D. pseudoobscura* (Rockwell, Cooke, and Harmsen, 1975).

• Sexual isolation is associated with different male courtship songs for the two species (Ewing, 1969). Males of *D. pseudoobscura* produce two songs controlled by the wings, a song with a low repetition rate consisting of trains of 525-Hz pulses at 6 per second, and a song with a high repetition rate in which 250-Hz pulses are repeated 24 times per second. In *D. persimilis* the low-repetition-rate song is absent or occurs in a very abbreviated form, and the high-repetition-rate song consists of 525-Hz pulses repeated 15 times per second.

The first four of these factors are not entirely effective because both species have been found feeding side by side on the same slime flux on the black oak *Quercus kellogii* (Carson, 1951). This suggests that the absence of interspecific matings in natural habitats is largely due to ethological isolation. In any case, when interspecific matings take place in the laboratory, fewer sperm are transferred than in intraspecific crosses, the F_1 males are sterile and the F_1 females have reduced vigor.

In the laboratory, hybridization occurs relatively readily. Virgins were about 4 days old in many experiments (see Section 8-4 about this specific age). However, when flies of both sexes were placed together a few hours after emergence, the proportion of hybrids was lower. Spieth (1958) has suggested that this higher level of isolation may be due to individuals of both species maturing together, enabling them to discriminate between the species before sexual maturity. Furthermore, a *D. persimilis* female once having mated with a *D. persimilis* male will not subsequently accept a *D. pseudoobscura* male. This

Table 13-4 Number of *D. pseudoobscura* and *D. persimilis* Collected Morning and Evening in Yosemite Region of California

Month	Morning		Evening	
	D. pseudoobscura	*D. persimilis*	*D. pseudoobscura*	*D. persimilis*
June	68	111	682	432
July	210	297	694	446
August	65	75	681	443

Source: Dobzhansky et al. (1956).

indicates that the high level of isolation in the wild may not be completely innate but partly learned. More evidence on the effect of experience on mating behavior is presented in Section 8-4, where it is noted that *Drosophila* females prefer to mate with the type of male previously accepted by them.

Laboratory experiments indicate other relevant variables. The degree of isolation has been found to be temperature-dependent (Mayr and Dobzhansky, 1945), being relatively low for flies raised at 16.5°C. However, the level of sexual isolation can be increased or decreased by selection (Koopman, 1950; Kessler, 1966), showing that the degree of sexual isolation itself is under genetic control. Further discussions of the genetics of sexual isolation are part of Section 5-3.

We have in this section concentrated on interspecific differences for habitat preference. Intraspecific (between genotype) habitat selection would of course be expected—but more difficult to detect. However, in *D. persimilis* Taylor and Powell (1977) studied the effects of a heterogeneous environment made up of various vegetational types associated with differing moisture regimes. They found differing allozyme and inversion frequencies among habitats and argued for habitat selection after eliminating the possibilities of natural selection through differential survivorship, random genetic drift, and nonrandom dispersal rates of the various genotypes. In addition, flies were captured from different habitats, marked according to origin, and released from a common site. The flies exhibited a tendency to return to the habitats of initial collection. There may be behavioral differences among genotypes whereby different genotypes seek out different parts of the environment. Given the demonstrated importance of temperature and humidity in *Drosophila* population biology, it is tempting to surmise that the environmental factors relate to such variables.

Other intraspecific evidence for possible habitat selection comes from odor and/or metabolite perception studies. Differences in larval reactions to alcohol in *D. melanogaster* were discussed in Section 8-5. In addition, adults and larvae of different geographical strains vary somewhat in attraction to alcohol, as well as to acetic acid, DL-lactic acid, and ethyl acetate (Fuyama, 1976; Parsons, 1979a).

Manning (1967b) found that adult *D. melanogaster* reared on a medium containing geraniol (a peppermint smell) showed reduced aversion to its odor; this may be a form of habituation. It is reasonable to postulate that there might be genetic assimilation for such learned behavior. Extrapolating to the wild, if a niche entered is characterized by some sort of habitat alteration that can be assimilated genetically, the possibility of evolutionary change occurs. Various races and species of the Mediterranean honeybee (summary in Lindauer, 1975) were trained with 17 different odors. Naive bees of the different races assess and learn these odors with varying degrees of ease; each race learns with the greatest facility the scents characteristic of its own native flora. The odor of orange was not learned. Oranges, however, originated in southeast Asia and have been cultivated in Mediterranean countries for only 300 years. Evidently this is an insufficient length of time to allow western honeybees to become adapted to this addition to available food sources. The learning of odors in honeybees is therefore strongly race-specific and thus has a genetic component.

The possibility of a genetic component in food selection in *Drosophila* appears worthy of investigation especially in the generalist species. Stalker's (1976) observation of differing inversion frequencies between *D. melanogaster* breeding on windfall oranges compared with grapefruit could be a starting point. Other recent evidence of habitat preference includes a correlation between larval habitat difference (oak tree hole versus beech tree hole) and gene-frequency distribution at an esterase locus for the mosquito *Aedes triseriatus* (Saul et al., 1978) and the differential migration of amylase genotypes in the crustacean *Asellus aquaticus* (Isopoda) in a small pond where decaying beech leaves were the dominant food resource in one section and decaying willow leaves in another (Christensen, 1977). Our final piece of evidence concerns the settlement behavior of marine planktonic larvae, the polychaete *Spinorbus borealis*. Larvae settle and metamorphose on several algal species, with marked local variations in algal substrate preferences. Doyle (1976) found a tendency for "habitat loyalty," or a tendency of animals to rank a habitat higher on a scale of preferences including several types of habitats, if the parents came from that habitat. That is, the primary selective agent determining the probability of settlement is the suitability of that alga as a substrate (see also McKay and Doyle, 1978).

From just these isolated observations it is reasonable to speculate that food and habitat selection may well be of considerable significance in the formation of races within species, and ultimately in speciation. This may apply particularly to generalist species that are able to utilize a variety of resources.

13-4 HABITAT SELECTION: RODENTS

Selection of an optimal environment is important not only in *Drosophila* but in any organism that exists in nature over a wide spectrum of habitats. Temperature is one of the prime factors involved in many adaptations. At first it seems that those organisms possessing mechanisms of thermal adaptation have a reproductive advantage over those that do not. *Poikelotherms* (animals not having an internal mechanism for regulating body heat) adapt by immobility and reduced metabolic rate during periods of cold or by physiological and behavioral adaptations allowing maximum utilization of heat and protection from the cold. *Homeotherms,* on the other hand, have an internal mechanism for regulating body heat and can function efficiently within a wider range of temperatures. Even so, they still possess various special ways of combating extreme variations, and many of these ways are behavioral—burrowing, seeking shade, basking, shivering, migrating, and various locomotor activities.

The laboratory experiments described in Section 9-3 show that mice, when faced with a thermal gradient, select a preferred temperature that presumably denotes a physiological adjustment of importance for optimal functioning. The data discussed there show that the temperature preference of mice may have a fairly close association with various physiological and morphological traits, so that it can be regarded as an innate trait permitting the selection of a most favored habitat.

Studies of the deer mouse, *Peromyscus,* have shown behaviors predictable from the habitats occupied by them in nature. The prairie deer mouse of the Midwestern and Plains states of the United States, *P. maniculatus bairdii,* is a strictly field-dwelling subspecies that avoids all forested areas, unlike the closely related woodland form, *P. maniculatus gracilis.* Some work has been done on trying to identify the environmental cues that the deer mice follow in choosing a place to live. Harris (1952) presented individual prairie and woodland deer mice with a choice between field and wood environments re-created in the laboratory. Each type of mouse exhibited a clear preference for the artificial habitat most closely resembling its natural environment. Furthermore, laboratory-bred *Peromyscus* that had had no previous experience with either of the natural habitats exhibited a preference for the type of habitat it normally inhabits in nature. Habitat selection can be regarded as basically genetic in nature, under the control of natural selection for the field or wood environment.

Ogilvie and Stinson (1966) found the thermotactic optima of *P. maniculatus bairdii* and *P. maniculatus gracilis* to be 25.8°C and 29.1°C, respectively, in agreement with the warmer wooded environment of *P. maniculatus gracilis* and the cooler prairie environment of *P. maniculatus bairdii.* Another species, *P. leucopus,* which comes from an area where the ground temperature is 3 to 4°C higher than the woodland regions of *P. maniculatus gracilis,* preferred an even higher temperature, 32.4°C. It can be concluded that the animals tend to select those habitats that most closely resemble their natal homes and that this preference is influenced by genotypes. Wecker (1964), in experiments comparable to those of Harris (1952), argued for a behavioral feedback to the genotype that occurs simply by keeping a given population restricted to its habitat.

In the two *P. maniculatus* subspecies, genetic differences are apparent in their reaction to sand (King, 1967). Over a 24-hour period, the median amount of sand removed from a tunnel by *P. maniculatus gracilis* was 0.1 pound, while *P. maniculatus bairdii* removed 5.9 pounds. The differences are compatible with our knowledge of the life history of the deer mice, since *P. maniculatus gracilis* is semiarboreal and *P. maniculatus bairdii* is strictly terrestrial. Other differences are such that *P. maniculatus bairdii* matures more rapidly in locomotor responses than does *P. maniculatus gracilis,* and, conversely, *P. maniculatus gracilis,* the semiarboreal species, develops a greater clinging response. These results are also compatible with the life histories of the two subspecies. Various differences in developmental rates as assessed by morphological and physiological traits are found between the two subspecies, suggesting the likelihood of developmental differences in the morphology and biochemistry of their central nervous systems. Here then we have good evidence for an association of morphological, physiological, and behavioral traits that are assuredly related to habitat selection.

Evidence for the genetic control of habitat preference also emerges from an examination of the north to south distribution of *Peromyscus* in Canada and in the United States (King, Maas, and Weisman, 1964). From north to south the following taxa occur: *P. maniculatus gracilis, P. maniculatus bairdii, P. polionotus,* and *P. floridanus* (Figure 13-2). This distribution approximates the ranked differences in the net amount of material used by the four species and subspecies to construct nests in the laboratory. *P. maniculatus gracilis* uses more nest material than *P. maniculatus bairdii,* and both use more material than either *P. polionotus* or *P. floridanus.* Only *P. polionotus* and *P. floridanus* build

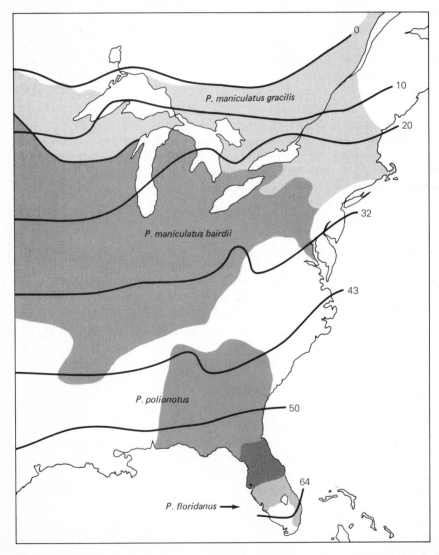

Figure 13-2 Geographical range of *Peromyscus. Dark lines* show January isotherms in degrees Fahrenheit. (*From King, Maas, and Weisman, 1964.*)

nests of about equal size. In the north, large nests are built presumably to provide insulation from cold weather, while in warmer climates nests of this size are unnecessary. The strains tested were descendants of mice collected in the wild and then bred in the laboratory. The differences are at least partly genetic in origin and so contribute to the confirmation of behavioral adaptations as having evolved through the influence of natural selection.

The work on *Peromyscus* points up the need to study wild house mouse populations in more detail, since genetic studies can be readily done in house mice. Work on laboratory strains (Section 9-3) certainly indicates the likelihood of variations between wild strains for traits similar to those described in *Peromyscus*. Recent studies (Lynch and Hegmann, 1972) have revealed differences in nesting behavior in house mice, assessed by the utilization of cotton in nest building, among five inbred strains. Furthermore, Lynch and Hegmann (1973) found that differences between two strains, BALB/cJ and C57BL/6J, were greater for animals tested at 5°C than for those tested at 26°C. Data ought to be collected over a series of environments, since this result implies the existence of genotype-environment interactions relatable to habitat selection.

Bruell (1970), in a perceptive article on behavioral population genetics in mice wrote:

> The phenomenon of substrate selection is only one of many that illustrate behavioral adaptations to local conditions. Most successful species occupy not one, but many different environments, each environment requiring special kinds of morphologic, physiologic, and behavioral adaptations. Indeed, the success of a species is measured by its ability to adapt to a variety of environments. This raises the question whether successful species colonize differing environments with highly adaptable but genetically identical populations, or whether such species consist of many genetically different populations, each adapted by natural selection to its own particular habitat.

He also considered that "one of the ambitious aims of the population-genetic study of the mouse could be to obtain 'behavioral profiles' of the various races and subraces of mice." This still applies, given the enormous diversity of habitats and strains of wild mice.

It is clear that the behavioral differences between *P. maniculatus bairdii* and *P. maniculatus gracilis* are maintained by natural selection. Stocks maintained in the laboratory for 12 to 20 generations showed no preference when offered a choice of field or forest habitat. However, if the laboratory-bred prairie mice were raised in the field, they selected the field habitat significantly more frequently than the forest (Wecker, 1964). Recently captured prairie mice raised in the laboratory, however, definitely selected the field. Therefore, a genetic change occurred among the mice raised in the laboratory for several generations. The inherent tendency to select the field habitat was diminished, but it could be regained by early exposure to the environment that had previously imposed selection for this trait. This effect shows that both heredity and experience can play a role in determining the preference of the prairie deer mouse for the field habitat. Presumably in the wild there is behavioral evolution from the learned behavior to an innate response. The learned behavior that initially arose becomes innate and hence under genetic control by natural selection (Wecker, 1964). Evolutionary changes that increase hereditary control are advantageous because they tend to limit the number of possible ways an organism can respond to a particular environmental stimulus (Waddington, 1957). This is beneficial in that natural selection favors those responses conducive to survival. As long as the environment remains stable, the population as a whole eventually becomes adjusted to the ecological situation it is best able to exploit.

Habitat selection in birds too is partly a genetic trait. This probably accounts for a slow response by some birds to changes in the environment. Many old birds return year after year to the same nesting site even if the habitat at that site is deteriorating. Experimental analyses are few, although Klopfer (1963) has shown that chipping sparrows (*Spizella passerina*) raised in the laboratory preferred to spend their time in pine rather than oak leaves just as wild birds do. However, laboratory birds reared with oak leaves showed a decreased preference for pine as adults; in other words, the innate preference for staying in pine could be somewhat modified by early experience.

It is not surprising that changes occur in a laboratory regimen, since natural selection for the habitat preference is relaxed as the laboratory habitat differs from the natural. For traits important in habitat selection, it is expected under natural conditions that stabilizing selection occurs to keep them within relatively narrow limits. Animals showing behavior away from the norm in a certain population are unlikely to breed with other members. Furthermore, animals occupying the most suitable habitat in a heterogeneous environment have less need to utilize the physiological and behavioral adaptations an animal possesses to combat imperfect environments. Another advantage of occupying a suitable habitat is that the chances of interbreeding by individuals with similar genotypes are greatly enhanced, thereby ensuring their continuity. Ultimately, such a process could lead to reproductive isolation among populations, a thing that has occurred frequently in evolution. However, Doyle (1976) found difficulties in interpreting his data on habitat selection by marine planktonic larvae in this way, and felt that he was dealing with a fitness trait having an optimum above the observed mean. One is forced to the conclusion that while stabilizing selection may be of predominant importance, directional selection may frequently occur in response to changing environments. Additional work on the genetic bases of habitat selection will be awaited with interest. Indeed we may not have to wait long if an excellent review by Partridge (1978) on habitat selection is any guide.

13-5 POPULATION DYNAMICS

Clearly, behavioral mechanisms are important as evolutionary forces leading to changes in the gene pool of a species, as illustrated by the discusssions on habitat selection. Concerning the dynamics of populations in general, our knowledge is much more restricted. For rodents, definitive measures are re-quired of genetic changes resulting from population parameters such as migra-tion, aggression, mating systems, differential fertility, and differential mortality. In recent years the importance of behavior as a force in evolution has become particularly evident from studies on the population structure of the house mouse and the vole.

Earlier ecological work showed that the home range of house mice is relatively small. Southern and Laurie (1946) demonstrated that the home range of house mice occupying corn ricks must be about 50 square feet, with vertical movement being less than lateral movement. Similar evidence has been found for wild house mice in Canada and the United States. Petras (1967) studied house mice in the buildings of six neighboring farms in southeast Michigan over a 4-year period. Small breeding units seem usual with low population sizes. In fact, estimates of "effective" population size (defined in Li, 1955, and based on the number of breeding individuals) range from 6 to 80. Such esti-mates were obtained from both genetic and ecological data. The genetic data were based on the frequency of two loci controlling biochemical polymorphisms—the esterase-2 (Es-2) locus and the hemoglobin (Hb) locus. These loci are polymorphic but showed a deficiency of heterozygotes, which can be explained by a subdivision of the population into a number of separate small breeding units in which there is an excess of homozygotes compared with heterozygotes, based on an expectation of random mating (Li, 1955). In other words, this subdivision into small breeding units leads to a situation equivalent to inbreeding. The theory is too complex for a general text of this nature. The first detailed population study that could be explained only by assuming small breeding units was described by Lewontin and Dunn (1960) with reference to data for a polymorphism at the T (for tail) locus in mice, which controls certain aspects of development of axial structures in the caudal region of the spine.

Ecological data cited by Petras (1967) are in aggreement, showing strong territoriality, leading to the subdivision of populations into small breeding units (demes), several of which existed within single farm buildings. Migration rates were low. Petras quoted work showing that not more than 5 percent of all mice in a farm building move their nests either in or out of isolated buildings. Be-tween farms, migration is probably negligible, since populations separated by land lacking ground cover have an extremely restricted exchange of genes. Crowcroft (1966 and earlier papers) reported experiments with wild mice placed into a large enclosure (250 square feet). The mice dispersed into spatially distinct breeding areas or territories. Aggressive behavior was rare within a family group, where social hierarchies were presumably established, but when unrelated mice were introduced into the family group, markedly aggressive behavior followed, often resulting in the destruction of the intruder.

Reimer and Petras (1967) used both wild and laboratory stocks in a study of the breeding structure of the house mouse in a population cage. Mice were released into a cage consisting of a series of nest boxes connected by runways. The mice formed small breeding colonies, each consisting of one dominant male, several females, and several subordinate males, due primarily to male territoriality. Between demes migration was rare and occurred primarily due to migrating females. Such breeding colonies were found to be stable over several generations. Therefore, it seems that mice arrange themselves into small breeding units due to territoriality on the part of the males. Experiments described by DeFries and McClearn (1972) and performed in the laboratory have shown evidence for an association between the social dominance of males of different genotypes as assessed by fighting and Darwinian fitness as assessed by the proportion of litters sired by dominant males. Thus, even though artificial laboratory conditions were used, the data fit with those obtained under more natural conditions and, furthermore, indicate a genotypic basis for social dominance.

Selander (1970) reported on the biochemical genetics of wild house mice by analyzing allele variations at various genetic loci for hemoglobin and esterase variants. Between regions in Texas, there are gradients of allele frequencies (clines). However, a marked heterogeneity of allele frequencies was found when samples of wild mice were collected from different barns within the same region. These differences were found even for barns a few yards apart, agreeing with the behavioral and ecological evidence already cited. Within the same barn a complex mosaic pattern was found for each locus, with small regions of high and low allele frequencies. The clustering of similar genotypes is regarded as a direct outcome of the demes found for wild mice. The population structure is therefore a mosaic of small breeding units (demes) where the effective population size may be extremely small. This means that chance plays an extremely important role in determining gene frequencies at the very local levels. Because the small effective population sizes are greatly dependent on social behavior, that behavior must have important effects on the genetic composition of mouse populations.

A feature of small rodents is the population cycle that occasionally leads to increases of colossal proportions followed by declines to very low numbers. The population cycle of small rodents has long been a classic problem in population ecology. There are two opposing schools of thought about what stops population increase in small rodents. One school considers that extrinsic agents such as food supply, predators, or disease stop populations from increasing. The other school, which is favored as the more important, looks to intrinsic factors—the effects of one individual upon another. Krebs et al. (1973) reviewed evidence in the field vole, *Microtus*.

First consider demographic changes during a population cycle. In *Microtus pennsylvanicus* (Figure 13-3), a typical cycle may, once begun, continue through the winter. The peak phase begins with a spring decline in numbers; then a summer or fall increase restores the population to its former level. The decline phase is most variable and may begin in the fall of the peak year or may be delayed until the next spring. The decline may be very rapid, as in Figure 13-3, but often it is gradual and extends over a year or more. Then follows a phase of low numbers about which little is known. This pattern is characteristic of several species of voles. The immediate cause of the fluctuations resides in

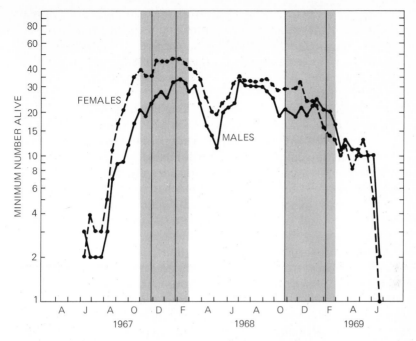

Figure 13-3 Population density changes in the vole, *M. pennsylvanicus,* on a grassland area in south Indiana. Shaded areas indicate winter months. (*From Krebs et al., 1973; copyright 1973 by the American Association for the Advancement of Science.*)

changes in birth and death rates. The percentage of adult females captured that are visibly lactating is reduced in both the peak phase and the decline phase, and in these phases, the death rate of juvenile animals increases dramatically. By contrast, the death rate of subadult and adult animals is not increased in the peak phase but is increased in the decline phase, along with that of the juveniles. Thus, in a peak population, if an animal survives through the early juvenile phase, it has a good chance for adult survival. However, declining populations are characterized by a low birth rate and high mortality rate for both juveniles and adults.

A population experiment was carried out with fenced and unfenced *M. pennsylvanicus.* On two adjacent 0.8-hectare fields, vole population sizes increased rapidly, but in the early peak phase there was a sharp divergence such that the fenced population increased to 310 animals—about three times the unfenced population. The overpopulation of the fenced group led to habitat destruction and overgrazing, followed by a sharp decline with symptoms of starvation—a situation that did not occur in the unfenced population. The same result was found in *M. ochrogaster.* The conclusion is that fencing a *Microtus* population destroys the regulatory machinery that normally prevents overgrazing and starvation. Dispersal is the obvious process prevented by a fence, especially as there was no indication that predation pressure was changed by the fence.

Two ways are envisaged in which dispersal may operate in support of population regulation. First, dispersal may be related to population density so that more animals emigrate in the peak and decline phases. These animals are at a great disadvantage from environmental hazards such as other voles, predators, and weather. Second, the quality of the dispersers may be more important than the numbers; if animals of only a certain genotype can tolerate high densities, dispersal may be a mechanism for sorting out these individuals. An experiment was done in which two areas were kept free of *Microtus* by removing all animals by trapping for 2 days every 2 weeks. Between the episodes of trapping, voles were free to colonize the areas. Dispersal was most common in the increase phase of a population fluctuation and least common in the decline phase. In fact, Krebs et al. (1973) consider that much of the loss rate in increasing populations is due to emigration. Conversely, little of the heavy loss in declining population is due to dispersal, so most losses must result from death in situ.

For the polymorphic serum protein loci *Tf* (transferrin) and *LAP* (leucine aminopeptidase), evidence has been found of large changes in gene and genotype frequencies in association with population changes. The LAP^s allele frequency (distinguished by slow electrophoretic mobility) dropped about 25 percent in *Microtus* males beginning at the time of high losses and 4 to 6 weeks later declined an equal amount in females. This type of observation strongly supports the hypothesis that demographic losses are genetically selective and that losses are not equally distributed over all genotypes.

The transferrin genotype frequencies of dispersing *Microtus* females are compared in Figure 13-4 with resident or nondispersing females. Clearly heterozygous females Tf^C/Tf^E (Tf^C and Tf^E are alleles of the transferrin gene) are more common in dispersing than in resident populations. In fact, 89 percent of the loss of heterozygous females from the resident populations during the population increase was due to dispersal. Certain genotypes show a tendency to disperse, a possibility suggested in the literature (e.g., Lidicker, 1962) but not previously demonstrated in animal populations. Therefore, intense selection pressure must be occurring. If the intrinsic factor of behavioral interactions among individual voles is the primary mechanism, then the behavioral characteristics of individuals should change during the cycle. This has been tested for males of *M. pennsylvanicus* and *M. ochrogaster*. Laboratory studies showed

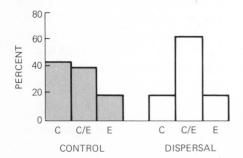

Figure 13-4 Transferrin genotypes during increase phase of the vole, *Microtus pennsylvanicus*, in the fall of 1969, for dispersing females compared with control or resident females. *C* and *E* are alleles. (*From Krebs et al., 1973. Copyright 1973 by the American Association for the Advancement of Science.*)

significant changes in aggressive behavior as assessed by paired round-robin encounters in the laboratory during the population cycle (Myers and Krebs, 1971): individuals in peak populations were the most aggressive. Furthermore, males of *M. pennsylvanicus* that dispersed during peak periods tended to be even more aggressive than the residents. This needs to be looked at further in the field, as Krebs et al. (1973) pointed out.

The results are consistent with the hypothesis of genetic and behavioral effects proposed by Chitty (1967) and displayed in Figure 13-5. Chitty's model of population regulation is based on a "behavioral polymorphism" in which there are socially tolerant and intolerant animals with respect to crowding, such that changing density acts as a selective force on the behavioral types. Accord-

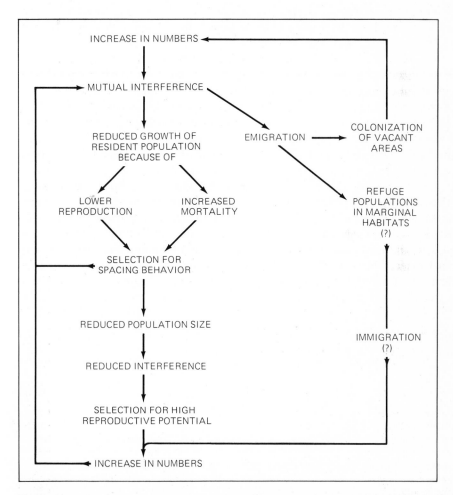

Figure 13-5 Population regulation in small rodents. Model modified from Chitty's original proposal. (*From Krebs et al., 1973. Copyright 1973 by the American Association for the Advancement of Science.*)

ing to this hypothesis, as the population size increases so too does mutual interference and selection for aggressive behavior. Emigration is one of the ways through which selection takes place. As a selecting force this occurs to the greatest extent during the increase phase. There is now clear evidence (see above) for genetic differences between the emigrants and resident animals. As Krebs et al. (1973) and Fairbairn (1978) point out, the greatest gap in our knowledge is in the behavior-genetics area. For example, there is no knowledge about the heritability or genetic architectures of aggression in natural populations of small rodents. The vole example is given here, not because the story is complete, but because it shows the likely importance of genetic and behavioral changes associated with population cycles—a result that, if generalized, seems of far-reaching significance, perhaps even for our own species.

Vale, Vale, and Harley (1971) studied 45- to 55-day old male house mice from five inbred strains in population numbers of 2, 4, and 8 animals per cage. Aggressive behavior and social grooming were recorded over a period of 10 days, at the end of which adrenals, testes, and seminal vesicles were removed and weighed. There were differences among all strains for five behavioral measures and the three weight measurements (Table 13-5). Population number was associated with significant effects in two of the behavioral measures, number of chases and number of attacks, and in adrenal weight. There was a positive association between agonistic behavior and adrenal weight, as has been reported elsewhere (references in Vale, Vale, and Harley, 1971), and also a positive association between social grooming and gonad weight, both revealed by principal component analysis. For the two variables, number of attacks and adrenal weight, there were genotype–population number interactions indicating that not all strains act in an identical way with respect to aggression as population numbers are increased. That is, population number increase does not necessarily lead to increased aggression or adrenal weight in all genotypes. This agrees with Chitty's (1967) model of socially tolerant and socially intolerant animals with respect to crowding; it begins to close the gap between behavior and genetics.

Table 13-5 Summary of Results of Analyses of Eight Variables in Males of Five Inbred Strains of House Mice

Variables	Effects		
	Strain	Population number	Interaction
Number of chases	$P < 0.01$	$P < 0.05$	NS
Number of attacks	$P < 0.0001$	$P < 0.025$	$P < 0.01$
Number of fights	$P < 0.01$	NS	NS
Number of social grooms	$P < 0.0001$	NS	NS
Number of tail pulls	$P < 0.0001$	NS	NS
Adrenal weight	$P < 0.0001$	$P < 0.005$	$P < 0.005$
Testis weight	$P < 0.0001$	NS	NS
Seminal vesicle weight	$P < 0.0001$	NS	NS

NS, not significant.
Source: Vale, Vale, and Harley (1971).

In recent years, reports have been appearing on territoriality in many organisms, although analyses at the level of sophistication needed for a text on behavior genetics are as yet few. O'Donald (1976, 1977, and earlier papers) has, however, been carrying out extensive studies on the arctic skua. This is a polymorphic seabird with three phenotypes in plumage: pale, intermediate, and dark. On average, dark males have a greater reproductive success than intermediate males who have a greater reproductive success than pale males, the effect being most pronounced when the male has no previous breeding experience. The observations can be explained by assuming that females have mating preferences for the darker males.

Each breeding pair defends a territory. O'Donald reports that in some cases territories are defended very strongly with males attacking intruders viciously, while others rarely attack. Males with higher levels of the hormones gonadotrophin and androgen maintain larger territories and court females more actively and persistently than do other males. A model finding some support from the data is that many females have a lower threshold of response toward the dark or intermediate males. Dark or intermediate males tend to have the larger sized territories; however, pale birds have an overall advantage because they are younger when they first breed.

In conclusion, a polymorphism in the arctic skua is maintained by a combination of breeding age, mating preferences, and sexual selection differences associated with territory size variations among three phenotypes. Detailed work of the type being carried out by O'Donald is clearly a necessary concomitant to our understanding of the population structures in territorial species exhibiting obvious polymorphisms.

13-6 GENETIC AND CULTURAL DIVERGENCES IN HUMAN TRIBES

Howells (1966) analyzed population structure in Bougainville in the Solomon Islands, considering 18 ethnic groups representing most of the Islands' territory (Figure 13-6). Linguistic and cultural evidence makes it clear that the existing ethnic differentiation has not been purely a local process, since at least three distinct immigrant groups are involved. The physical environment varies from coastal beaches to hill slopes and ridges to mountainous areas along the central spine. Based on data collected on over 1,300 males, a number of "distances" between the 18 ethnic groups were computed, as well as correlation coefficient between the 153 possible pairings among the 18 groups. The distances relevant for our discussion are:

• Geographic distance (GEOG): Measured between centers of group areas.
• Linguistic distance (LING): A measure taking into account the number of shared words between languages.
• Size distance (SIZE): Penrose's (1954) size distance, which measures general size independent of proportional differences. It was based on eight

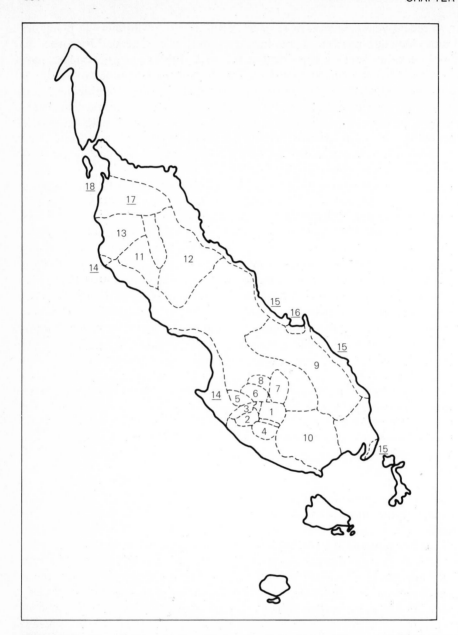

Figure 13-6 Map of Bougainville showing location of populations studied. *Underlined groups,* 14 to 18, are Melanesian speaking. *(From Howells, 1966.)*

measurements including sitting height, arm length, chest breadth, head length, and four other head measurements.

• Shape distance (SHAPE): Penrose's (1954) shape distance, which measures differences in proportions keeping size constant. It was based on the same eight measurements as SIZE.

• Morphological observations (SCOPIC): Anthroposcopic or nonmetric observations of a number of traits including hair form, hair texture, head hair color, eye color, height of eye opening, forehead slope, nasal root breadth, nasal tip inclination, frontal nostril visibility, and integumental lip thickness. A generalized mean distance was obtained as described in Howells (1966).

The correlation coefficients among the measures are all positive (Table 13-6). Of particular interest to us is the association between the linguistic distance and the three biological distances (SIZE, SHAPE, SCOPIC), since the biological distances can be regarded as under strong genetic control. High and significant correlations were found between these three biological measures, especially SHAPE and SCOPIC, and LING. In other words, there is an association between biological (genetic) and linguistic divergence on Bougainville; or, there is an association between biological and cultural divergence using language as a measure of culture. Such linguistic divergence is presumably due to a combination of migration of differing groups to Bougainville and the occurrence of linguistic drift taking place in an initially genetically homogeneous population, which subsequently leads to the isolation and linguistic differentiation of the respective speakers. Especially if, for some reason such as terrain, populations speaking the same language were partially isolated, this can easily occur. Once having developed, a linguistic difference can discourage social contact and acts as a barrier to gene flow so that genetic divergence evolves. Indeed, Friedlander et al. (1971), in a consideration of 18 villages in Bougainville, found that intermarriage among villages was largely confined to the same linguistic group; there was little migration among linguistic groups. In their research they used blood group polymorphisms and anthropometric data as measures of biological distance and came to conclusions qualitatively similar to Howell's (Table 13-6).

Table 13-6 Coefficients of Correlation Among Various Distance Measures in Two Studies

Measure*	GEOG	LING	SIZE	SHAPE
LING	0.58			
SIZE	0.13	0.31		
SHAPE	0.24	0.43	0.36	
SCOPIC	0.22	0.42	0.45	0.28

Measure†	1	2	3
1 Geographical			
2 Linguistic	0.506		
3 Serological	0.406	0.565	
4 Anthropometric	0.170	0.547	0.416

* Modified from Howells (1966).
† Modified from Friedlander et al. (1971).
In both examples $P = 0.01$ for a deviation from zero when the correlation coefficient is 0.22.

The other measure in Table 13-6 (GEOG) is of less relevance for the behavior geneticist. For Howells's data there is a high correlation with LING, but lower correlations with the three biological measures (SIZE, SHAPE, SCOPIC). An association of geographical distance with the other four variables seems, in any case, inevitable. However, the correlations of LING with the biological measures are higher in all cases than for GEOG. Friedlander et al. (1971) reported results that are essentially similar (Table 13-6): a high correlation between geographical and linguistic distances and the correlation of the linguistic distances with the two biological distances exceeding those with geographical distance. Taken together, the two surveys therefore agree in showing an association between biological and linguistic divergence, which means an association between genetic and linguistic divergence.

Studies made of preindustrial people in other parts of the world are in general agreement with the findings of Howells and Friedlander, for example, in Ruanda-Urundi and Kivu in central Africa (Hiernaux, 1956), in New Guinea (Livingstone, 1963), and the Yanomama Indians in the tropical regions of South America (Spielman, Migliazza, and Neel, 1974; Neel et al., 1977). The same conclusions hold for Australian aboriginal tribes in the Northern Territory of Australia (White and Parsons, 1973). Here sociocultural divergence also was found to be associated with linguistic and genetic divergence. A particularly interesting example based on genetic distances from blood groups comes from the relatively close relation between the Yolngu tribe, who live in northeast Arnhem Land (the northern section of the Northern Territory), and the Aranda tribe, who live in the arid country in the center of Australia (Figure 13-7). Dermatoglyphic data are in agreement. Clearly, this result is at variance with expectation based on geographical distance. The probable explanation is that the Aranda represent a relatively recent migration of northern individuals southward (Birdsell, 1950). Linguistic data support this explanation, since the Yolngu and Aranda languages are more alike than are the languages of the Aranda and the desert tribes adjacent to them in Central Australia.

In Australia as in Bougainville, gene flow between tribes is assumed to be low (Tindale, 1953). This assumption supports the probability that language and associated sociocultural factors discourage social contact and therefore act as a barrier to gene flow. Before European settlement the Aborigines were subdivided into tribes that to some extent were genetically discrete, the isolation being maintained by low migration rates between tribes. Although it is difficult to know what actually happened in the past, it is clear from such tribal groups that remain that linguistic and genetic heterogeneity occurs among tribes and that the two are associated.

It could be proposed that linguistic drift can occur within a genetically homogeneous population, which subsequently leads to the isolation and genetic differentiation of the respective speakers. Such drift would function as a positive feedback system and, as such, gains support from high correlations between linguistic and geographic distance (White and Parsons, 1973). Indeed there is a parallel between linguistic diversity in the Arnhem Land region

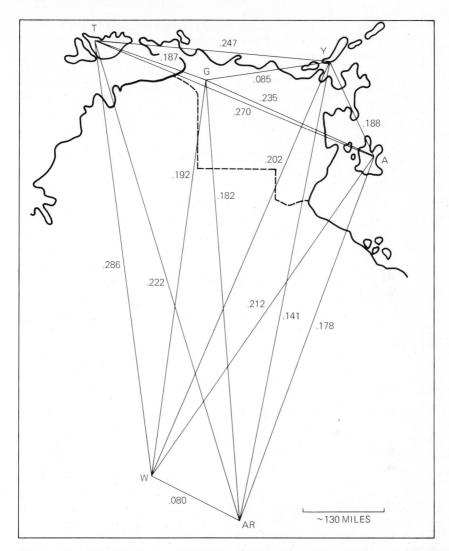

Figure 13-7 Genetic distances between Australian aboriginal tribes based on blood groups and serum protein systems. Tribes are (A) Andilyaugwa, (AR) Aranda, (G) Gunwinggu, (Y) Yolngu, (T) Tiwi, (W) Walbiri. Note the location of the two tribes discussed in the text—Yolngu in northeast Arnhem Land and Aranda in Central Australia. *Dashed line* indicates division between Arnhem Land and Central Australia. *(From White and Parsons, 1973.)*

especially on the coast and genetic heterogeneity—findings in accord with a drift hypothesis (White, 1979). As White and Parsons (1973) wrote

> It is difficult to know exactly what is the cause and effect relationship here, i.e. do initial language differences imply different gene pools, or does linguistic drift occur within a given genetically homogeneous population, which subsequently leads to the isolation and genetic differentiation of the respective speakers? Perhaps the

latter is the more satisfactory explanation and implies that language itself tends to drift with isolation, such that a linguistic difference once having developed may discourage social contact and act as a barrier to gene flow.

The resultant population structure is in any case a mosaic of tribes with somewhat differing overall gene pools.

White and Parsons (1976) then proceeded to consider intratribal variation in northeast Arnhem Land (Figure 13-8) within the Yolngu tribe. Genetic heterogeneity again was found at this level and was related to linguistic and social components especially marriage clusters. The marriage clusters were also associated with drainage divisions such as mountain ranges, which act as natural barriers to communication. The major drainage divisions of northeast Arnhem Land that overlap into the Yolngu territory are shown in Figure 13-8. The conclusion is that the Yolngu function as a marriage network of endogamous units within a broad cultural and linguistic complex. At both the tribal and regional level, these endogamous units are closely related to the drainage division. Indeed, the Yolngu society can be regarded as a confederacy of several tribes, at least when the term *tribe* is used in a biological sense, that is, as a unit largely isolated reproductively from neighboring units.

Birdsell (1973) discussed in some detail tribal size among the Australian aborigines. Many tribes having a population averaging 500 are dialectical tribes, that is, those with no political organization or authority and so without leadership. An exception is the Aranda, whose total number was estimated at about 1,500 at the time of historical contact. The Aranda tribe consists of three subgroups: the Northern Aranda, the Western Aranda, and the Southern Aranda. They recognize themselves as belonging to the same dialectical community, yet are also aware of regional differences in speech. At the time of historical contact, then, it seems likely that this oversized tribe was in the process of differentiating into three new dialectical tribal units that would have averaged about 500. As Birdsell (1973) put it, "The tendency for oversized dialectical tribes to fail to maintain homogeneous speech throughout their bands is not unexpected, and may be explained in terms of the concept of density of communication." This division into new tribes must then be assumed to be associated with sociocultural, linguistic, and genetic differentiation; furthermore, the evidence obtained so far argues for positive correlations among all three factors. As noted above, White and Parsons (1976) have obtained detailed field evidence on this phenomenon for the Yolngu, which currently number 2,400 individuals.

The social structure of the Australian aborigines at the time of historical contact consisted of a number of separate breeding units, so that across (and even within) tribes a mosaic of allele frequencies is expected; even so, broad clines in allele frequencies are apparent on an Australia-wide basis (Kirk, 1966), as found by Selander (1970) in his study of mice over the state of Texas. The expected level of heterogeneity can be assumed to be lower than in wild mice where effective breeding units are normally less than 100. The relative differences in population size can possibly be attributed to the development of language as a method of communication in the Australian Aborigines.

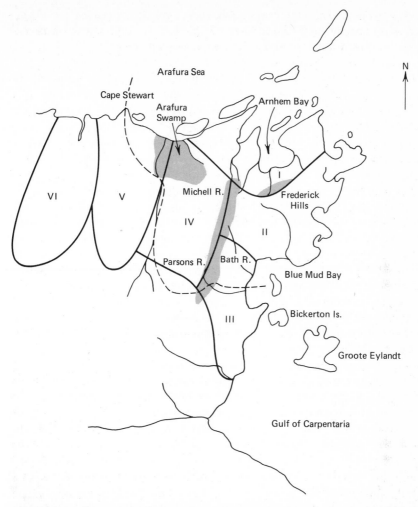

Figure 13-8 Map of northeast Arnhem Land, Northern Territory, showing drainage divisions (I–VI) and some topographical features (shaded). Solid lines indicate major river systems, and the dashed line is the approximate boundary of the Yolngu. (*From White and Parsons, 1976.*)

13-7 EVOLUTION OF BEHAVIOR IN *HOMO SAPIENS*

The evolution of human beings has been described in many texts (Mayr, 1963; Dobzhansky, 1964; Harrison et al., 1964); here our aim is to mention some of those behavioral changes that are important in this evolutionary process. Contemporary *Homo sapiens* is the end product of a long evolutionary history, as was realized by Darwin (1871). From the fossil record it appears that the Hominidae (human) and Pongidae (ape) families separated during the Eocene geological period and the differentiation was well established by the Miocene and Pliocene epochs about 10 million to 15 million years ago. In the later part of the Pleistocene epoch, about 2 million years ago, the first definite hominid appeared, *Australopithecus.* This remarkable form was characterized by: (1) enlarged cranium and hence increased neural tissue, although his brain was only slightly larger than that of the modern chimpanzee and hardly more than one-third the size of that of *Homo sapiens;* (2) bipedal locomotion; (3) the use of tools, which is permitted by the development of bipedal locomotion; (4) the ability to communicate and to hunt in bands; and (5) the development of a carnivorous diet. This form existed until well into the Ice Age, perhaps up to 700,000 years ago.

The next major step was the appearance of *Homo erectus* about 600,000 years ago. This species was characterized by a brain size of about 1,000 cubic centimeters, approximately double that of *Australopithecus* and 75 percent of that of *Homo sapiens.* Fossils of *H. erectus* often coincide with stone tools, including axes, that were made and used. Since the fossil sites contain bones of large animals that the hunters apparently killed, the existence of well-organized bands is implied. The existence of both stone tools and organized bands implies a form of speech among members of this species, a much more developed level of communication than that found among apes and other animals. Communication is quite general in the nonhuman primates, but not verbal communication. The genetically determined neural substrate is presumably not sufficient to support speech behavior (DeVore, 1965). Evidence from fossil sites indicates that the behavioral traits of *H. erectus* were closer to those of *Homo sapiens* than to those of *Australopithecus.*

The first people indistinguishable from ourselves evolved about 35,000 to 40,000 years ago during the last advance of the glaciers. Primitive forms of *Homo* disappeared, but as yet there are inadequate fossils to determine whether this occurred by evolution, hybridization, or extinction (Washburn, 1978). Their appearance was accompanied by a rapid expansion, diversification, and improvement of culture. They buried their dead with flowers and implements carefully laid around the body, so it is not unreasonable to assume that they believed in an afterlife and had some form of religion. This was *Homo sapiens*—modern people.

The evolutionary trend is toward the development of intellectual capacity—the feature that makes human beings' position unique. Morphological trends such as an increase in brain size from about 500 cubic centimeters in *Australopithecus* to about 1400 cubic centimeters in *H. sapiens* and the development of bipedalism, plus behavioral trends such as the development of the ability to communicate and make tools, all are in agreement with the developing capacity to engage in highly coordinated cooperative actions. In modern *Homo*, in addition to the advances pioneered by *H. erectus,* we have: (1) advanced toolmaking, (2) elaborate cultural organization, (3) additional increase in brain size, (4) prolonged childhood and adolescence providing a longer period in which cultural achievements can be assimilated, (5) a degree of control of the environment by means of advances in medicine and technology.

The increase of brain size is an example of directional selection that in terms of the paleontological time scale was very rapid. Since it was associated with the development of progressively more advanced intellectual capacity, there must have been a selective advantage in more efficient communication, perhaps related to tool development, the use of fire, and hunting in bands. Quite likely the period when *Homo's* brain was increasing most rapidly coincided with the evolution of the ability to invent and use language for communication. Speech is not only essential for these behavioral developments, it also is basic to the development of ideas and plans for the future. When we look at the latest evolutionary strides, the importance of behavioral changes in initiating new evolutionary phenomena is undeniable. It can reasonably be assumed that initially there were only minor modifications at the structural level but that the evolution of a morphological change followed a permanent behavioral change. For example, the perfection of bipedalism may well have been speeded up by the preoccupation of the anterior extremities with manipulation, a behavioral characteristic. Similarly, the significant increases in brain size were associated with the development of an efficient system of communication, speech. It is remarkable that the development of language was accompanied by tremendous linguistic diversity and isolation. This was associated with genetic diversity as observed in those regions where tribes of *H. sapiens* can be still studied (see Section 13-6).

Quite likely the breeding structure of primitive people affected their evolutionary rate. If the male leader of a group had several wives (polygyny), he contributed a greater than average share to the genetic composition of the next generation of the group. This tremendous reproductive advantage would favor the characteristics of the particular man, since for leadership certain physical and mental traits would inevitably be favored. These attributes would depend to a considerable extent on the genotype of the individual, so that the reproductive advantage would ultimately make a maximal contribution to the fitness of the entire group. The actual evidence for polygyny is difficult to obtain. But it is observed in a few living primitive tribes and may be an original condition. It is more or less developed in nearly all anthropoid apes (Bartholomew and Birdsell, 1953).

Information available on the population structure of hunters and gatherers is scarce. However, White (1979) tabulated demographic data for the Yolngu tribe of northeast Arnhem Land (Figure 13-8) collected by himself and earlier workers. The average number of surviving offspring per female did not vary greatly, ranging from 2.3 to 3.5. These numbers are consistent with Salzano's 1972 data showing that hunters and gatherers may produce lower means than agriculturalists. This can be explained in terms of child spacing as a response to the inability to carry and suckle two children. An overall male fertility value of 8.1 ± 6.1 living offspring compared with 3.8 ± 2.9 for females was found for the Donydji, a community within the Yolngu tribe studied in detail by White (1979). Women are uniformly exposed to the risk of pregnancy and rarely fail to reproduce, while men are characterized by an appreciably higher variance in their reproductive performance. The basis for this difference is polygyny associated with a relatively late age at which a male obtains his first wife. Concomitantly, many men of reproductive age are without wives. Therefore, some men will reproduce very little and others will make large, often dramatic contributions to the gene pool. We can conclude that male fertility differentials may offer a greater opportunity for selection than those for females. In addition, these data strongly argue for polygyny as an original condition of great importance during the hunter-gatherer stage of human evolution.

Modern humankind, appearing 35,000 to 40,000 years ago, was anatomically indistinguishable from ourselves: there has been little selective pressure since then for altered anatomical features. Has natural selection ceased? The answer must be negative. We have changed from a species living in small hunting communities to a species of which many members live in large, highly organized communities. Up to this century the rate of population increase in the human species was quite slow because of an extrinsic factor—disease. There must have been a high premium placed upon genes for resistance to specific diseases. Some of these diseases must have become important because of the human way of life. For example, Livingstone (1958) traced malaria to the slash-and-burn agriculture that opened the forest floor to stagnant pools and brought us into contact with insect vectors and hence malaria. One consequence of malaria was gene pool change. Because of their greater resistance to malaria, carriers of genes for sickle cell anemia, thalassemia, and glucose-6-phosphate dehydrogenase deficiency were favored, and this led to polymorphisms in regions where malaria was present. (See Bergsma, 1979, for details about these hereditary conditions.)

Technological advances have brought people into contact with other diseases (Omenn and Motulsky, 1972). Rodents, attracted to settled populations, brought epidemic diseases. The practice of single-crop agriculture brought nutritional risks (since each cereal has its own limiting amino acids) and a propensity to protein undernutrition and endemic dysentery. Most bizarre is the culturally based occurrence of kuru (Section 11-1), a disease contracted by the cannibalistic practice of eating the brains of the dead. Today, contagious disease has receded as a major problem, but a concomitant effect of the control of disease is a rate of population increase reminiscent of voles during their increase phase (Section 13-5). In the vole situation, emigration of certain genotypes reduced the rate of increase. Until recently emigration has been a factor in human populations, but for us this phase too is nearly over.

The extreme territoriality of small rodents has been noted. Beginning with population sizes of 32 and 56 Norway rats in different experiments in the presence of adequate food and water, Calhoun (1962) showed that rats confined in a quarter-acre enclosure reached a population size of about 150 at 27 months. One would expect, from the very low adult mortality in uncrowded conditions, a population of 5,000 at 27 months. However, at this level infant mortality is high and pathological behavior is evident in both sexes, associated with social dominance of certain males. It seems that much of the abnormal behavior observed was a consequence of the disruption of territories and peck orders by the presence of excess numbers. As a consequence, normal behavior patterns broke down. The predicted population growth curve did not occur because of insufficient space to develop normal social behavioral patterns. Based upon studies of rodents, it can be argued that space may well become a progressively more limiting factor for *Homo sapiens* (Hoagland, 1966); and as it becomes more limiting, various behavioral changes in populations may occur that may be to some extent under genetic control.

Is there evidence for stress syndromes in human populations leading to a reduction of the growth rate? The human pituitary responds to stress in a way similar to the response of other mammals, and there is indirect evidence that inmates of concentration camps experienced acute stress syndromes that accounted for their deaths. Concentration camps may be more analogous to highly congested animal populations than city slums, since even in very crowded cities the poor have some mobility, although the occurrence of street gangs and juvenile delinquency characteristic of crowded cities is a form of social pathology. The increased incidence of atherosclerosis and other cardiovascular pathology associated with urban living and its competitive stresses may be enhanced by crowding. In underdeveloped countries where high birth rates and recently lowered death rates produce a population growth of 2 percent or more per year, the use of health measures increasing life expectancies masks any growth-retarding effect of a stress syndrome. Therefore, direct comparisons with rodents are hardly possible. Although our societies are far from the rat colonies, space limitations are likely to become progressively more important; indeed, they are already apparent in the poorer sections of some of the world's largest cities and the tendency is likely to increase. The development of technology reduces the number of people needed to produce food and encourages the drift of people toward overcrowded conditions, a tendency that was accelerated by the industrial revolution and that is continuing today.

Demographic, medical, and technologic changes associated with social change are now occurring so rapidly that populations have difficulty adapting. Consequently, improved understanding of human behavior and its genetic basis is essential. There may be, for two examples, differences among human genotypes in tolerances to crowding and perhaps to noise levels enhanced by crowding. In these ways humankind may still be evolving in directions different from those favored in the past.

13-8 SOCIAL STRUCTURES: THE EVOLUTIONARY SIGNIFICANCE OF BEHAVIOR

In this book we have mainly considered natural selection in terms of individuals in a population—and the action on genotypes through the differential reproductive success of individuals. For the most part, species' characteristics involve physical and behavioral properties that serve to increase the chances of each individual passing on his genes to the next generation. In Darwinian terms, fitness may be viewed as a particular individual's success in achieving this goal. However, in this chapter from Section 13-4 on, we have considered various forms of selection at the level of the local population. This is most generally *interdeme* selection, which is the differential reproduction of different local populations. When the populations involved are social groups the term *group selection* is frequently used. When selection is based upon the kinship as the social group, the term *kin selection* is used. According to Wilson (1973), pure kin selection and pure interdeme selection form the opposite ends in the spectrum of all conceivable cases of group selection. (See Maynard Smith, 1976, for further discussions.)

Population biologists have argued extensively about the reality of interdeme selection. Williams (1966), Dawkins (1976), and others have argued that modifications of individual selection are capable of bringing about the effects attributed to population selection "except under the most restrictive of conditions" (Maynard Smith, 1976). On the other hand, Wright (1960) and others have considered that interdeme selection is a real process in nature. Wright argues that the optimal population structure is one where there is considerable heterogeneity brought about by subdivision into nearly isolated "island" populations. The optimum is due to a delicate balance between intensity of selection, amount of migration, and local effective population size. If to this we add the increasing evidence for habitat selection discussed in Sections 13-3 and 13-4 then the possibility of interdeme selection becomes a biological possibility. Under interdeme selection, the less fit "island" populations will dwindle and become extinct perhaps because of a subtle habitat change, while the better-adapted colonies will expand and replace them throughout the population range. This is followed by the development of a new subdivided population system with a new set of adaptive characteristics as the result of initial selection within colonies followed by the differential reproduction of different colonies. Even so, as Dawkins (1976) cogently pointed out, such population structures are almost certainly reducible to the selection of fitter at the expense of less fit individuals, that is, individual selection.

Those considering kin selection have looked for traits benefiting the population but disadvantageous to the individuals possessing them. These are referred to as *altruistic* traits and must necessarily be considered both in terms of the whole population and the individual. Interdeme selection is not ruled out since kin selection in many cases may come close to deme selection. Examples of altruistic traits include the sharing of food, giving warning signals, and the cooperative efforts found in social insect colonies. In the case of the honeybee, *Apis mellifica*, the unit of selection definitely appears to be the colony as a whole; indeed, the social bees represent an extreme of social integration in the animal world. In this case the functions of food getting, defense and maintenance of the colony, and rearing of the broods are carried out by the workers.

However since the worker bees are sexually neuter, they cannot reproduce as individuals and consequently have no opportunities for passing on genes determining their adaptive characteristics across generations—a function performed by the queens and drones. It follows that if the queens and drones do not carry genes that produce biologically fit workers, then the hive will not thrive and may be eliminated by competition from other hives with better adapted workers. Such replacements have been observed in recent years (Michener, 1975) showing that the unit of selection is indeed the colony.

Space does not permit a long discussion of altruism—a recent review is given by Maynard Smith (1978). As he points out, R. A. Fisher, Sewall Wright, and J. B. S. Haldane were all aware of the problem and came close to solving it. The current understanding of the evolution of altruistic behavior is traceable to Hamilton (1964). Consider an individual donor who performs an act that reduces its own Darwinian fitness or expected number of surviving offspring by a cost C but increases the beneficiary's fitness by a benefit B. For a pair of alleles A and a, let the presence of A make an individual more likely to perform the act. Then the change in frequency of gene A in the population after the act, depends upon the coefficient of relationship (defined in Section 6-9) between the donor and the recipient. With certain approximations, Hamilton showed that the frequency of gene A will increase because of the donor's act if the coefficient of relationship r is greater than C/B. For example, consider full sibs where $r = \frac{1}{2}$, then if the cost $C = 1$ and the benefit $B > 2$, then indeed $r > C/B$. From Hamilton's work, it can be predicted that altruistic and cooperative behavior will be found more frequently in interactions of related rather than unrelated individuals. Obviously from the above, the more remote the relationship between two individuals, the greater the benefit relative to the cost for the possibility of the evolution of altruistic behavior. Social insects provide excellent examples of cooperative behavior. (Note however that relationships are rather more complex in *Hymenoptera* because of their haplo-diploid modes of inheritance.) In recent years, Hamilton's ideas have been increasingly applied to the social life of higher animals, especially birds and mammals. For elaboration of these models and of evolutionary strategies in animal contests, see Dawkins (1976) and Maynard Smith (1978). Unfortunately, the genetics of such contests is not well known (Parsons, 1967); it is an area worthy of more attention from behavior geneticists.

Because social bees represent an extreme of social integration, the effects of interdeme selection are pronounced. In less socialized animals, such as many mammals and birds, we would expect lesser effects of interdeme selection but some characteristics of such selection, as discussed in Section 13-5. In the primates there is much species-to-species variation in social structure (E. O. Wilson, 1975), although the effects on the genetics of population have been little studied. Eisenberg, Muchenhirn, and Rudran (1972) classified models of social organization in primates as presented below incorporating the modifications of Grant (1977):

1 Maternal family consisting of mother and offspring. Adult females and males have separate centers of activity. Found in the aye-aye and some species of lemur and loris.
2 Biparental family. The family group consists of a female, a male, and their young. Found in a species of woolly lemur, and in some marmosets, New World monkeys, and gibbons.

3 One-male troop. The troop consists of several maternal families and one adult male in contact with them all. The male has a strong intolerance of other mature or maturing males. Found in some New World monkeys (including the howler monkey) and Old World monkeys (including Hamadryas baboons and Gelada baboons).

4 Age-graded multiple-male troop. A cohesive group consisting of several females, several males, and young. There is an intermediate degree of male tolerance. This degree of tolerance permits the coexistence of several males of varying ages, mostly young males, with the dominance order corresponding to the age of the males. Found in some New World monkeys (including, again, the howler monkey) and Old World monkeys (including macaques), and the gorilla.

5 True multiple-male troop. As in mode 4, but with a high degree of male tolerance, permitting the coexistence of several adult males. These are co-dominant and cooperative and maintain a flexible oligarchy in the troop. Found in a species of lemur, a species of woolly lemur, in some Old World monkeys (including baboons and macaques), and in the chimpanzee.

The above series of types of social organization probably corresponds approximately to evolutionary trends where (1) is primitive, and (4) and (5) derived. In addition, along with social structure changes is a trend toward larger groups. From considerations presented in this chapter, it is perhaps not surprising that social organization types correspond to habitats. One-male troops (3) are mainly found in species with an arboreal foraging habitat, while true multiple-male troops (5) tend to occur mainly among those species with semiterrestrial foraging habitats. The advantages of social groups reside in holding a territory, defense against predators, sharing of experiences, and, most important, communication. As we have seen, these adaptive features apply equally well to hominids. These trends are developed maximally in ourselves as shown by an overall increase in the size of the social group, the complexity of social organization, the power of technology, the sophistication of language, intelligence, and the length of the immature stages of the individual's development. The uniqueness of man's social behavior, derivative from that of primates, has led to the hybrid "discipline" of sociobiology (E. O. Wilson, 1975). This is not the place, in a book on the genetics of behavior, to discuss the full ramifications of sociobiology. However, the behavior genetics of human beings is clearly an underlying component of sociobiology. Therefore, application of the methods and approaches we have discussed should assist in the scientific study of social structures and hence sociobiology. The difficulties in the scientific study of animal populations from this point of view show that real progress in analogous human population studies is not likely to be rapid.

Even so, the emphasis on sociobiology has since Wilson's 1975 book led to a resurgence of activities on the study of mammalian populations, which is an approach through which real progress is likely to be made. As an example, the discussion of Clutton-Brock and Harvey (1978) can be cited. They considered four aspects of mammalian biology—population density, group size, breeding systems, and sexual dimorphism. All four are related and they develop the thesis (1) that variation in the distribution, density, and quality of food supplies is largely responsible for species differences in the density and distribution of mammalian populations, and (2) that differences in population density and distribution play an important part in determining differences in breeding systems and that these in themselves affect a wide variety of morphological and physiological traits. We conclude by using their quote from Aristotle, which unmistakably relates sociobiology with habitat selection:

> For it is differences in feeding habits that make some animals live in herds and others scattered about; some are carnivorous, some vegetarian, others will eat anything. So, in order to make it easier for them to get these nutrients, nature has given them different ways of life. Again, since animals do not all like the same food but have different tastes according to their nature, so the ways of living of carnivorous animals differ among their different kinds, as do also those of the vegetarian.

More generally, one of the emerging findings of sociobiology is that many aspects of an animal's social organization can be predicted on the basis of a limited set of environmental variables. Emlen and Oring (1977) propose that certain environmental factors determine the degree to which mates can be defended or monopolized; that is, ecological constraints impose limits upon the degree to which sexual selection can operate. They characterize mating systems on the basis of the ecological and behavioral potential to monopolize mates. The greater the potential for individuals to monopolize resources or mates, the greater the intensity of sexual selection and the greater the environmental potential for polygamy. Emlen and Oring consider examples mainly from birds, but the models are of much wider application. One consequence is that assuming ecological variables do indeed influence or constrain the intensity of sexual selection, lability in mating systems is to be expected between different populations of a given species in different environmental and/or density situations. Such variations should be predictable. For example, there are situations in a wide range of species including insects suggestive of lek behavior at high population densities but shifting to simpler forms of polygyny at lower population densities.

SUMMARY

Behavior genetics is a central discipline in the study of evolution. In *Drosophila*, male mating behavior is an extremely important component of fitness in laboratory experiments, and field examples are observed in many species.

In nature, the interplay of behavior genetics and ecological genetics has led to the interdisciplinary topic of habitat selection. In *Drosophila*, this means the study of oviposition and hence resource utilization under permissive physical conditions of temperature, light, and humidity. Closely related species often diverge considerably for such variables. In the deer mouse, *Peromyscus*, complex behavioral phenotypes have been described relating to field and woodland habitats in the wild.

Although our knowledge of habitat selection is limited, knowledge of population dynamics is even more restricted. It is, however, genetically important that small rodents such as mice are often subdivided in the wild into small breeding units or demes, and that during the dispersal phases of voles, certain genotypes migrate more than others. The understanding of population structure in territorial species exhibiting obvious polymorphisms is not yet great, but a start has been made in the arctic skua.

Contemporary people are the end-point of a long evolutionary history involving major behavioral changes that incorporated the development of intellectual capacity. Population structures analyzed in hunter–gatherer societies such as Australian Aborigines who are organized into tribes, show an association of genetic divergence with sociocultural divergence and with ecological boundaries. Such structures have broken down in urban dwellers, but are important for our understanding of genetic heterogeneity in modern peoples.

In recent years, the evolutionary significance of behavior has attracted great attention leading to the hybrid area of sociobiology. The behavior genetics of humans and experimental animals is a critically important component underlying sociobiology.

GENERAL READINGS

Barash, D. P. 1977. *Sociobiology and Behavior*. New York: Elsevier. One of the "second generation" accounts of the rapidly developing field of sociobiology.

Dawkins, R. 1976. *The Selfish Gene*. Oxford: Oxford University Press. A formulation of sociobiology in terms of the genetical theory of natural selection, stressing the importance of individual selection.

Dobzhansky, T., F. J. Ayala, G. L. Stebbins, and J. W. Valentine. 1977. *Evolution*. San Francisco: Freeman. A modern, comprehensive book on evolution.

Grant, V. 1977. *Organismic Evolution*. San Francisco: Freeman. An excellent, succinct modern book on evolution putting the study of social structures into an evolutionary framework.

Krebs, J. R., and N. B. Davies (eds.). 1978. *Behavioural Ecology: An Evolutionary Approach*. Oxford: Blackwell Scientific Publications. Several commendable articles pertain to the theme of this chapter.

Mayr, E. 1963. *Animal Species and Evolution*. Cambridge, Mass.: Harvard University Press. A classic account of evolution in which the importance of behavior is firmly stressed.

Parsons, P. A. 1973. *Behavioural and Ecological Genetics: A Study in Drosophila*. Oxford: Oxford University Press. A general account based on *Drosophila* research and concluding with a consideration of evolutionary implications.

Washburn, S. L., and R. Moore. 1974. *Ape into Man*. Boston: Little, Brown. A good example of a discussion of human evolution.

Wilson, E. O. 1975. *Sociobiology: The New Synthesis*. Cambridge, Mass.: Belknap Press. Irrespective of approaches to human studies, this is an excellent account of social structures in other animals.

Epilog

14-1 FROM MECHANISMS TO EVOLUTION

In this book, we began by considering the phenotype of an organism in the widest sense, which includes the totality of physiological, anatomical, and behavioral components of an individual, and confirmed that the mechanism of the inheritance of behavior is as for other phenotypic components. Indeed, the behavioral phenotype we observe is the end-product of many genetic pathways interacting among themselves, and with the environment. The study of this phenotype becomes progressively more difficult as we move up the phylogenetic series, culminating in our species for which environments cannot be controlled and breeding experiments cannot be done. Yet a full appreciation of human behavior genetics comes from a deep understanding of methods and techniques used in experimental animals. This is critically important, since armed with knowledge of the genetic bases of human characteristics, it is possible to offer advice to individuals on their chances of having a disorder appear in themselves or in their immediate family. As our knowledge of the genetical and especially molecular bases of human disorders increases, so will the accuracy and predictive value of such genetic counseling increase.

Extending our coverage to quantitative traits, we found that all are controlled by genotype and environment, with the possible exception of laterality. Although the demonstration of a genetic component in experimental animals is simple enough, doing so is much more difficult for people because of complex interactions between genotypes and environments. However, if we assume that measurable behavioral traits have genetic components even if difficult to pinpoint, then behavior genetics becomes especially important in understanding biology especially at the population and the evolutionary levels. At the applied level, the expected and concomitant behavioral and genetic changes consequent upon domestication have not been elaborated in the literature but they obviously become important as agricultural production procedures become more intensive on a worldwide basis.

Even though we have attempted to consider the field from the points of view of mechanism and evolution, there remains today a deep split between these two areas. Concerning a slightly narrower sense, the contrast between mendelian and quantitative genetics, Caspari (1977) recently wrote:

> This is a remnant of the violent fight between two schools which dominated research in genetics in the first two decades of the 20th century, the Mendelians and the biometricians. The history of this fight has been described by Provine (1971) who concludes that it was ended by the theory of R. A. Fisher: the laws of biometrical genetics can be understood as the effects of a large number of Mendelian genes with small additive effects. While this is correct in principle, in practice the two areas of genetics have remained separated, both in technique and in the questions they ask. This division has become even sharper due to the merger of Mendelian genetics with molecular biology.

The field of behavior genetics illustrates this perhaps better than any other, because the merger of mendelian genetics with molecular genetics has led to the field of neurogenetics, which is rapidly developing. Neurogenetics can be regarded as an extension of developmental genetics, which had its origins many years ago. The genetic basis of the behavioral phenotype, both at the mendelian and quantitative genetic levels, is now of central importance. Indeed the behavioral phenotype both in the laboratory and in natural populations, is the foundation material upon which we work.

In Chapter 1 three particular concerns for behavior genetics were mentioned: (1) difficulty of environmental control, (2) difficulty of objective measurement, and (3) the study of learning and reasoning. These are discussed many times in the text. Fundamental to the study of behavior genetics in this way is quantitative genetics, which allows the disentangling of genetic and environmental influences, particularly in nonhuman animals. In man this is enormously difficult but the problems must be appreciated for progress to occur. It must be stressed, however, that despite the difficulties, progress is being made. In recent years there has been steadily increasing feedback from behavior genetics to evolutionary studies. Some of these will now be discussed here.

14-2 POPULATION GENETICS: DOES RANDOM MATING OCCUR?

Random mating is commonly assumed in natural populations—and indeed the theoretical foundations of population genetics are based on this assumption. It must be admitted that this is done in part to permit simple mathematical modeling. In practice, random mating rarely occurs. This statement seems to apply whether laboratory or nonlaboratory populations are considered. It is certainly true that assortative mating is the rule for *Homo sapiens* for a variety of reasons ranging from relative attractions based on varying somatotypes or intelligence levels to cultural elements such as religion. Evidence for assortative mating and nonrandom mating in general is readily obtainable for humans, but interpreting the patterns thus found is quite another issue. In that most-studied laboratory species, *Drosophila melanogaster,* extrapolations to the field are essential, but at least the evidence for rapid male mating speed as a component of fitness is overwhelming in laboratory studies. It is difficult to avoid the conclusion that random mating may be a rare phenomenon in natural populations of *Drosophila,* especially if we add the rare-male mating advantage phenomenon. Particularly unclear is the extent to which deviations from random mating occur in the *Drosophila* species showing lek behavior, as is found in higher organisms showing such behavior. What are the genetic consequences of such differences? Indeed our information on the genetic consequences of territoriality is limited, although where studied in rodents and birds they are real enough and include nonrandom mating.

Such considerations lead us to sexual selection and sexual isolation—phenomena under repeated discussion in this book. Thoughts on sexual selection lead to consideration of those secondary sexual characteristics conferring some social benefits on their bearers. Two main types of benefits can be envisaged. Either their bearers have superior competitive ability against others of the same sex (e.g., antlers in deer) or they possess increased sexual appeal in competition for the opposite sex. In the former case, selection is within one sex (*intrasexual* selection), and in the latter, selection is performed by the opposite sex (*intersexual* selection). Intrasexual selection includes selection for social rank, especially if dominant individuals contribute more of certain genes—for example, for size and aggressiveness—to future generations than those that are not socially dominant. Sexual selection has been considered several times in this book mainly in *Drosophila* but also in mice, fish, and chickens and other birds. Writing on guppies, Farr (1976) predicts that ''closer examination will reveal that interstrain competition and resulting dimorphisms or polymorphisms are present in many species'' and that ''once we have begun to discover more diverse examples, we will be able to define the evolutionary and ecological prerequisites for maintenance of polymorphisms by sexual selection.'' Although there are difficulties in distinguishing intrasexual from intersexual selection, female preferences occur in mating. For example, when *Drosophila* flies meet on food masses, each female has the opportunity of choosing among competing males and because she effectively rejects some suitors, intersexual selection can be inferred; indeed Spieth (1974a) concludes that, in nature, *Drosophila* females are responsible for sexual selection. This is supported by Crossley's 1975 experimental work on selection for ethological isolation in *D. melanogaster* (Section 5-3). The reconciliation of this observation with laboratory data on male mating speed as a component of fitness is clearly needed.

We have seen that levels of sexual isolation in *Drosophila* are determined by a combination of genotype and environment. In addition, as noted in Section 13-2, a *D. persimilis* female once having mated with a *D. persimilis* male will not subsequently accept a male of its sibling species, *D. pseudoobscura*. Similarly in *D. pseudoobscura*, females tend to mate with the type of male previously accepted by them (Section 8-4). These and other data indicate a learned component in *Drosophila* mating; certainly it is a phenomenon of potential significance in natural populations, and of unknown importance for the structure of populations.

Referring mainly to observations on vertebrates, especially birds, it appears that when choosing a mate, members of a polymorphic species may (see Brown, 1975):

- choose mates of similar phenotype to their own phenotype—positive assortative mating
- choose mates differing from their own phenotype—negative assortative mating
- choose mates of the same phenotype as their mother or father—a process of *imprinting* on the mother or father respectively
- mate at random (unlikely)

Imprinting has been referred to in Chapter 10 in fish, chickens, and blue snow geese and has been reported or inferred in many other species including *Homo sapiens* (Section 7-5). Indeed it has been suggested that examples of assortative mating in natural populations may often be due to imprinting.

Imprinting is of particular interest to us since it involves a learning process, whereby sexual selection (and isolation) is affected by early experience. The house mouse provides another example of early experience on mating preference. Sexually mature female laboratory mice, *Mus musculus domesticus*, reared with both parents approach males of their own species more readily than males of the distantly related subspecies *M.m. bactrianus*, while females reared by their mothers alone showed no discrimination (Mainardi, 1963). Mainardi suggested that infant female mice are imprinted on their fathers and that this experience affects later sexual preferences. An intriguing positive association with parental traits was found when young females of *M.m. domesticus* were reared with artificially perfumed parents (using the perfume Violetta di Parma) and later allowed to choose between perfumed and ordinary males in a special cage measuring social attraction rather than actual matings (Mainardi, Marsan, and Pasquali, 1965). More recently, Doty (1974) has reviewed evidence suggesting the importance of olfaction on mating behavior of rodents, especially for

female preferences, and Yamazaki et al. (1976) found that the mating preferences of male mice are influenced by genetic differences in the major histocompatibility (H-2) complex. These results reinforce the possibility suggested by
Ehrman and her colleagues (Section 8-4) that mating in *Drosophila* may depend
on a characteristic "bouquet"—clearly such a bouquet may be determined
both genetically and environmentally. In the human context, such preference
may be a culturally inherited trait. Or to repeat a question posed in Section 7-5,
to what extent is nonrandom mating of human beings due to an imprinting
process based on the traits of one's relatives? Questions of this nature are just
beginning to appear in the mouse literature (Gilder and Slater, 1978).

In the terminology of Mayr's (1974) closed and open genetic programs (in
Chapter 1), sexual isolation in *Drosophila* is considered to be largely a closed
program, with experiential components enhancing isolation as already discussed (Parsons, 1977d). Open systems will be favored in animals with prolonged parental care; imprinting is clearly an example of an open program.
Generally, open programs are expected to be more prominent in the higher
vertebrates, mammals and birds, this being particularly important in human
behavior, including mating patterns.

We therefore conclude that the study of mating patterns suggests the
nonexistence of random mating within species and within populations within
species. Most evidence is at the gross morphological level or in association with
a particular visual polymorphism. Genetic effects will clearly depend upon the
traits and genetic systems involved, together with correlated traits and, in some
instances, linkage relationships. More subtle behavioral effects occur when
mating preferences are affected by early experience; if these involve chemical
cues, mating patterns could be modified extremely frequently.

14-3   BEHAVIOR AND SPECIATION

For an understanding of speciation, the study of behavior is critically important. Interspecies differences are effective in ensuring reproductive isolation in
nature. Mayr (1963) wrote, "The shift into a new niche or adaptive zone is
almost without exception, initiated by a change in behavior." Studies in
Drosophila and the deer mouse, *Peromyscus,* show substantial behavioral variations among the various taxa that can be related to habitats, with some associated morphological and physiological changes. A dramatic example of associated behavioral and morphological changes is the evolution of *Homo sapiens.*
Indeed if one constructs the phylogeny of a group of animals such as ducks and
pigeons based on behavioral traits, that phylogeny is exceedingly similar to one
independently designed on the basis of strictly morphological traits. This paral-
lelism signifies that both sets of characters are the product of the same genotype
(Mayr, 1974). Perhaps the best insect example is Spieth's (1952) *Drosophila*
study which encompassed 101 species and subspecies representing 21 species
groups; generally the evolution of mating behavior paralleled the morphological
evolution of the group. Spieth concluded that divergence of the mating behavior

among species occurred first at the physiological and behavioral levels and that the visually observable morphological changes arose much later. Brown (1965) quantified differences among 11 species of the *D. obscura* group for behavioral and morphological characters using a measure of "mean character differences" based on 20 behavioral and 24 morphological traits, from which a high correlation between behavioral and morphological divergence emerged. Considering these data, plus those for the sibling species *D. melanogaster* and *D. simulans,* it is clear that both behavioral and morphological differences between mutants within species are slight, those between sibling species are greater, and those between nonsibling species in the same division greater still (Parsons, 1977d). The same applies to the level of subgenera, which show major differences in behavior and morphology. Spieth (1974b) reiterated this general view after working on the Hawaiian Drosophilidae.

Of vital significance in our understanding of the role of behavior in speciation are those quite rare taxa at the species boundary. We have seen in Sections 5-3 and 8-4 that *D. paulistorum* consists of six semispecies or incipient species among which many degrees of sexual isolation exist even though the species are extremely similar morphologically. Clear evidence for natural selection comes from the higher degrees of isolation in sympatric compared with allopatric situations (Ehrman, 1965); this result is generalized to frogs in Section 10-10. From direct observations we know that a fly of a given semispecies once having mated within its semispecies prefers subsequent matings even more strongly within that semispecies—that is, isolation is enhanced and eventually this would be fixed into the genome (Pruzan et al., 1979). In these ways levels of sexual isolation are increased by natural selection. We therefore see that isolation is initially augmented by a programmed learning process and often precedes morphological changes. It almost certainly follows for species complexes where there are no apparent behavioral isolating mechanisms that they occur and have not yet been detected by the study methods available. It can be concluded that taxa of the *D. paulistorum* type are of great value in our understanding of mechanisms of speciation, since not only direct assessments of mating behavior can be made but also the genetic bases of sexual isolation levels can be investigated. The one gap in *Drosophila* speciation is extrapolation in nature. This poses problems, but since the genuine study of *Drosophila* ecology is recent, we hope they are not insurmountable. As far as mating behavior is concerned, the lek species discussed in Table 13-2, which use the undersides of bracket fungi as courting territories and can therefore be filmed in nature, may provide a starting point especially as two species *D. mycetophaga* and *D. polypori* frequently occur under the same fungus.

Comments have already been made on the paucity of studies on wild mice—studies of the type Mainardi has carried out in species complexes such as *Mus* and *Peromyscus*. Incorporating olfactory components would seem an area ripe for investigation as indeed Parsons (1967a) suggested when writing on "The Genetic Analysis of Behavior."

It can be concluded that under suitable conditions behavioral differences in courtship *within* species may become the prime traits differentiating closely related species, and eventually—possibly after a considerable period of time—morphological differences would become apparent. What is needed however, is to see how such differences could have arisen during the course of evolution and how behavior serves to adapt the animal to its environment. In the more general sense, the question has been more rarely posed, although an investigation through habitats selected (Parsons, 1978b) would seem a method to consider. Parsons's conclusion from the genus *Drosophila* is that "there is an association between resource utilization divergence, habitat selected, and taxonomic divergence," and conversely that "closely related species will select differing habitats assessible by a consideration of resources utilized or by some other behavioral difference." Section 13-2 argues that food and habitat selection may be of significance in the formation of races within species and, ultimately, in speciation. One observation of note in this regard is Manning's (1967b) finding that *Drosophila* reared on a medium containing geraniol (a peppermint smell) showed reduced aversion to the odor when adult, which Manning regards as a form of habituation. Since certain *Drosophila* species feed on fungi in nature, another perhaps more significant observation is that adults of *D. guttifera* reared on a mushroom medium tend to oviposit on it (Cushing, 1941). It is reasonable that there could be genetic assimilation for such behaviors over a number of generations, as suggested by Moray and Connolly (1963). In passing it should be noted that habituation to chemical gradients occurs in the nematode, *Caenorhabditis elegans* (Section 10-3), which could relate in some ways to environmental cues in nature. This is reasonable as *C. elegans* normally eats soil bacteria. Habitat selection in response to physicochemical gradients with no evidence (as yet) of any form of habituation or learning seems the rule in lower organisms with simple nervous systems such as *Paramecia* (Gould, 1974).

Extrapolating to the wild, if a newly-entered habitat is characterized by some sort of resource difference that can be assimilated genetically, then evolutionary change is possible in response to the habitat change. Studies to distinguish learning from habituation are of obvious significance in this regard. Finally, to return to Mayr's terminology of closed and open programs, it would appear that an open program is advantageous in adapting to habitat changes, as compared with the advantage of a closed program with its associated stereotyped behavior in mating and communication.

Further discussions of behavior in *Drosophila* speciation are given in Parsons (1980), where it is argued that the combination of resource utilization heterogeneity followed by habitat selection, and assortative mating within habitats (Maynard Smith, 1966) could in theory lead to speciation without geographic isolation. The process could be accelerated by learning followed by genetic assimilation during the adaptation to new resources. An integrated approach involving aspects of the ecology, behavior, and genetics of natural populations is needed for an understanding of speciation in these circumstances. This has also been stressed by Bush (1969, 1975) in his studies of the true fruit flies (Tephritidae) where species of *Rhagoletis* have evolved in North America over a relatively short period of time to parasitize domesticated European fruits such as cherries and apples. Indeed, these considerations suggest that ecological independence preceded reproductive isolation, which evolved as a consequence of differing preferred habitats. Therefore it is essential to relate results from the laboratory to natural populations and vice versa for an understanding of the genetics of speciation.

14-4 QUANTITATIVE TRAITS: GENOTYPES, ENVIRONMENTS, INTERACTIONS, AND CORRELATIONS

The behavioral phenotype is made up of heredity and environment. Genotypes differ in response to a common environment, and conversely one genotype varies in response to different environments, as illustrated by McKenzie's mating speed data in Section 6-2. For the behavior geneticist interested in laboratory populations of *Drosophila* this is not a serious problem, since such genotype-environmental interactions (GE) can be estimated by appropriate statistical designs, as can the main effects of genotypes and environments. It is also possible, in appropriately designed experiments, to obtain heritability estimates.

When we turn to rodents, complications are more likely. We refer to Henderson's (1970) analysis of learning as assessed by a food-seeking task in mice (Section 9-3), where in essence he found that as an environment becomes more favorable, the degree of genetic expression of a character also increases. That is, in an enriched or more advantageous environment, genetic effects tend to be expressed more fully than in a situation approaching environmental impoverishment. Also, while improvement is differential among different genotypes, the performance of all groups increases as a function of environmental improvement. This is an animal demonstration of the reaction range concept of GE interaction described by Gottesman (1963) for human behavior. This concept suggests that a shift to a more favorable environment increases individual differences through an increase in genetic variation. Even so, it is important to note that it is often impossible to generalize "environmental improvement" across different groups.

Since the expression of a genotype may then be correlated with the environment, one of the basic assumptions of many quantitative genetic models, namely that of no GE correlations, breaks down. If there is a GE interaction, total phenotypic variance is increased, but on the other hand if genotypic and environmental contributions are correlated, the variance can be either increased or decreased. Such covariance could be manifested by different genotypes being differentially distributed across environmental conditions; an obvious example is the ecological distribution of species, subspecies, races, and even genotypes, to those habitats to which they are best adapted. Viewed in this way, habitat selection could be regarded as a form of GE correlation; the discussions in the previous chapter should be assessed in this light. We can then regard GE correlation as resulting from selection based upon earlier GE interactions. It is difficult to think of mutually exclusive examples of GE correlation and GE interaction. It is highly probable that both types of relationships between genetic and environmental factors are continually operating.

One of the consequences of GE correlation is that the heritability is affected, and heritability estimates will vary according to whether the correlation is positive, negative, or zero. In other words, following Moran (1973), "for characteristics such as human intelligence in which genetic and environmental components are correlated, 'heritability' cannot be defined." Indeed, most estimates of heritability of intelligence usually include the covariance in the proportion of the variance due to heredity. In fact, as noted in Chapters 7 and 12, correlations in the range $+0.2$ to $+0.3$ are repeatedly obtained between intelligence and socioeconomic status, beginning at 6 years of age or earlier. We are now entering the difficult field of cultural inheritance, with the problems of distinguishing between the two types of inheritance when confronted with a set of data (Section 12-9).

The understanding of gene-environment relationships is thus one of the features of great concern for behavior genetics. Partly resulting from the studies of behavior geneticists and their statistically minded colleagues, this concern is beginning to feed back a methodology that may slowly modify our approaches to quantitative genetics. Quantitative genetics, in estimating the degree of genetic control of traits, frequently minimizes environmental variation, usually by making it optimal. Behavior genetics must regard both the genotype and the environment as equally important, since genotypes may vary in response to a fixed environment and, conversely, one genotype may have varying responses in different environments. It is important to study behavior under both situations, perhaps simultaneously if the experiment itself does not then become excessively unwieldy.

It has been argued several times in this book that phenotypes controlled by directional selection for extreme expression of a trait in hybrids will have been subject to natural selection for this extreme. The genotype is then evolved to maintain the superiority of the extreme phenotype as an optimum. Mating speed has been discussed in detail, and learning in rodents and IQ in human beings are other examples showing this at a somewhat lesser level. In some cases, as pointed out by Henderson (1978), it is possible from a knowledge of the organism and its environment to predict genetic architectures. Reference has been made to the work of Quinn, Harris, and Benzer (1974) which shows that *Drosophila* can readily acquire a complex odor discrimination shock avoidance response (see also Section 10-5 for similar results in blowflies). Fulker (1979) argued that the ability to learn a complex discrimination task of this nature would have small adaptive value to fast-breeding, short-lived species such as *Drosophila*. Yet a 9×9 diallel cross among wild-type strains indicated consistent dominance for high level of performance, so that avoidance learning is apparently an important component of reproductive fitness in *Drosophila*. Why should this be so? Is it in relation to avoiding various possible noxious substances? When, however, we consider Hay's (1975) demonstration of learning for left or right turns in a maze after flies are initially forced to turn left and right, and his subsequent demonstration (Hay, 1979) that directional dominance for high performance is shown by this trait, there is a real dilemma in attempting to extrapolate to nature. Still, Hay (1972) did find directional dominance for high reactivity to disturbance after mechanical stimulation. This was associated with greater mortality among the less reactive flies, strongly indicating that high reactivity is a fitness trait. Presumably after disturbance it is adaptive in nature to be able to move rapidly to avoid predators.

Henderson's predictions concerning genetic architectures in mice are based upon assessments of the potential adaptive significance of a trait. An artificial test situation that was only peripheral to fitness was expected to and was found to have a largely additive genetic architecture, while directional dominance was found for obvious fitness-associated traits (Section 9-3). In other words, according to the test situation and the environment generally, the genetic architecture can vary markedly. It is often stressed for morphological traits that estimates of parameters providing indications of genetic architecture are dependent upon the array of genotypes tested (e.g., Falconer, 1960). Since most experiments are done under relatively optimal environments rather than an array of realistic environments, the environment is usually imperfectly adjudicated. However, in this book we have shown the need to consider an array of meaningful environments for behavioral traits. It should be noted, too, that for environmental stresses of potential ecological significance, the genetic architecture may indeed vary markedly according to stress levels (Parsons, 1979b). The realization of potentially large effects of environment and its effect on genetic architecture, together with potential effects on GE interactions and correlations, is an area of necessary concern for the quantitative geneticist. It is a concern that has primarily emerged from studies in the field of behavior genetics, but it is subject to universal application.

14-5 FUTURE TRENDS

Predicting future trends in any area of science is tricky; nevertheless it should be attempted. Some of our previous predictions (Ehrman and Parsons, 1976) are now shown fulfilled in the body of this text; this indeed is an indication of rapid progress. Many of the possibilities already mentioned in the book are included in the following list simply because we feel that they may be emphasized further in the future:

1 As we have seen, many studies are based on a restricted series of genotypes in a restricted set of environments. The calculated heritability values are frequently based upon studies of one population in a single environment. When the number of genotypes and environments affecting the expression of a trait is considered, the problem of generalization is enormous. It becomes incredibly complex if the environment is defined in the broadest sense to include not only the physical environment but also effects of previous experience.

Future endeavors will have to come to grips with this problem and with the associated difficulty of GE correlation.

2   Related to point 1 is the study of behavioral responses of different genotypes to stress. The study of the great variety of neural, endocrine, and cardiovascular structures and functions relevant to stress in a number of animals will be of importance (see Emlen, 1973, for references). Therefore a comprehensive understanding of genetic factors relevant to stress responses may be aided by behavior-genetics studies.

3   The effect of drugs on behavior, particularly on learning, is well known. The use of differing genotypes should help provide more precise information of use in drug therapy and psychotherapy. Ultimately, progress with regard to learning and its genetic and biochemical bases may be predicted. Recent studies on learning in a variety of organisms including *Drosophila* are probing in this direction. Now that this field is termed psychopharmacogenetics (Section 9-7), the dignity of a name, long as it is, may provide an impetus to further work.

4   For complex behaviors, the trend of looking at components of a behavioral trait and studying their genetic bases is likely to develop further. Such an approach is described in Chapter 12 for primary mental abilities and may be expected to accelerate, especially now that vast amounts of data can be processed by computer. This approach ought to be particularly valuable in studying human beings. In some cases, complex traits with ambiguous hereditary mechanisms may be broken down into subunits, each showing distinct and simple hereditary mechanisms. Psychoses may be amenable to this approach. Complex social behavior in animals including contests may be another candidate, but extreme care in sampling techniques and experimental design is essential for meaningful results (Fuller and Hahn, 1976, offer some discussion of this topic).

5   The tendency of some behavioral traits to vary over time merits more emphasis. The behavior geneticist necessarily needs to know about cyclic variations, notably diurnal and/or seasonal. Furthermore, changes over a life span need further study, a point that undoubtedly applies to traits modified by learning. At present, the competent experimenter controls age or time of day at measurement in carrying out a behavior-genetics analysis. However, the lability of behavioral traits over time has been little explored.

6   More generally, it seems that behavior genetics will play a major part in the slowly growing rapprochement between sociology and the biological sciences (particularly in the study of human variation). For example, studies of early experience and prenatal environment should lead to an increased participation by social scientists in behavior-genetics research. Behavior genetics should play a central role in future integrated research projects involving social, behavioral, and biological scientists. For example, Eckland (1972) wrote:

> At least in sociology, we never seem to run out of social and psychological hypotheses for almost any phenomenon. Yet, although our theories have much to recommend them, our tools and our findings leave much to be desired. In fact, when weighed in balance, the evidence on a number of points tends to push the search for explanation back to biology.

This is a point worthy of note by any person interested in sociobiology.

7 Increased study of social structures, mating patterns, and contests (agones) in human beings and other animals is likely, with the aim of assessing their effects on the gene pools of populations. The significance of this topic in the evolutionary biology of species has already been emphasized. Ecological components are likely to assume progressively more importance with time, since a social structure must interact with its habitat.

8 The genetic basis of behavioral factors involved in determining the niche of organisms has been little considered. As Emlen (1973) pointed out, there are numerous definitions of *niche,* but in brief the niche is the set of conditions, physical and biotic, in which an organism can live. In a broad niche, a greater range of behavioral responses is more likely than in a narrow niche. Conversely in a stable habitat behavior is likely to be more stereotyped and responses rigid. Closely related species of *Drosophila* and rodents provide excellent experimental material to add to those studies already published. The assessment of behavioral factors in determining a niche will involve the extremely difficult extrapolation of laboratory studies of behavioral traits to natural populations. In other words, it will involve the transition from a relatively determinate environment to one that, except for rare special circumstances, is essentially indeterminate. Indications of genetics of habitat selection as discussed in Chapter 13 offer a promising start to this field.

9 Following on from this is the investigation of behavioral changes under domestication. This has been discussed especially in Chapter 10, but we consider it to be an extremely important future applied aspect of behavior genetics in its evolutionary context, and so it is stressed here. As a final important example, sterile male screwworm flies are factory bred in enormous quantities as part of a screwworm control program in Texas; Bush (1978) has shown that factory conditions may unintentionally promote the genetic selection of noncompetitive flies, which among other things do not become active until early afternoon, while wild flies are active throughout the day. In other words, mating between wild flies may be completed before the factory flies become active enough for sexual activity. The differences appear to be associated with forms of the enzyme α-glycerol phosphate dehydrogenase, which is involved in insect flight. Therefore the factory is an artificial environment leading to a genetic change so that insect control becomes ineffective. This confirms that basic biological information, especially at the behavioral level, is essential for an understanding of domestication in its many forms and as such is an area of undoubted importance in the future.

10 Mayr (1974) asked, "Under what circumstances is a closed genetic program favored and under what others an open one?" His answers were: "Selection should favor the evolution of a closed program when there is a reliable relationship between a stimulus and only one correct response" and that "non-communicative behavior leading to an exploitation of natural resources should be flexible, permitting an opportunistic adjustment to rapid changes in the environment and also permitting an enlargement of the niche as well as a shift into a new niche." In the latter case an open program would be favored since such flexibility would be impossible if such behavior were too rigidly determined genetically. The observed facts discussed in this book, especially in this chapter, by Mayr (1974), and by Parsons (1977d) conform with these expectations. Even so, behavior constantly interacts with both the living and the inanimate environment and thus is constantly the target of natural selection. Sometimes therefore it will be advantageous for optimal responses to be largely closed and in other cases selection will favor an open behavior program. We conclude with Mayr (1974): "There is a wide and largely un-explored field of research to determine the selective advantages of various possible options in different organisms and under different conditions."

Finally, we regard behavioral changes as belonging to the most important components in evolutionary processes, whether considering aspects of behavior concerned with mating or those improving adaptedness to a new environment. A common approach, as presented in parts of this book—to study only the heritable components of traits—does not contribute greatly to the comprehension of the role of behavior in evolutionary processes simply because the extent to which the traits are relevant in determining the population continuity of genotypes control-ling them is unknown. Behaviors relevant in population continuity need to be isolated and studied under a multiplicity of environments. In other words, the question of the connection of a trait to fitness must be posed. And replies to questions of these types are now becoming feasible in *Drosophila* with its largely genetically closed behavior program. This will surely aid in the study of the more complex paths between genes, behavior, and evolutionary processes that must occur in vertebrates such as mice and human beings, where genetic programs controlling behavior are certainly more open.

Bibliography

Adler, J. 1969. Chemoreceptors in bacteria. *Science 166*:1588–1597.

Adler, J. 1976. The sensing of chemicals by bacteria. *Sci. Am. 234*:40–47.

Adler, J., G. C. Hazelbauer, and M. M. Dahl. 1973. Chemotaxis toward sugars in *Escherichia coli. J. Bacteriol. 115*:824–847.

Alawi, A., V. Jennings, J. Grossfield, and W. L. Pak. 1972. Phototransduction mutants of *Drosophila melanogaster*. In *The Visual System*. G. B. Arden (ed.). New York: Plenum, pp. 1–21.

Alexander, R. D. 1962. The role of behavioral study in cricket classification. *Syst. Zool. 11*(2):53–72.

Alexander, R. D. 1968. Arthropods. In *Animal Communication,* T. Seboek (ed.). Bloomington: Indiana University Press, pp. 167–216.

Allard, R. W., S. K. Jain, and P. L. Workman. 1968. The genetics of inbreeding populations. *Adv. Genet. 14*:55–131.

Allee, W., A. Emerson, O. Park, T. Park, and K. Schmidt. 1949. *Principles of Ecology*. Philadelphia: Saunders.

Amark, C. A. 1951. A study in alcoholism: clinical, social-psychiatric and genetic investigations. *Acta Psychiat. Neurol. Scand. Suppl. 70*.

Anderson, V. E. 1977. Genetic counseling for epilepsy. In *Plan for Nationwide Action on Epilepsy,* Vol. 2, Part 1, Pub. No. (NIH) 78–311. Washington, D.C.: U.S. Department of Health, Education and Welfare, pp. 141–162.

Anderson, W. W., and L. Ehrman 1969. Mating choice in crosses between geographic populations of *Drosophila pseudoobscura. Am. Midland Nat. 81*:47–53.

Anderson, W. W., and P. McGuire. 1978. Mating patterns and mating success of *Drosophila pseudoobscura* karyotypes in large experimental populations. *Evolution* *32*:416–423.

Andrewartha, H. G., and L. C. Birch, 1954. *The Distribution and Abundance of Animals*. Chicago: University of Chicago Press.

Anisman, H. 1975. Task complexity as a factor in eliciting heterosis in mice: aversively motivated behaviors. *J. Comp. Physiol. Psychol. 8*:976–984.

Annett, M. 1978. Genetic and nongenetic influences on handedness. *Behav. Genet. 8*:227–249.

Archer, J. 1973. Tests for emotionality in rats and mice: a review. *Anim. Behav. 21*:205–235.

Arehart-Treichel, J. 1978. Enzymes: medicine's new gold mine. *Sci. News 114*:58–60.

Aristotle. 1962. *The Politics*. Translator: J. A. Sinclair. Harmondsworth: Penguin.

Ashburner, M., and E. Novitski (eds.). 1976. *Genetics and Biology of* Drosophila. Vol. 1. New York: Academic.

Ashburner, M., and Wright, T. (eds.). 1978. *Genetics and Biology of* Drosophila, Vols. 2a, 2b. New York: Academic.

Åslund, S. E. 1977. Mating behaviour as a fitness component in maintaining allozyme polymorphism in *Drosophila melanogaster*. II. Selection in the leucine aminopeptidase-A allozyme system. *Hereditas 87*:261–270.

Åslund, S. E., and M. Rasmuson. 1976. Mating behaviour as a fitness component in maintaining allozyme polymorphism in *Drosophila melanogaster*. *Hereditas 82*:175–178.

Asmundson, V. S., and F. W. Lorenz. 1955. Pheasant-turkey hybrids. *Science 121*:307–308.

Atkinson, W., and B. Shorrocks. 1977. Breeding site specificity in the domestic species of *Drosophila*. *Oecologia 29*:223–232.

Averhoff, W. W., and R. H. Richardson. 1974. Pheromonal control of mating patterns in *D. melanogaster*. *Behav. Genet. 4*:207–225.

Averhoff, W. W., and R. H. Richardson. 1976. Multiple pheromone system controlling mating in *D. melanogaster*. *PNAS 73*:591–593.

Bakwin, H. 1973. Reading disability in twins. *Dev. Med. Child Neurol. 15*:184–187.

Bamber, R., and E. Herdman. 1932. A report on the progeny of a tortoise-shell male cat, together with a discussion of his gametic constitution. *J. Genet. 26*:117–128.

Barash, D. P. 1977. *Sociobiology and Behavior*. New York: Elsevier.

Barker, J. S. F. 1971. Ecological differences and competitive interaction between *Drosophila melanogaster* and *Drosophila simulans* in small laboratory populations. *Oecologia 8*:139–156.

Barlow, G. W. 1973. Competition between color morphs of the polychromatic midas cichlid *Cichlasoma citrinellum*. *Science 179*:806–807.

Barr, M. L. 1959. Sex chromatin and phenotype in man. *Science 130*:679–685.

Barrows, E. M., W. J. Bell, and C. D. Michener. 1975. Individual odor differences and their social functions in insects. *PNAS 72*:2824–2828.

Barrows, S. L. 1945. "The Inheritance of the Ability to Taste Brucine." Master's thesis, Stanford University.

Bartholomew, G. A., Jr., and J. B. Birdsell. 1953. Ecology and the protohominids. *Am. Anthropol. 55*:481–498.

Bastock, M. 1956. A gene mutation which changes a behavior pattern. *Evolution* *10*:421–439.

Bastock, M. 1967. *Courtship: An Ethological Study*. Chicago: Aldine.

Bateman, A. J. 1948. Intra-sexual selection in *Drosophila*. *Heredity 2*:349–368.

Bateman, A. J. 1949. Analysis of data on sexual isolation. *Evolution 3*:174-177.

Beach, F. A. 1950. The snark was a boojum. *Am. Psychol. 5*:115–124.

Beadle, M. 1977. *The Cat*. New York: Simon and Schuster.

Beardmore, J. A. 1965. A genetic basis for lateral bias. In *Mutation in Populations*. Prague: Symposium on the Mutational Process, pp. 75–83.

Beardmore, J. A. 1970. Ecological factors and the variability of gene-pools in *Drosophila*. In *Essays in Evolution and Genetics in Honor of Theodosius Dobzhansky*. M. K. Hecht and W. C. Steere (eds.). New York: Appleton, pp. 299–314.

Becker, H. J. 1970. The genetics of chemotaxis in *Drosophila melanogaster*: selection for repellent insensitivity. *Mol. Gen. Genet. 107*:194–200.

Beckett, P., and T. Bleakley (eds.). 1968. *A Teaching Program in Psychiatry*, Vol. 1. Detroit: Wayne State University Press, p. 73.

Beckman, L. 1962. Assortative mating in man. *Eugenics Rev. 54*:63–67.

Begg, M., and L. Hogben. 1946. Chemoreceptivity of *Drosophila melanogaster*. *Proc. Roy. Soc. Lond. Ser. B. 133*:1–19.

Belmont, L., and F. A. Marolla. 1973. Birth order, family size, and intelligence. *Science 182*:1096–1101.

Bentley, D. R. 1971. Genetic control of an insect network. *Science 174*:1139–1141.

Bentley, D. R., and R. R. Hoy. 1972. Genetic control of the neuronal network generating cricket (*Teleogryllus gryllus*) song patterns. *Anim. Behav. 20*:478–492.

Bentley, D. R., and W. Kutsch. 1966. The neuromuscular mechanism of stridulation in crickets (Orthoptera: Gryllidae). *J. Exp. Biol. 45*:151–164.

Benzer, S. 1971. From the gene to behavior. *JAMA 218*:1015–1026.

Benzer, S. 1973. Genetic dissection of behavior. *Sci. Am. 229*:24–37.

Bergsma, D. (ed.). 1973. *Birth Defects: Atlas and Compendium*. Baltimore: Williams & Wilkins.

Bergsma, D. (ed.). 1979. *Birth Defects Compendium*, 2d ed. New York: A. R. Liss.

Bergsma, D. R., and K. S. Brown, 1971. White fur, blue eyes, and deafness in the domestic cat. *J. Heredity 62*:171.

Berry, R. J. 1963. Epigenetic polymorphism in wild populations of *Mus musculus*. *Genet. Res. 4*:193–220.

Berry, R. J. 1970. The natural history of the house mouse. *Field Stud. 3*:219–262.

Bigelow, R. S. 1960. Interspecific hybrids and speciation in the genus *Acheta* (Orthoptera: Gryllidae). *Can. J. Zool. 38*:509–524.

Bignami, G. 1965. Selection for high rates and low rates of avoidance conditioning in the rat. *Anim. Behav. 13*:221–227.

Birdsell, J. B. 1950. Some implications of the genetical concept of race in terms of spatial analysis. *Cold Spring Harbor Symp. Quant. Biol. 15*:259–314.

Birdsell, J. B. 1973. A basic demographic unit. *Curr. Anthropol. 14*:337–356.

Blair, W. F. 1974. Character displacement in frogs. *Am. Zool. 14*:1119–1125.

Blake, R., and J. Camisa. 1978. Is binocular vision always monocular? *Science 200*:1497–1498.

Blewett, D. B. 1954. An experimental study of the inheritance of intelligence. *J. Ment. Sci. 100*:922–933.

Blinkov, S. M. and I. I. Glezer. 1968. *The Human Brain in Figures and Tables.* New York: Plenum.

Bliss, E. (ed.). 1962. *Roots of Behavior.* New York: Hoeber-Harper.

Blizard, D. A. 1971. Autonomic reactivity in the rat: effects of genetic selection for emotionality. *J. Comp. Physiol. Psychol. 76*:282–289.

Blizard, D. A., and D. W. Fulker. 1979. Interactions of strain of male and female on the mating behavior of the rat. *Behav. & Neur. Biol. 25*:99–114.

Bodmer, W. F. and L. L. Cavalli-Sforza. 1970. Intelligence and race. *Sci. Am. 223*(4):19–29.

Bodmer, W. F., and L. L. Cavalli-Sforza. 1976. *Genetics, Evolution and Man.* San Francisco: Freeman.

Böök, J. A. 1953. A genetic and neuropsychiatric investigation. *Acta Genet. 4*:3–100.

Borisov, A. I. 1970. Disturbances of panmixia in natural populations of *Drosophila funebris* polymorphous by the inversion II-1. *Genetika 6*:61–67.

Bösiger, E. 1957. Sur l'activité sexuelle des males de plusieurs souches de *Drosophila melanogaster. C. R. Acad. Sci (D) (Paris) 244*:1419–1422.

Bösiger, E. 1967. La signification evolutive de la selection sexuelle chez les animaux. *Scientia 102*:207–223.

Bossert, W. H., and E. O. Wilson. 1963. The analysis of olfactory communication among animals. *J. Theor. Biol. 5*:443.

Brenner, S. 1973. The genetics of behavior. *Br. Med. Bull. 29*:269–271.

Bridges, C. B., and K. Brehme. 1944. *The Mutants of* Drosophila melanogaster. *Carnegie Institution Pub.* No. 552. Washington, D. C.: Carnegie Institution.

Brncic, D., and S. Koref-Santibañez. 1963. Life cycle and mating activity as criteria of heterosis in heterokaryotypes in *Drosophila pavani. Genetics Today 1*:157–158.

Brncic, D., and S. Koref-Santibañez. 1964. Mating activity of homo- and heterokaryotypes in *Drosophila pavani. Genetics 49*:585–591.

Broadhurst, P. L. 1960. Experiments in psychogenetics: applications of biometrical genetics to the inheritance of behaviour. In *Experiments in Personality,* Vol. 1, *Psychogenetics and Psychopharmacology.* H. J. Eysenck (ed.). London: Routledge, pp. 1–102.

Broadhurst, P. L. 1967. The biometrical analysis of behavioural inheritance. *Sci. Progr. 55*:123–129.

Broadhurst, P. L., and J. L. Jinks. 1961. Biometrical genetics and behaviour: reanalysis of published data. *Psychol. Bull. 58*:337–362.

Broadhurst, P. L., and J. L. Jinks. 1963. The inheritance of mammalian behavior re-examined. *J. Hered. 54*:170–176.

Broadhurst, P. L., and R. H. J. Watson. 1964. Brain cholinesterase, body build and emotionality in different strains of rats. *Anim. Behav. 12*:42–51.

Brown, J. L. 1975. *The Evolution of Behavior.* New York: Norton.

Brown, R. G. B. 1965. Courtship behaviour in the *Drosophila obscura* group. Part II. Comparative studies. *Behaviour 25*:281–323.

Bruell, J. H. 1970. Behavioral population genetics and wild *Mus musculus.* In *Contributions to Behavior-Genetic Analysis: The Mouse as a Prototype.* G. Lindzey and D. D. Thiessen (eds.). New York: Appleton-Century-Crofts, pp. 437–543.

Bruere, A., R. Marshall, and D. Ward. 1969. Testicular hypoplasia and XXY sex chromosome complement in two rams: the ovine counterpart of Klinefelter's syndrome in man. *J. Reprod. Fertil. 19*:103–108.

Burt, C. 1966. The genetic determination of differences in intelligence: a study of monozygotic twins reared together and apart. *Br. J. Psychol. 57*:137–153.

Bush, G. L. 1969. Sympatric host race formation and speciation in fungivorous flies of the genus *Rhagoletis* (Diptera, Tephritidae). *Evolution 23*:237–251.

Bush, G. L. 1975. Modes of animal speciation. *Ann. Rev. Ecol. Syst. 6*:339–364.

Bush, G. L. 1978. Planning a rational quality control program for the screwworm fly. In *The Screwworm Problem*. R. H. Richardson (ed.). Austin: University of Texas Press, pp. 37–47.

Calhoun, J. 1962. Population density and social pathology. *Sci. Am. 206*(2):139–148.

Campbell, M., S. Wolman, H. Breurer, F. Miller, and B. Perlman. 1972. Klinefelter's syndrome in a three-year-old severely disturbed child. *J. Autism Child Schizo. 2*:34–48.

Carpenter, F. W. 1905. The reactions of the pomace fly (*Drosophila melanogaster* Loew) to light, gravity and mechanical stimulation. *Am. Nat. 39*:141–171.

Carson, H. L. 1951. Breeding sites of *Drosophila pseudoobscura* and *Drosophila persimilis* in the transition zone of the Sierra Nevada. *Evolution 5*:91–96.

Carson, H. L. 1973. The genetic system in parthenogenetic strains of *Drosophila mercatorum*. *PNAS 70*:1772–1774.

Carson, H. L., D. E. Hardy, H. T. Spieth, and W. S. Stone. 1970. The evolutionary biology of the Hawaiian Drosophilidae. In *Essays in Evolution and Genetics in Honor of Theodosius Dobzhansky*. M. K. Hecht and W. C. Steere (eds.). New York: Appleton, pp. 437–543.

Carter, C. O. 1965. The inheritance of common congenital malformations. *Progr. Med. Genet. 4*:59–84.

Carter, J. E. L. 1970. The somatotypes of athletes: a review. *Hum. Biol. 42*:535–569.

Caspari, E. W. 1951. On the biological basis of adaptedness. *Am. Sci. 39*:441–451.

Caspari, E. W. 1955. On the formation in the testis sheath of *Rt* and *rt Ephestia kuhniella*. *Zell. Biol. Z. 74*:585–602.

Caspari, E. W. 1958. Genetic basis of behavior. In *Behavior and Evolution*. A. Roe and G. G. Simpson (eds.). New Haven: Yale University Press, pp. 103–127.

Caspari, E. W. 1961. Implications of genetics for psychology. Review of *Behavior Genetics* by J. L. Fuller and W. R. Thompson. *Contemp. Psychol. 6*:337–339.

Caspari, E. W. 1967. Introduction to part I and remarks on evolutionary aspects of behavior. In *Behavior-Genetic Analysis*. J. Hirsch (ed.). New York: McGraw-Hill, pp. 3–9.

Caspari, E. W. 1977. Genetic mechanisms and behavior. In *Genetics, Environment and Intelligence*. A. Oliverio (ed.). New York: Elsevier North-Holland, pp. 3–22.

Caspari, E. W., and F. J. Gottlieb. 1959. On a modifier of the gene *a* in *Ephestia kuhniella*. *Z. Vererbungslehre 90*:263–272.

Castellion, A. W., E. A. Swingard, and L. S. Goodman. 1965. The effect of maturation on the development and reproducibility of audiogenic and electroshock seizures. *Exp. Neurol. 13*:206–217.

Cattell, R. B. 1965. Methodological and conceptual advances in evaluating hereditary and environmental influences and their interaction. In *Methods and Goals in Human Behavior Genetics*. S. G. Vandenberg (ed.). New York: Academic, pp. 95–139.

Cavalli-Sforza, L. L. 1975. Quantitative genetic perspectives: implications for human development. In *Developmental Human Behavior Genetics*. K. W. Schaie, V. E.

Anderson, G. E. McClearn, and J. Money (eds.). Lexington, Mass.: Lexington Books, pp. 123–149.

Cavalli-Sforza, L. L., and W. F. Bodmer. 1971. *The Genetics of Human Populations.* San Francisco: Freeman.

Cavalli-Sforza, L. L., and M. W. Feldman, 1973. Cultural versus biological inheritance: phenotypic transmission from parent to children. *Am. J. Hum. Genet. 25*:618–637.

Chabora, P. C., and Kessler, M. E. 1977. Genetic and light intensity effects on the activity and emigratory behavior of the house fly, *Musca domestica* (L). *Behav. Genet. 7*:281–290.

Chang, S., and C. Kung, 1973. Temperature-sensitive pawns: conditional behavioral mutants of *Paramecium aurelia. Science 180*:1197–1199.

Chang, S., J. van Houten, L. Robles, S. S. Lui, and C. Kung. 1974. An extensive behavioural and genetic analysis of the pawn mutants of *Paramecium aurelia. Genet. Res. 23*:165–173.

Chetverikov, S. S. 1926. On certain features of the evolutionary process from the viewpoint of modern genetics. *J. Exp. Biol.* (Russian) *2*:3–54.

Chiarelli, B. 1963. Sensitivity to PTC (phenyl-thio-carbamide) in primates. *Folia Primatol. 1*:103–107.

Child, I. L. 1950. The relation of somatotype to self-ratings on Sheldon's temperamental traits. *J. Pers. 18*:440–453.

Chitty, D. 1967. The natural selection of self-regulatory behaviour in animal populations. *Proc. Ecol. Soc. Aust. 2*:51–78.

Christensen, B. 1977. Habitat preference among amylase genotypes in *Asellus aquaticus* (Isopoda, Crustacea). *Hereditas 87*:21–26.

Clark, E., L. R. Aronson, and M. Gordon. 1954. Mating behavior patterns in two sympatric species of xiphophorin fishes: their inheritance and significance in sexual isolation. *Bull. Am. Mus. Nat. Hist. 103*:135–226.

Cloninger, C. R., K. Christiansen, T. Reich, and I. Gottesman. 1978. Implications of sex differences in the prevalences of antisocial personality, alcoholism, and criminality for familial transmission. *Arch. Gen. Psychiat. 35*:941–951.

Clutton-Brock, T. H., and P. H. Harvey. 1978. Mammals, resources and reproductive strategies. *Nature* (London) *273*:191–195.

Cole, B. L. 1970. The colour blind driver. *Aust. J. Optometry 53*:261–269.

Cole, B. L. 1972. The handicap of abnormal colour vision. *Aust. J. Optometry 58*:304–310.

Coleman, D. L. 1960. Phenylalanine hydroxylase activity in dilute and nondilute strains of mice. *Arch. Biochem. Biophys. 91*:300–306.

Coleman, D. L. 1979. Obesity genes: beneficial effects in heterozygous mice. *Science 203*:663–665.

Coleman, D. L., and K. P. Hummel 1973. The influence of genetic background on the expression of the obese (*ob*) gene in the mouse. *Diabetologia 9*:287–293.

Coleman, J. S., E. Q. Campbell, C. J. Hobson, J. McPartland, A. M. Mood, F. D. Weinfeld, and R. L. York. 1966. *Equality of Educational Opportunity.* Washington, D. C.: GPO.

Collins, R. L. 1964. Inheritance of avoidance conditioning in mice: a diallel study. *Science 143*:1188–1190.

Collins, R. L. 1968. On the inheritance of handedness. I. Laterality in inbred mice. *J. Hered. 59*:9–12.

Collins, R. L. 1977a. Origin of the sense of asymmetry: Mendelian and non-Mendelian models of inheritance. *Ann. N.Y. Acad. Sci. 299*:283–305.

Collins, R. L. 1977b. Toward an admissible genetic model for the inheritance of the degree and direction of asymmetry. In *Lateralization in the Nervous System.* S. Harvard, R. L. Doty, L. Goldstein, J. Jaynes, and G. Krauthamer (eds.). New York: Academic, pp. 137–150.

Comings, D. C. 1979. Pc 1 Duarte, a common polymorphism of a human brain protein and its relationship to depressive disease and multiple sclerosis. *Nature* (London) *277*:28–32.

Connolly, K. 1966. Locomotor activity in *Drosophila.* 2. Selection for active and inactive strains. *Anim. Behav. 14*:444–449.

Connolly, K. 1968. The social facilitation of preening behavior in *Drosophila melanogaster. Anim. Behav. 16*:385–391.

Connolly, K., B. Burnet, and D. Sewell. 1969. Selective mating and eye pigmentation: an analysis of the visual component in the courtship behavior of *Drosophila melanogaster. Evolution 23*:548–559.

Conterio, F., and L. L. Cavalli-Sforza. 1959. Evolution of the human constitutional phenotype: an analysis of mortality effects. *Ricerca Sci. (Suppl.) 29*:3–14.

Cook, W. T., P. Siegel, and K. Hinkelmann. 1972. Genetic analyses of male mating behavior in chickens. 2. Crosses among selected and control lines. *Behav. Genet. 2*:289–300.

Cooke, F. 1978. Early learning and its effect on population structure: studies of a wild population of snow geese. *Z. Tierpsychol. 46*:344–358.

Cooke, F., G. H. Finney, and R. F. Rockwell. 1976. Assortative mating in lesser snow geese (*Anser caerulescens*). *Behav. Genet. 6*:127–140.

Cooke, F., and C. M. McNally. 1975. Male selection and colour preferences in lesser snow geese. *Behaviour 53*:151–170.

Cooke, F., and P. J. Mirsky. 1972. A genetic analysis of lesser snow goose families. *Auk 89*:863–871.

Cooper, R. M., and J. P. Zubek. 1958. Effects of enriched and restricted early environments on the learning ability of bright and dull rats. *Can. J. Psychol. 12*:159–164.

Cornsweet, T. N. 1971. *Visual Perception.* New York: Academic.

Corson, S., E. Corson, R. Becker, B. Ginsburg, A. Trattner, R. Conner, L. Lucas, J. Panksepp, and J. P. Scott. 1979. Interaction of genetics and separation in canine hyperkinesis and in differential responses to amphetamine. *Pavlovian J. Bio. Sci. 15*:5–11.

Cotter, W. B. 1951. "The Genetic and Physiological Analyses of the Silk-Spinning Behavior of *Ephestia kühniella.*" Master's thesis, Wesleyan University.

Cowan, B. D., and W. M. Rogoff. 1968. Variation and heritability of responsiveness of individual male house flies, *Musca domestica,* to the female sex pheromone. *Ann. Entomol. Soc. Am. 61*:1215–1218.

Craig, G. 1965. The role of genetics in mosquito control. *Proc. and Papers of the 33rd Annual Conf. of the Calif. Mosquito Control Assoc. 33*:80–82.

Craig, G., and R. VandeHey. 1962. Genetic variability in *Aedes aegypti.* 1. Mutations affecting color pattern. *Ann. Entomol. Soc. Am. 55*:58–67.

Crossley, S. A. 1975. Changes in mating behavior produced by selection for ethological isolation between ebony and vestigial mutants of *Drosophila melanogaster. Evolution 28*:631–647.

Crow, J. F. 1976. *Genetics Notes,* 7th ed. Minneapolis: Burgess.

Crow, J. F., and J. Felsenstein. 1968. The effect of assortative mating on the genetic composition of a population. *Eugen. Q. 15*:85–97.

Crowcroft, P. 1966. *Mice All Over*. Chester Springs, Pa.: Dufour.

Cunningham, D. L., and P. B. Siegel. 1978. Response to bidirectional and reverse selection for mating behavior in Japanese quail *Coturnix coturnix japonica. Behav. Genet. 8*:387–397.

Cunningham, E. P., 1976. Genetic studies in horse populations. In *Proceedings of the International Symposium on Genetics and Horse-Breeding*. Dublin, Ireland, September 17–18, 1975. Dublin: Royal Dublin Society, pp. 2–8.

Cushing, J. E. 1941. An experiment on olfactory conditioning in *Drosophila guttifera. PNAS 17*:496–499.

Darwin, C. 1859. *The Origin of Species by Means of Natural Selection*. London: Murray.

Darwin, C. 1871. *The Descent of Man and Selection in Relation to Sex*. London: Murray.

Dawkins, R. 1976. *The Selfish Gene*. Oxford: Oxford University Press.

Dawson, W. M. 1932. Inheritance of wildness and tameness in mice. *Genetics 17*:296–326.

DeFries, J. C. 1964. Prenatal maternal stress in mice: differential effects on behavior. *J. Hered. 55*:289–295.

DeFries, J. C., M. C. Gervais, and E. A. Thomas. 1978. Response to 30 generations of selection for open-field activity in laboratory mice. *Behav. Genet. 8*:3–13.

DeFries, J. C., and J. P. Hegmann. 1970. Genetic analysis of open-field behavior. In *Contributions to Behavior-Genetic Analysis: The Mouse as a Prototype*. G. Lindzey and D. D. Thiessen (eds.). New York: Appleton, pp. 23–56.

DeFries, J. C., J. P. Hegmann, and M. W. Weir. 1966. Open-field behavior in mice: evidence for a major gene effect mediated by the visual system. *Science 154*:1577–1579.

DeFries, J. C., and G. E. McClearn. 1972. Behavioral genetics and the fine structure of mouse populations: a study in microevolution. *Evol. Biol. 5*:279–291.

DeFries, J. C., and R. Plomin. 1978. Behavioral genetics. *Ann. Rev. Psychol. 29*:473–515.

DeFries, J. C., E. A. Thomas, J. P. Hegmann, and M. W. Weir. 1967. Open-field behavior in mice: analysis of maternal effects by means of ovarian transplantation. *Psychon. Sci. 8*:207–208.

DeFries, J. C., S. G. Vandenberg, and G. E. McClearn. 1976. Genetics of specific cognitive abilities. *Ann Rev. Genet. 10*:179–207.

Deol, M. S. 1975. Genes affecting behaviour and the inner ear in the mouse. In *The Genetics of Behavior*. J. H. F. Van Abeelen (ed.). New York: North-Holland.

Desforges, M. F., and D. G. M. Wood-Gush. 1975a. A behavioural comparison of domestic and mallard ducks. Habituation and flight reactions. *Anim. Behav. 23*:692–697.

Desforges, M. F., and D. G. M. Wood-Gush. 1975b. A behavioural comparison of domestic and mallard ducks. Spatial relationships in small flocks. *Anim. Behav. 23*:698–705.

Desforges, M. F., and D. G. M. Wood-Gush. 1976. Behavioural comparison of Aylesbury and mallard ducks: sexual behaviour. *Anim. Behav. 24*:391–397.

De Souza, H. M. L., A. B. Da Cunha, and E. P. Dos Santos. 1970. Adaptive polymorphism of behavior evolved in laboratory populations of *Drosophila willistoni. Am. Nat. 104*:175–189.

Dethier, V. G. 1976. *The Hungry Fly*. Cambridge, Mass.: Harvard University Press.

De Vore, I. (ed.). 1965. *Primate Behavior: Field Studies of Monkeys and Apes*. New York: Holt.

Dewsbury, D. A. 1975. A diallel cross analysis of genetic determinants of copulatory behavior in rats. *J. Comp. Physiol. Psychol. 88*:713–722.

Dilger, W. 1962a. Behavior and genetics. In *Roots of Behavior*. E. Bliss (ed.). New York: Hoeber-Harper.

Dilger, W. 1962b. The behavior of lovebirds. *Sci. Am. 206*:88–98.

Dimond, S. J., and D. A. Blizard (eds.). 1977. Evolution and lateralization of the brain. *Ann. N.Y. Acad. Sci. 299*:1–501.

Dingman, H. 1968. Psychological test patterns in Down's syndrome. In *Progress in Human Behavior Genetics*. S. Vandenberg (ed.). Baltimore: Johns Hopkins University Press, pp. 19–25.

Dobzhansky, T. 1937. *Genetics and the Origin of Species*. New York: Columbia University Press.

Dobzhansky, T. 1940. Speciation as a stage in evolutionary divergence. *Am. Nat. 74*:312–321.

Dobzhansky, T. 1950. Genetics of natural populations. 19. Origin of heterosis through natural selection in populations of *Drosophila pseudoobscura*. *Genetics 35*:288–302.

Dobzhansky, T. 1951. *Genetics and the Origin of Species*, 3d ed. New York: Columbia University Press.

Dobzhansky, T. 1964. *Heredity and the Nature of Man*. New York: Harcourt.

Dobzhansky, T. 1968. On some fundamental concepts of Darwinian biology. *Evol. Biol. 2*:1–34.

Dobzhansky, T. 1970. *Genetics of the Evolutionary Process*. New York: Columbia University Press.

Dobzhansky, T., F. J. Ayala, G. L. Stebbins, and J. W. Valentine. 1977. *Evolution*. Freeman: San Francisco.

Dobzhansky, T., D. M. Cooper, H. J. Phaff, E. P. Knapp, and H. L. Carson. 1956. Studies on the ecology of *Drosophila* in the Yosemite region of California. 4. Differential attraction of species of *Drosophila* to different species of yeasts. *Ecology 37*:544–550.

Dobzhansky, T., and C. Pavan. 1950. Local and seasonal variations in relative frequencies of species of *Drosophila* in Brazil. *J. Anim. Ecol. 19*:1–14.

Dobzhansky, T., and O. Pavlovsky. 1962. A comparative study of the chromosomes in the incipient species of the *Drosophila paulistorum* complex. *Chromosoma 13*:196–218.

Dobzhansky, T., and J. R. Powell. 1975. The *willistoni* group of sibling species of *Drosophila*. In *Handbook of Genetics*, Vol. 3. R. C. King (ed.). New York: Plenum, pp. 589–622.

Dobzhansky, T., and B. Spassky. 1962. Genetic drift and natural selection in experimental populations of *Drosophila pseudoobscura*. *PNAS 48*:148–156.

Dobzhansky, T., and B. Spassky. 1969. Artificial and natural selection for two behavioral traits in *Drosophila pseudoobscura*. *PNAS 62*:75–80.

Doose, H., H. Gerken, K. F. Hien-Volpel, and E. Völzke. 1969. Genetics of photosensitive epilepsy. *Neuropädiatrie 1*:56–73.

Doty, R. L. 1974. A cry for the liberation of the female rodent: courtship and copulation in Rodentia. *Psychol. Bull. 81*:159–171.

Doyle, R. S. 1976. Analysis of habitat loyalty and habitat preference in the settlement behavior of planktonic marine larvae. *Am. Nat. 110*:719–730.

Dudai, Y., Y. N. Jan, D. Byers, W. G. Quinn, and S. Benzer. 1976. Dunce, a mutant of *Drosophila* deficient in learning. *PNAS 73*:1684–1688.

Dunham, H. W. 1965. *Community and Schizophrenia: An Epidemiological Analysis.* Detroit: Wayne State University Press.

Dunlop, K. 1943. Mental maladjustment and color vision. *Science 98*:470–472.

Dustman, R. E., and E. C. Beck. 1965. The visually evoked potential in twins. *Electroencephalogr. Clin. Neurophysiol. 19*:570–575.

Eaton, S. W., and R. S. Weil. 1955. *Culture and Mental Disorders.* Glencoe, Ill.: Free Press.

Eaves, L. J. 1973. Assortative mating and intelligence: an analysis of pedigree data. *Heredity 30*:199–210.

Eaves, L. J. 1976. The effect of cultural variation of continuous transmission. *Heredity 37*:41–57.

Eaves, L. J., and Eysenck, H. J. 1977. A genotype-environment model for psychoticism. *Adv. Behav. Res. Ther. 1*:5–26.

Eckert, R. 1972. Bioelectric control of ciliary activity. *Science 176*:473–481.

Eckland, B. K. 1972. Comments on school effects, gene-environment covariance, and the heritability of intelligence. In *Genetics, Environment and Behavior: Implications for Educational Policy.* L. Ehrman, G. S. Omenn, and E. Caspari (eds.). New York: Academic, pp. 297–306.

Edwards, J. H. 1958. Congenital malformations of the central nervous system in Scotland. *Br. J. Prev. Soc. Med. 12*:115–130.

Edwards, J. H. 1960. The simulation of mendelism. *Acta Genet. Statist. Med. 10*:63–70.

Ehrman, L. 1960a. The genetics of hybrid sterility in *Drosophila paulistorum. Genetics 46*:212–223.

Ehrman, L. 1960b. A genetic constitution frustrating the sexual drive in *Drosophila paulistorum. Science 131*:1381–1382.

Ehrman, L. 1961. The genetics of sexual isolation in *Drosophila paulistorum. Genetics 46*:1025–1038.

Ehrman, L. 1964. Courtship and mating behavior as a reproductive isolating mechanism in *Drosophila. Am. Zool. 4*:147–153.

Ehrman, L. 1965. Direct observation of sexual isolation between allopatric and between sympatric strains of the different *Drosophila paulistorum* races. *Evolution 19*:459–464.

Ehrman, L. 1966. Mating success and genotype frequency in *Drosophila. Anim. Behav. 14*:332–339.

Ehrman, L. 1968. Frequency dependence of mating success in *Drosophila pseudoobscura. Genet. Res. 11*:135–140.

Ehrman, L. 1969. Genetic divergence in M. Vetukhiv's experimental populations of *Drosophila pseudoobscura.* 5: a further study of rudiments of sexual isolation. *Am. Midland Nat. 82*:272–276.

Ehrman, L. 1970a. The mating advantage of rare males in *Drosophila. PNAS 65*:345–348.

Ehrman, L. 1970b. A release experiment testing the mating advantage of rare *Drosophila* males. *Behav. Sci. 15*:353–365.

Ehrman, L. 1971. The small gilded fly does lecher in my sight. *Psychology Today* (July) 5(2):62–82.

Ehrman, L. 1972. Genetics and sexual selection. In *Sexual Selection and the Descent of Man.* B. Campbell (ed.). Chicago: Aldine, pp. 105–135.

Ehrman, L. 1979. Still more on natural selection for the origin of reproductive isolation. *Am. Nat. 113*:148–150.

Ehrman, L., W. Anderson, and L. Blatte. 1977. A test for rare male mating advantage at an enzyme locus. *Behav. Genet. 7*:427–432.

Ehrman, L., G. S. Omenn, and E. Caspari (eds.). 1972. *Genetics, Environment, and Behavior: Implications for Educational Policy.* New York: Academic.

Ehrman, L., and J. Probber. 1978. Rare *Drosophila* males: the mysterious matter of choice. *Am. Sci. 66*(2):216–222.

Ehrman, L., B. Spassky, O. Pavlovsky, and T. Dobzhansky. 1965. Sexual selection, geotaxis and chromosomal polymorphism in experimental populations of *Drosophila pseudoobscura. Evolution 19*:337–346.

Ehrman, L., and M. Strickberger. 1960. Flies mating: a pictorial record. *Nat. Hist. 69*:28–33.

Ehrman, L., J. Thompson, I. Perelle, and B. Hisey. 1978. Some approaches to the question of *Drosophila* laterality. *Genet. Res. 32*:231–238.

Eiduson, S., E. Geller, A. Yuwiller, and B. T. Eiduson. 1964. *Biochemistry and Behavior.* Princeton: Van Nostrand.

Eisenberg, J. F., N. A. Muchenhirn, R. Rudran. 1972. The relation between ecology and social structure of primates. *Science 176*:863–874.

Eisenberg, L. 1973. Psychiatric intervention. *Sci. Am. 229*:116.

Eisner, V., L. L. Pauli, and S. Livingstone. 1959. Hereditary aspects of epilepsy. *Johns Hopkins Hosp. Bull. 105*:245–271.

Eldridge, F., and Y. Suzuki, 1976. A mare mule—dam or foster mother? *J. Heredity 67*:353–360.

Eleftheriou, B. 1975. *Psychopharmacogenetics.* New York: Plenum.

Elens, A. A., and J. M. Wattiaux. 1964. Direct observation of sexual isolation. *Drosophila Inf. Serv. 39*:118–119.

El-Helw, M. R., and A. M. M. Ali. 1970. Competition between *Drosophila melanogaster* and *D. simulans* on media supplemented with *Saccharomyces* and *Schizosaccharomyces. Evolution 24*:531–537.

Emlen, J. M. 1973. *Ecology: An Evolutionary Approach.* Reading, Mass.: Addison-Wesley.

Emlen, S. T., and L. W. Oring. 1977. Ecology, sexual selection, and the evolution of mating systems. *Science 197*:215–223.

Epps, S., and R. W. Parnell. 1952. Physique and temperament of women delinquents compared with women undergraduates. *Br. J. Med. Psychol. 25*:249–255.

Eriksson, C. J. P. 1973. Ethanol and acetaldehyde metabolism in rat strains genetically selected for their ethanol preference. *Biochem. Pharm. 22*:2283–2292.

Erlenmeyer-Kimling, L. 1968. Studies on the offspring of two schizophrenic parents. In *The Transmission of Schizophrenia.* D. Rosenthal and S. Kety (eds.). New York: Pergamon, p. 75.

Erlenmeyer-Kimling, L. 1972. Gene-environment interactions and the variability of behavior. In *Genetics, Environment, and Behavior: Implications for Educational Pol-*

icy. L. Ehrman, G. S. Omenn, and E. Caspari (eds.). New York: Academic, pp. 181–208.

Erlenmeyer-Kimling, L. 1978a. Genetic approaches to the study of schizophrenia: the genetic evidence as a tool in research. *Birth Defects: Original Article Series* *XIV*:59–74.

Erlenmeyer-Kimling, L. 1978b. Fertilité des psychotiques-demographie. *Confrontations Psychiatriques 16*:47–81.

Erlenmeyer-Kimling, L., and L. F. Jarvik. 1963. Genetics and intelligence: a review. *Science 142*:1477–1479.

Erlenmeyer-Kimling, L., and W. Paradowski. 1966. Selection and schizophrenia. *Am. Nat. 100*:651–665.

Erway, L., L. S. Hurley, and A. Fraser, 1966. Neurological defect: manganese in phenocopy and prevention of a genetic abnormality of inner ear. *Science 152*:1766–1768.

Ewing, A. W. 1963. Attempts to select for spontaneous activity in *Drosophila melanogaster. Anim. Behav. 11*:369–377.

Ewing, A. W. 1969. The genetic basis of sound production in *Drosophila pseudoobscura* and *D. persimilis. Anim. Behav. 17*:555–560.

Eysenck, H. J. 1956. The inheritance of extraversion. *Acta Psychol. 12*:95–110.

Eysenck, H. J. 1964. *Crime and Personality.* London: Routledge.

Eysenck, H. J. 1967. Intelligence assessment: a theoretical and experimental approach. *Br. J. Educ. Psychol. 37*:81–98.

Eysenck, H. J., and P. L. Broadhurst, 1964. Experiments with Animals: Introduction. In *Experiments in Motivation.* H. J. Eysenck (ed.). Elmsford, N.Y.: Pergamon.

Eysenck, H. J., and D. B. Prell. 1951. The inheritance of neuroticism: an experimental study. *J. Ment. Sci. 97*:441–465.

Fairbairn, D. J. 1978. Behaviour of dispersing deer mice (*Peromyscus maniculatus*). *Behav. Ecol. Sociobiol. 3*:265–282.

Falconer, D. S. 1960. *Introduction to Quantitative Genetics.* Edinburgh: Oliver & Boyd.

Falconer, D. S. 1965. The inheritance of liability to certain diseases, estimated from the incidence among relatives. *Ann. Hum. Genet. 29*:51–71.

Falconer, D. S. 1967. The inheritance of liability to diseases with variable age of onset, with particular reference to diabetes mellitus. *Ann. Hum. Genet. 31*:1–20.

Falek, A., and S. Britton. 1974. Phases in coping: the hypothesis and its implications. *Soc. Bio. 21*:1–7.

Falk, C., and L. Ehrman. 1975. Random mating revisited. *Behav. Genet. 3*:91–95.

Farr, J. A. 1976. Social facilitation of male sexual behavior, intrasexual competition, and sexual selection in the guppy, *Poecilia reticulata* (Pisces: Poeciliidae). *Evolution 30*:707–717.

Farr, J. A. 1977. Male rarity or novelty, female choice behavior, and sexual selection in the guppy, *Poecilia reticulata* Peters (Pisces: Poeciliidae). *Evolution 31*:162–168.

Feldman, M. W., and L. L. Cavalli-Sforza. 1976. Cultural and biological evolutionary processes, selection for a trait under complex transmission. *Theor. Pop. Biol. 9*:239–259.

Fenna, D., L. Mix, and O. Schaefer. 1971. Ethanol metabolism in various racial groups. *Can. Med. Assoc. J. 105*:472–475.

Fischer, M. 1973. Genetic and environmental factors in schizophrenia. *Acta Psychiat. Scand. Suppl. 238.*

Fischer, R., and F. Griffin. 1960. Factors involved in the mechanism of "taste-blindness." *J. Hered. 51*:182–183.

Fischer, R., F. Griffin, S. England, and S. M. Garen. 1961. Taste thresholds and food dislikes. *Nature* (London) *191*:1328.

Fisher, R. A. 1930. *The Genetical Theory of Natural Selection.* Oxford: Clarendon Press.

Fisher, R. A. 1958. Cancer and smoking. *Nature* (London) *182*:596.

Fisher, R. A., E. B. Ford, and J. H. Huxley. 1939. Taste testing the anthropoid apes. *Nature* (London) *144*:750.

Fisher, R. A., and F. Yates. 1967. *Statistical Tables for Biological, Agricultural and Medical Research,* 6th ed. Edinburgh: Oliver & Boyd.

Franck, D. 1970. Verhaltengenetische Untersuchungen an Artbastarden der Gattung *Xiphophorus* (Pisces). *Zeitschrift für Tierpsychologie 27*:1–34.

Friedlander, J. S., L. A. Sgaramella-Zonta, K. K. Kidd, L. Y. C. Lai, P. Clark, and R. J. Walsh. 1971. Biological divergences in south-central Bougainville: an analysis of blood polymorphism gene frequencies and anthropometric measurements utilizing tree models and a comparison of these variables with linguistic, geographic, and migrational "distances." *Am. J. Hum. Genet. 23*:253–270.

Fuhrmann, W., and F. Vogel. 1969. *Genetic Counseling.* New York: Springer-Verlag.

Fulker, D. W. 1966. Mating speed in male *Drosophila melanogaster:* a psychogenetic analysis. *Science 153*:203–205.

Fulker, D. W. 1970. Maternal buffering of rodent genotypic responses to stress: a complex genotype-environment interaction. *Behav. Genet. 1*:119–124.

Fulker, D. W. 1972. Applications of a simplified triple-test cross. *Behav. Genet. 2*:185–198.

Fulker, D. W. 1979. Some implications of biometrical genetical analysis for psychological research. In *Theoretical Advances in Behavior Genetics.* J. R. Royce and L. Mos (eds.). Alphen on den Rign, Netherlands; Germantown, Maryland: Suthoff and Nordhoff, pp. 337–376.

Fulker, D. W., J. Wilcock, and P. L. Broadhurst. 1972. Studies in genotype-environment interaction. 1. Methodology and preliminary multivariate analysis of a diallel cross of eight strains of rats. *Behav. Genet. 2*:261–287.

Fuller, J. L., and R. L. Collins. 1968. Temporal parameters of sensitization for audiogenic seizures in SJL/J mice. *Dev. Psychobiol. 1*:185–188.

Fuller, J. L., and Hahn, M. E. 1976. Issues in the genetics of social behavior. *Behav. Genet. 6*:391–406.

Fuller, J. L., and W. R. Thompson. 1960. *Behavior Genetics.* New York: Wiley.

Fuller, J. L., and W. R. Thompson. 1978. *Foundations of Behavior Genetics.* St. Louis: Mosby.

Fuyama, Y. 1976. Behavior genetics of olfactory responses in *Drosophila* I. Olfactometry and strain differences in *Drosophila melanogaster. Behav. Genet. 6*:407–420.

Gajdusek, D. C. 1964. Factors governing the genetics of primitive human populations. *Cold Spring Harb. Symp. Quant. Biol. 29*:121–135.

Gajdusek, D. C. 1977. *Unconventional Viruses and the Origin and Disappearance of Kuru.* Stockholm: Nobel Foundation.

Garn, S. M. (ed.). 1961. *Human Races.* Springfield, Ill.: Charles C Thomas.

Garside, R. F., and D. W. K. Kay. 1964. The genetics of stuttering. In *The Syndrome of Stuttering.* G. Andrews and M. M. Harris (eds.). London: Heinemann.

Gayral, L., M. Barraud, J. Carrie, and C. Candebat. 1960. Pseudohermaphrodisme à type de "testicule feminisant": 11 cas. *Toulouse Med. 9*:637–647.

Gershon, E. S. 1979. Genetics of the affective disorders. *Hosp. Prac. March:*117–122.

Gershon, E. S., W. E. Bunney, Jr., J. F. Leckman, M. Van Eerdewegh, and B. A. DeBauche. 1976. The inheritance of affective disorders: a review of data and hypotheses. *Behav. Gent. 6*:227–261.

Gershon, E. S., S. Targum, L. Kessler, C. Mazure, and W. Bunney, Jr. 1977. Genetic studies and biologic strategies in the affective disorders. *Prog. Med. Genet.* New Series, *2*:101–164.

Gibbs, C. J., and D. C. Gajdusek. 1978. Virus-induced subacute slow infections of the brain associated with a cerebellar-type ataxia. *Adv. Neurol. 21*:359–372.

Gibson, J. B., and C. G. N. Mascie-Taylor. 1973. Biological aspects of a high socio-economic group. 2. I.Q. components and social mobility. *J. Biosoc. Sci. 5*:17–30.

Gibson, J. B., and J. M. Thoday. 1962. Effects of disruptive selection. 9. Low selection intensity. *Heredity 19*:125–130.

Gilder, P. M., and P. J. B. Slater. 1978. Interest in mice in conspecific male odours is influenced by degree of kinship. *Nature* (London) *274*:364–365.

Ginsburg, B. E. 1967. Genetic parameters in behavioral research. In *Behavior Genetic Analysis*. J. Hirsch (ed.). New York: McGraw-Hill, pp. 135–153.

Glueck, S., and E. Glueck. 1956. *Physique and Delinquency*. New York: Harper & Row.

Godoy-Herrera, R. 1977. Inter- and intrapopulational variation in digging in *Drosophila melanogaster* larvae. *Behav. Genet. 7*:433–439.

Goodenough, D. R., E. Gandini, I. Olkin, L. Pizzamiglio, D. Thayer, and H. A. Witkin. 1977. A study of X chromosome linkage with field dependence and spatial visualization. *Behav. Genet. 7*:373–387.

Goodenough, U., and R. P. Levine. 1974. *Genetics*. New York: Holt.

Goodwin, D. W., F. Schulsinger, L. Harmanson, S. B. Guze, and G. Winokur. 1973. Alcohol problems in adoptees raised apart from alcoholic biological parents. *Arch. Gen. Psychiat. 28*:238–243.

Gottesman, I. I. 1963. Genetic aspects of intelligent behavior. In *The Handbook of Mental Deficiency: Psychological Theory and Research*. N. Ellis (ed.). New York: McGraw-Hill, pp. 253–296.

Gottesman, I. I. 1965. Personality and natural selection. In *Methods and Goals in Human Behavior Genetics*. S. G. Vandenberg (ed.). New York: Academic, pp. 63–80.

Gottesman, I. I. 1978. Schizophrenia and genetics: Where are we? Are you sure? In *The Nature of Schizophrenia*. L. C. Wynne (ed.). New York: Wiley, pp. 59–70.

Gottesman, I. I., and L. Heston. 1972. Human behavioral adaptations: speculations on their genesis. In *Genetics, Environment, and Behavior: Implications for Educational Policy*. L. Ehrman, G. S. Omenn, and E. Caspari (eds.). New York: Academic, pp. 105–122.

Gottesman, I. I., and J. Shields. 1966. Schizophrenia in twins: 16 years' consecutive admissions to a psychiatric clinic. *Dis. Nerv. Syst. 27*:11–19.

Gottesman, I. I., and J. Shields. 1971. Schizophrenia: geneticism and environmentalism. *Hum. Hered. 21*:517–522.

Gottesman, I. I., and J. Shields. 1972. *Schizophrenia and Genetics: A Twin Study Vantage Point*. New York: Academic.

Gottesman, I. I., and J. Shields. 1973. Genetic theorizing and schizophrenia. *Br. J. Psychiatry 122*:15–30.

Gould, J. L. 1974. Genetics and molecular ethology. *Tierpsychol. 35*:267–292.

Goy, R. W., and J. S. Jakway. 1959. The inheritance of patterns of sexual behavior in female guinea pigs. *Anim. Behav.* 7:142–149.

Gramberg-Danielson, B. 1962. Investigation of traffic accident frequency of persons with defective color sense. *Klin. Monatsbl. Augenheilkd. 139*:677–682.

Grant, B., G. A. Snyder, and D. L. Glessner, 1974. Frequency-dependent mate selection in *Mormoniella vitripennis. Evolution 28*:259–264.

Grant, V. 1977. *Organismic Evolution.* San Francisco: Freeman.

Green, E. L. 1966. Breeding systems. In *Biology of the Laboratory Mouse,* 2d ed. E. L. Green (ed.). New York: McGraw-Hill, pp. 11–22.

Gregory, R. L. 1966. *Eye and Brain: The Psychology of Seeing.* London: Weidenfeld & Nicolson.

Griffing, B. 1956. Concept of general and specific combining ability in relation to diallel crossing systems. *Aust. J. Biol. Sci. 9*:463–493.

Griffiths, D. R. 1970. Assessment of personality. In *Psychological Assessment of Mental and Physical Handicaps.* P. Mittler (ed.). London: Methuen, pp. 83–117.

Grossfield, J. 1971. Geographic distribution and light-dependent behavior in *Drosophila. PNAS 68*:2669–2673.

Grossfield, J. 1975. Behavioral mutants of *Drosophila.* In *A Handbook of Genetics,* Vol. 3, *Invertebrates of Genetic Interest,* R. C. King (ed.). New York: Plenum, pp. 679–701.

Grüneberg, H. 1963. *The Pathology of Development: A Study of Inherited Skeletal Disorders in Animals.* Oxford: Blackwell.

Guhl, A. M. 1962. The behaviour of chickens. In *The Behaviour of Domestic Animals,* E. S. E. Hafez (ed.). London: Ballière pp. 491–530.

Guhl, A. M., Craig, J. V., and Mueller, C. D. 1960. Selective breeding for aggressiveness in chickens. *Poult. Sci. 39*:970–980.

Guillery, R. W. 1974. Visual pathways in albinos. *Sci. Am. 230*:44–54.

Guillery, R. W., and J. H. Kass. 1973. Genetic abnormality of the visual pathways in a "white" tiger. *Science 180*:1287–1289.

Gunderson, J. G., and L. R. Mosher. 1975. The cost of schizophrenia. *Am. J. Psychiatry 132*:901–906.

Guttman, R. 1974. Genetic analysis of analytical spatial ability: Raven's progressive matrices. *Behav. Genet. 4*:273–284.

Gwadz, R. 1970. Monfactorial inheritance of early sexual receptivity in the mosquito, *Aedes atropalpus. Anim. Behav. 18*:358–361.

Hadler, N. M. 1964. Heritability and phototaxis in *Drosophila melanogaster. Genetics 50*:1269–1277.

Hafez, E. S. E. 1968. *Reproduction in Farm Animals,* 2d ed. Philadelphia: Lea & Febiger.

Hafez, E. S. E. (ed.). 1969. *The Behavior of Domestic Animals,* 2d ed. Baltimore: Williams & Wilkins.

Hafez, E. S. E. (ed.). 1975. *The Behaviour of Domestic Animals,* 3d ed. London: Baillière Tindall.

Hafez, E. S. E., M. Williams, and S. Wierzbowski, 1969. The behaviour of horses. In *The Behavior of Domestic Animals.* 2d ed. E. S. E. Hafez (ed.). Baltimore: Williams & Wilkins, pp. 391–416.

Haldane, J. B. S. 1946. The interaction of nature and nurture. *Ann. Eugen. 13*:197–205.

Hale, E. B. 1969. Domestication and the evolution of behaviour. In *The Behavior of Domestic Animals*. 2d ed. E. S. E. Hafez (ed.). Baltimore: Williams & Wilkins, pp. 22–42.

Hall, C. S. 1951. The genetics of behavior. In *Handbook of Experimental Psychology*. S. S. Stevens (ed.). New York: Wiley, pp. 304–329.

Hall, C. S., and G. Lindzey. 1957. *Theories of Personality*. New York: Wiley.

Hamerton, J. L. 1976. Human population cytogenetics: dilemmas and problems. *Am. J. Hum. Genet. 28*:107–122. Reprinted and discussed in N. Korn (ed.). 1978. *Human Evolution*, 4th ed. New York: Holt, pp. 46–60.

Hamilton, W. D. 1964. The genetical evolution of social behavior. *J. Theor. Biol. 7*:1–52.

Hancock, J. 1954. *Studies in Monozygotic Cattle Twins*. Pub. No. 63, Washington, D.C.: U.S. Department of Agriculture.

Harnard, S., R. W. Doty, L. Goldstein, J. Jaynes, and G. Krauthamer (eds.). 1977. *Lateralization in the Nervous System*. New York: Academic.

Harper, P. S. 1977. Mendelian inheritance or transmissible agent?—the lesson of kuru and the Australia antigen. *J. Med. Genetics 14*:389–398.

Harrell, R. F., E. R. Woodyard, and A. I. Gates. 1956. The influence of vitamin supplementation of the diets of pregnant and lactating women on the intelligence of their offspring. *Metabolism 5*:555–562.

Harris, H. 1959. *Human Biochemical Genetics*. London: Cambridge University Press.

Harris, V. T. 1952. An experimental study of habitat selection by prairie and forest races of the deermouse *Peromyscus maniculatus*. *Contrib. Lab. Vertebrate Biol. Univ. Michigan 56*:1–53.

Harrison, G. A., R. J. Morton, and J. S. Weiner, 1959. The growth in weight and tail length of inbred and hybrid mice reared at two different temperatures. *Philos. Trans. R. Soc. Lond. (Biol. Sci.) 242*:479–516.

Harrison, G. A., J. S. Weiner, J. M. Tanner, and N. A. Barnicot. 1964. *Human Biology: An Introduction to Human Evolution, Variation and Growth*. Oxford: Clarendon Press.

Harvald, B., and M. Hauge. 1965. Hereditary factors elucidated by twin studies. In *Genetics and the Epidemiology of Chronic Diseases*. J. V. Neel, M. W. Shaw, and W. J. Schull (eds.). Washington, D.C.: U.S. Department of Health, Education and Welfare, pp. 61–76.

Hatt, D., and P. A. Parsons. 1965. Associations between surnames and blood groups in the Australian population. *Acta Genet. 15*:309–318.

Hauser, A. W., and L. T. Kurland. 1975. The epidemiology of epilepsy in Rochester, Minnesota, 1935 through 1967. *Epilepsia 16*:1–66.

Hay, D. A. 1972. Recognition by *Drosophila melanogaster* of individuals from other strains or cultures: support for the role of olfactory cues in selective mating. *Evolution 26*:171–176.

Hay, D. A. 1975. Strain differences in the maze-learning abilities of *D. melanogaster*. *Nature* (London) *257*:44–46.

Hay, D. A. 1979. Genetic validation of a *Drosophila* learning task. *Experientia 35*:310–311.

Hayman, B. I. 1958. The theory and analysis of diallel crosses. *Genetics 43*:63–85.

Hayman, R. H. 1964. Exercise of mating preference by a Merino ram. *Nature* (London) *203*:160–162.

Henderson, N. D. 1968. The confounding effects of genetic variables in early experience research: Can we ignore them? *Dev. Psychobiol. 1*:146–152.

Henderson, N. D. 1970. Genetic influences on the behavior of mice can be obscured by laboratory rearing. *J. Comp. Physiol. Psychol. 72*:505–511.

Henderson, N. D. 1973. Brain weight changes resulting from enriched rearing conditions: a diallel analysis. *Dev. Psychobiol. 6*:367–376.

Henderson, N. D. 1976. Short exposures to enriched environments can increase variability of behavior in mice. *Dev. Psychobiol. 9*:549–553.

Henderson, N. D. 1978. Genetic dominance for low activity in infant mice. *J. Comp. Physiol. Psychol. 92*:118–125.

Henderson, N. D. 1979. Adaptive significance of animal behavior: The role of gene-environment interaction. In *Theoretical Advances in Behavior Genetics*. J. R. Royce and L. Mos (eds.). Alphen on den Rign, Netherlands; Germantown, Maryland: Suthoff and Nordhoff, pp. 243–284.

Henry, K. R. 1967. Audiogenic seizure susceptibility induced in C57BL/6J mice by prior auditory exposure. *Science 158*:938–940.

Herschel, M. 1978. Dyslexia revisited. *Human Genetics 40*:115–134.

Herskowitz, I. 1973. *Principles of Genetics*. New York: Macmillan.

Heston, L. L. 1966. Psychiatric disorders in foster home reared children of schizophrenic mothers. *Br. J. Psychol. 112*:819–825.

Heston, L. L. 1970. The genetics of schizophrenic and schizoid disease. *Science 167*:249–256.

Heston, L. L. 1972. Discussion. In *Genetics, Environment, and Behavior: Implications for Educational Policy*. L. Ehrman, G. S. Omenn, and E. Caspari (eds.). New York: Academic, pp. 99–102.

Hiernaux, J. 1956. Analyse de la variation des caractères physiques humains en une region de l'Afrique Centrale: Ruanda-Urundi et Kivu. *Anthropologie 3*:1–131.

Higgins, J. V., E. W. Reed, and S. C. Reed. 1962. Intelligence and family size: a paradox resolved. *Eugen. Q. 9*:84–90.

Hill, K. G., J. J. Loftus-Hills, and D. F. Gartside. 1972. Premating isolation between the Australian field crickets, *Teleogryllus commodus* and *T. oceanicus*. *Aust. J. Zool. 20*:153–163.

Hirsch, J. 1962. Individual differences in behavior and their genetic basis. In *Roots of Behavior*. E. L. Bliss (ed.). New York: Hoeber-Harper, pp. 3–23.

Hirsch, J. 1963. Behavior genetics and individuality understood. *Science 142*:1436–1442.

Hirsch, J. (ed.). 1967a. *Behavior-Genetic Analysis*. New York: McGraw-Hill.

Hirsch, J. 1967b. Intellectual functioning and the dimensions of human variation. In *Genetic Diversity and Human Behavior*. J. N. Spuhler (ed.). Chicago: Aldine, pp. 19–31.

Hirsch, J., and J. Boudreau. 1958. Studies in experimental behavior genetics. The heritability of phototaxis in a population of *Drosophila melanogaster*. *J. Comp. Physiol. Psychol. 51*:647–651.

Hoagland, H. 1966. Cybernetics of population control. In *Human Ecology: Collected Readings*. J. B. Bresler (ed.). Reading, Mass.: Addison-Wesley, pp. 351–359.

Hodgson, R. E. 1935. An eight generation experiment in inbreeding swine. *J. Hered. 26*:209–217.

Hollaender, A. (ed.). 1971. *Chemical Mutagens: A Method for Their Detection*. New York: Plenum.

Hölmberg, L. 1972. Genetic studies in a family with testicular feminization, hemophilia A and color blindness. *Clin. Genet. 3*:253–257.

Holzberg, S., and J. H. Schroder. 1975. The inheritance of aggressiveness in the convict cichlid fish, *Cichlasoma nigrofasciatum* (Pisces: Cichlidae). *Anim. Behav. 23:*625–631.

Holzinger, K. J. 1929. The relative effect of nature and nurture influences on twin differences. *J. Educ. Psychol. 20:*241–248.

Honzik, M. P. 1957. Developmental studies of parent-child resemblance in intelligence. *Child Dev. 28:*215–228.

Hosgood, S. M. W., and P. A. Parsons, 1967a. Genetic heterogeneity among the founders of laboratory populations of *Drosophila melanogaster* II. Mating behavior. *Aust. J. Biol. Sci. 20:*1193–1203.

Hosgood, S. M. W., and P. A. Parsons, 1967b. The exploitation of genetic heterogeneity among the founders of laboratory populations of *Drosophila* prior to directional selection. *Experientia 23:*1066–1067.

Hotta, Y., and S. Benzer. 1972. Mapping of behavior in *Drosophila* mosaics. *Nature* (London) *240:*527–535.

Hotta, Y., and S. Benzer. 1973. Mapping of behavior in Drosophila mosaics. In *Genetic Mechanisms of Development.* F. H. Ruddle (ed.). New York: Academic, pp. 129–167.

Howe, W. L., and P. A. Parsons. 1967. Genotype and environment in the determination of minor skeletal variants and body weight in mice. *J. Embryol. Exp. Morphol. 17:*283–292.

Howells, W. W. 1966. Population distances: biological, linguistic, geographical, and environmental. *Curr. Anthropol. 7:*531–540.

Hoy, R. R. 1974. Genetic control of acoustic behavior in crickets. *Am. Zool. 14:*1067.

Hoy, R. R., J. Hahn, and R. Paul. 1976. Hybrid cricket auditory behavior: evidence for genetic coupling in animal communication. *Science 195:*82–84.

Hoy, R. R., and R. L. Paul. 1973. Genetic control of song specificity in crickets. *Science 180:*82–83.

Hungerford, D. 1971. Chromosome structure and function in man. 1. Pachytene mapping in the male, improved methods and general discussion of initial results. *Cytogenetics 10:*23–32.

Hungerford, D., G. U. LaBadie, and G. B. Balaban. 1971. Chromosome structure and function in man. 2. Provisional maps of the two smallest autosomes (chromosomes 21 and 22) at pachytene in the male. *Cytogenetics 10:*33–37.

Huxley, J., E. Mayr, H. Osmond, and A. Hoffer. 1964. Schizophrenia as a genetic morphism. *Nature* (London) *204:*220–221.

Ikeda, K., and W. D. Kaplan. 1970a. Patterned neural activity of a mutant *Drosophila melanogaster. PNAS 66:*765–772.

Ikeda, K., and W. D. Kaplan. 1970b. Unilaterally patterned neural activity of gynandromorphs: mosaic for a neurological mutant of *Drosophila melanogaster. PNAS 67:*1480–1487.

Imperato-McGinley, J., L. Guerrero, T. Gautier, and R. Peterson. 1974. 5α-reductase deficiency in man: an inherited male pseudohermaphroditism. *Science 186:*1213–1214.

Jacobs, P. A., M. Brunton, M. M. Melville, R. P. Brittain, and W. F. McClemont, 1965. Aggressive behaviour, mental subnormality, and the XYY male. *Nature* (London) *208:*1351–1352.

Jacobson, M. 1972. *Insect Sex Pheromones.* New York: Academic.

Jakway, J. S. 1959. Inheritance of patterns of mating behavior in the male guinea pig. *Anim. Behav.* 7:150–162.

Jay, B. 1974. Recent advances in ophthalmic genetics. *Br. J. Ophthalmol. 58*: 427–437.

Jennings, H. S. 1906. *Behavior of the Lower Organisms.* New York: Columbia University Press.

Jensen, A. R. 1972. Discusssion. In *Genetics, Environment, and Behavior: Implications for Educational Policy.* L. Ehrman, G. S. Omenn, and E. Caspari (eds.). New York: Academic, pp. 240–246.

Jensen, A. R. 1973. *Educability and Group Differences.* New York: Harper & Row.

Jensen, A. R. 1979. *Bias in Mental Testing.* New York: Free Press.

Jinks, J. L., and D. W. Fulker. 1970. Comparison of the biometrical genetical, MAVA, and classical approaches to the analysis of human behavior. *Psychol. Bull. 73*:311–349.

Judd, D. 1943. Color blindness and the detection of camouflage. *Science* 97:544–546.

Jude, A., and A. Searle. 1957. A fertile tortoiseshell tomcat. *Nature* (London) *179*:1087–1088.

Juel-Nielsen, N., and B. Harvald. 1958. The electroencephalogram in uniovular twins brought up apart. *Acta Genet.* 8:57–64.

Kaij, L. 1957. Drinking habits in twins. *Acta Genetica et Medica* 7:437–441.

Kalmus, H. 1965. *Diagnosis and Genetics of Defective Colour Vision.* Elmsford, N. Y.: Pergamon.

Kalmus, H. 1967. Sense perception and behavior. In *Genetic Diversity and Human Behavior.* J. N. Spuhler (ed.). Chicago: Aldine, pp. 73–87.

Kamin, L. J. 1974. *The Science and Politics of IQ,* New York: Halsted Press.

Kaplan, W. D., and W. E. Trout. 1974. Genetic manipulation of an abnormal jump response in *Drosophila. Genetics* 77:721–739.

Karlson, P., and A. Butenandt. 1959. Pheromones (ectohormones) in insects. *Ann. Rev. Entomol. 4*:39.

Kaul, D., and P. A. Parsons. 1965. The genotypic control of mating speed and duration of copulation in *Drosophila pseudoobscura. Heredity 20*:381–392.

Kaul, D., and P. A. Parsons, 1966. Competition between males in the determination of mating speed in *Drosophila pseudoobscura. Aust. J. Biol. Sci. 19*:945–947.

Kawanishi, M., and T. K. Watanabe. 1978. Difference in photo-preferences as a cause of coexistence of *Drosophila simulans* and *D. melanogaster* in nature. *Jap. J. Genet. 53*:209–214.

Keith, S. J., J. G. Gunderson, A. Reifman, S. Buchsbaum, and L. R. Mosher. 1976. Special report: Schizophrenia 1976. *Schiz. Bull. 2*:509–565.

Kempthorne, O. 1969. *An Introduction to Genetic Statistics.* Ames: Iowa State University Press.

Kennedy, W. A., V. Van De Riet, and J. C. White, Jr. 1963. A normative sample of intelligence and achievement of Negro elementary school children in the southeastern United States. *Monogr. Soc. Res. Child Dev. 28*(6):1–112.

Kessler, M. E., and Chabora, P. C. 1977. Light intensity and phototaxis in the house fly: photonegativity in a yellow-eyed mutant. *Behav. Genet.* 7:129–137.

Kessler, S. 1966. Selection for and against ethological isolation between *Drosophila pseudoobscura* and *Drosophila persimilis. Evolution 20*:634–645.

Kessler, S. 1968. The genetics of *Drosophila* mating behavior. *Anim. Behav. 16*:485–491.

Kessler, S. 1969. The genetics of *Drosophila* mating behavior. 2. The genetic architecture of mating speed in *Drosophila pseudoobscura. Genetics 62*:421–433.

Kety, S. S. 1967. Biochemical aspects of mental states. In *The Human Mind.* J. D. Roslansky (ed.). Amsterdam: North-Holland, pp. 141–152.

Kidd, K. K. 1977. A genetic perspective on stuttering. *J. Fluency Dis. 2*:259–269.

Kidd, K. K., and L. L. Cavalli-Sforza. 1973. An analysis of the genetics of schizophrenia. *Soc. Biol. 20*:254–265.

Kidd, K. K., J. R. Kidd, and M. A. Records. 1978. The possible causes of the sex ratio in stuttering and its implications. *J. Fluency Dis. 3*:13–23.

Kiker, J. T., P. B. Siegel, and K. Hinkelmann. 1976. Genetic analysis of behaviors related to the solution of a detour learning task. *Behav. Genet. 6*:315–325.

Kilgour, R. 1975. The open-field test as an assessment of the temperament of dairy cows. *Anim. Behav. 23*:615–624.

King, J. A. 1967. Behavioral modification of the gene pool. In *Behavior-Genetic Analysis.* J. Hirsch (ed.). New York: McGraw-Hill, pp. 22–43.

King, J. A., D. Maas, and R. Weisman. 1964. Geographic variations in nest size among species of *Peromyscus. Evolution 18*:230–234.

King, R. C. (ed.) 1974–1976. *Handbook of Genetics,* Vols. 1–5. New York: Plenum.

Kirk, R. L. 1966. Population genetic studies in Australia and New Guinea. In *The Biology of Human Adaptability.* P. T. Baker and J. S. Weiner (eds.). Oxford: Clarendon Press, pp. 395–430.

Klein, T. W., and J. C. DeFries. 1970. Similar polymorphism of taste sensitivity to PTC in mice and man. *Nature* (London) *225*:555–557.

Klopfer, P. 1963. Behavioural aspects of habitat selection: the role of early experience. *Wilson Bull. 75*:15–22.

Knight, G., A. Robertson, and C. Waddington. 1956. Selection for sexual isolation within a species. *Evolution 10*:14–22.

Koch, R. 1967. Tagesperidik der Activität und der Orientierung nach Wald und Feld von *D. subobscura* und *D. obscura. Z. Vergl. Physiol. 54*:353–394.

Konopka, R. J., and S. Benzer. 1971. Clock mutants of *Drosophila melanogaster. PNAS 68*:2112–2116.

Koopman, K. R. 1950. Natural selection for reproductive isolation between *Drosophila pseudoobscura* and *Drosophila persimilis. Evolution 4*:135–148.

Kovach, J. K. 1974. The behaviour of Japanese quail: review of literature from a bioethological perspective. *Appl. Anim. Ethol. 1*:77–102.

Kovach, J. K. 1977. Binomial assessment of behavioral-phenotypic variations: constancy of choices, trial effects, and social interaction effects in mass-screened color preferences of quail chicks (*Coturnix coturnix japonica). J. Comp. Physiol. 91*:851–857.

Kovach, J. K. 1978. Color preferences in quail chicks: generalization of the effects of genetic selection. *Behaviour 65*:263–269.

Kraepelin, E. 1896. *Psychiatrie,* 5th ed. Leipzig: Barth.

Krebs, C. J., M. S. Gaines, B. L. Keller, J. H. Myers, and R. H. Tamarin. 1973. Population cycles in small rodents. *Science 179*:35–41.

Krebs, J. R., and N. B. Davies (eds.). 1978. *Behavioural Ecology: An Evolutionary Approach.* Oxford: Blackwell.

Kretchmer, N. 1972. Lactose and lactase. *Sci. Am. 227*:70–78.

Krüger, J. 1972. Zur Unterscheidung zwischen multifaktoriellem Erbgang mit Schwellenwerteffekt und einfachem diallelem Erbgang. *Humangenetik 17*:182–252.

Kung, C. 1976. Membrane control of ciliary motions and its genetic modifications. In *Cold Spring Harbor Conference on Cell Proliferation,* Vol. 3. *Cell Motility,* R. Goldman, T. Pollard, and J. Rosenbaum (eds.). Cold Spring Harbor, N.Y.: Cold Spring Harbor Laboratory, pp. 941–948.

Kung, C. 1978. Behavioral genetics of Paramecium. XIV International Congress of Genetics, Moscow, Plenary Sessions Symposia Abstracts, p. 100.

Kung, C., S. Y. Chang, Y. Satow, J. Van Houten, and H. Hansma. 1975. Genetic dissection of behavior in *Paramecium. Science 188*:898–904.

Kung, C., and Y. Naitoh. 1973. Calcium-induced ciliary reversal in the extracted models of "pawn," a behavioral mutant of *Paramecium. Science 179*:195–196.

Landis, B., and E. S. Tauber (eds). 1972. *In the Name of Life: Essays in Honor of Erich Fromm.* New York: Holt.

Law, J. H., and R. E. Regnier. 1971. Pheromones. *Ann. Rev. Biochem. 40*:533–548.

Leader, R. 1967. The kinship of animal and human diseases. *Sci. Am. 216*:110–116.

Leader, R., and I. Leader. 1971. *Dictionary of Comparative Pathology and Experimental Biology.* Philadelphia: Saunders.

Lee, B. T. O., and P. A. Parsons, 1968. Selection, prediction and response. *Biol. Rev. 43*:139–174.

LeFrancois, G. R. 1972. *Psychological Theories and Human Learning: Kongor's Report.* Monterey, Cal.: Brooks/Cole.

Leonard, J., and L. Ehrman. 1976. Recognition and sexual selection in *Drosophila. Science 193*:693–695.

Leonard, J., L. Ehrman, and A. Pruzan. 1974. Pheromones as a means of genetic control of behavior. *Annu. Rev. Genet. 8*:179–193.

Leonard, J., L. Ehrman, and M. Schorsch. 1974. Bioassay of a *Drosophila* pheromone influencing sexual selection. *Nature* (London) *250*:261–262.

Lerner, I. M. 1968. *Heredity, Evolution and Society.* San Francisco: Freeman.

Lerner, I. M., and W. Libby. 1976. *Heredity, Evolution and Society,* 2d ed. San Francisco: Freeman.

Leroy, Y. 1964. Transmission du paramètre fréquence dans le signal acoustique des hybrides F_1 et $P \times P_1$, de deux Grillons: *Teleogryllus commodus* Walker et *T. oceanicus* Le Guillon (Orthoptères, ensifères). *C. R. Acad. Sci. 259*:892–895.

Levine, L. 1958. Studies on sexual selection in mice. *Am. Nat. 92*:21–26.

Levine, L. 1969. *Biology of the Gene.* St. Louis: Mosby.

Levere, R. D., and A. Kappas. 1973. The porphyric diseases of man. In *Medical Genetics,* V. McKusick, and R. Claiborne (eds.). New York: H. P. Publ. Co., pp. 113–121.

Levitan, M., and A. Montagu. 1971. *Textbook of Human Genetics.* New York: Oxford University Press.

Levy, J. 1977. The origins of lateral asymmetry. In *Lateralization in the Nervous System.* S. Harned, R. Doty, L. Goldstein, J. Jaynes, and G. Krauthamer. (eds.). New York: Academic, pp. 195–209.

Lewontin, R. C. 1959. On the anomalous response of *Drosophila pseudoobscura* to light. *Am. Nat. 93*:321–328.

Lewontin, R. C. 1972. The apportionment of human diversity. *Evolutionary Biology 6*:381–398.

Lewontin, R. C. 1974. *The Genetic Basis of Evolutionary Change.* New York: Columbia University Press.

Lewontin, R. C., and L. C. Dunn. 1960. The evolutionary dynamics of a polymorphism in the house mouse. *Genetics 45*:705–722.

Li, C. 1955. *Population Genetics*. Chicago: University of Chicago Press.

Li, C. 1976. *First Course in Population Genetics*. Pacific Grove, Calif.: Boxwood Press.

Lidicker, W. Z., Jr. 1962. Emigration as a possible mechanism permitting the regulation of population density below carrying capacity. *Am. Nat. 96*:29–33.

Lieber, C. S. 1972. Metabolism of ethanol and alcoholism: racial and acquired factors. *Annals of Internal Med. 76*:326–327.

Liley, N. R., and B. H. Seghers. 1975. Factors affecting the morphology and behaviour of guppies in Trinidad. In *Function and Evolution in Behaviour*. G. Baerends, C. Beer, and A. Manning (eds.). Oxford: Clarendon Press, pp. 92–116.

Lill, A. 1966. A review of nonrandom mating behaviour in domestic poultry. *Poultry Review 6*:51–56.

Lill, A. 1968. An analysis of sexual isolation in the domestic fowl. 1. The basis of homogamy in males. *Behaviour 30*:107–126.

Lindauer, M. 1975. Evolutionary aspects of orientation and learning. In *Function and Evolution in Behaviour*. G. Baerends, C. Beer, and A. Manning (eds.). Oxford: Clarendon Press, pp. 228–242.

Lindsay, D. R., D. G. Dunsmore, J. D. Williams, and G. J. Syme. 1976. Audience effects on the mating behaviour of rams. *Anim. Behav. 24*:818–821.

Lindzey, G. 1967. Behavior and morphological variation. In *Genetic Diversity and Human Behavior*. J. N. Spuhler (ed.). Chicago: Aldine, pp. 227–240.

Lindzey, G., J. Loehlin, M. Manosevitz, and D. Thiessen. 1971. Behavioral genetics. *Annu. Rev. Psychol. 22*:39–94.

Lindzey, G., and D. Thiessen (eds.). 1970. *Contributions to Behavior-Genetic Analysis: The Mouse as a Prototype*. New York: Appleton.

Littlefield, J., A. Milunsky, and L. Jacoby. 1973. Prenatal genetic diagnosis: status and problems. In *Ethical Issues in Human Genetics*. B. Hilton, D. Callahan, M. Harris, P. Condliffe, and B. Berkley (eds.). New York: Plenum, pp. 43–51.

Littlejohn, M. J. 1965. Premating isolation in the *Hyla ewingi* complex (Anura: Hylidae). *Evolution 19*:234–243.

Littlejohn, M. J. 1969. The systematic significance of isolating mechanisms. In *Systematic Biology Publ. No. 1692*, Washington, D. C.: National Academy of Sciences, pp. 459–482. C. Sibley, Chairman, and W. F. Blair, E. Nevo, W. Bossert, and others. Discussion, pp. 483–493.

Littlejohn, M. J., and R. S. Oldham. 1968. *Rana pipiens* complex: mating call structure and taxonomy. *Science 162*:1003–1005.

Livingstone, F. B. 1958. Anthropological implications of sickle cell gene distribution in West Africa. *Am. Anthropol. 60*:553–562.

Livingstone, F. B. 1963. Blood groups and ancestry: a test case from the New Guinea highlands. *Curr. Anthropol. 4*:541–542.

Loehlin, J. C., G. Lindzey, and J. N. Spuhler. 1975. *Race Differences in Intelligence*. San Francisco: Freeman.

Loehlin, J. C., S. Sharan, and R. Jacoby. 1978. In pursuit of the "spatial gene": a family study. *Behav. Genet. 8*:27–41.

Lykken, D., A. Tellegen, and K. Thorkelson. 1974. Genetic determination of EEG frequency spectra. *Biol. Psych. 1*:245–259.

Lynch, C. B., and J. P. Hegmann. 1972. Genetic differences influencing behavioral temperature regulation in small mammals. 1. Nesting by *Mus musculus. Behav. Genet. 2*:43–53.

Lynch, C. B., and J. P. Hegmann. 1973. Genetic differences influencing behavioral temperature regulation in small mammals. 2. Genotype-environment interactions. *Behav. Genet. 3*:145–154.

Lynch, H. T. 1969. *Dynamic Genetic Counseling for Clinicians.* Springfield, Ill.: Charles C Thomas.

Lyon, M. F. 1962. Sex chromatin and gene action in the mammalian X-chromosome. *Am. J. Hum. Genet. 14*:135–148.

Macalpine, I., and R. Hunter. 1969. Porphyria and George III. *Sci. Am. 221*:38–46.

MacBean, I. T., and P. A. Parsons. 1967. Directional selection for duration of copulation in *Drosophila melanogaster. Genetics 56*:233–239.

McBride, G. 1958. The environment and animal breeding problems. *Anim. Breed. Abstr. 26*:349–358.

McClearn, G. E. 1972. Genetic Determination of Behavior (Animal). In *Genetics, Environment, and Behavior: Implications for Educational Policy.* L. Ehrman, G. S. Omenn, and E. Caspari (eds.). New York: Academic, pp. 55–67.

McClearn, G. E., and J. C. DeFries. 1973. *Introduction to Behavior Genetics.* San Francisco: Freeman.

McClure, H., K. Belden, W. Pieper. 1969. Autosomal trisomy in a chimpanzee: resemblance to Down's syndrome. *Science 165*:1010–1011.

McCracken, R. 1971. Lactase deficiency: an example of dietary evolution. *Curr. Anthropol. 12*:479–517.

McDonald, J., and P. A. Parsons. 1973. Dispersal activities of the sibling species *Drosophila melanogaster* and *Drosophila simulans. Behav. Genet. 3*:293–301.

McGaugh, J. L. (ed.). 1972. *The Chemistry of Mood-Motivation and Memory.* New York: Plenum.

McGill, T. E. 1970. Genetic analysis of male sexual behavior. In *Contributions to Behavior-Genetic Analysis: The Mouse as a Prototype.* G. Lindzey and D. D. Thiessen (eds.). New York: Appleton, pp. 57–88.

McGuire, T. R., and J. Hirsch. 1977. Behavior-genetic analysis of *Phormia regina:* conditioning, reliable individual differences, and selection. *PNAS 74*:5193–5197.

McKay, H., L. Sinisterra, A. McKay, H. Gomez, and P. Lloreda. 1978. Improving cognitive ability in chronically deprived children. *Science 200*:270–278.

McKay, T. F. C., and R. W. Doyle. 1978. An ecological genetic analysis of the settling behaviour of a marine polychaete. I. Probability of settlement and gregarious behaviour. *Heredity 40*:1–12.

McKenzie, J. A. 1974. The distribution of vineyard populations of *Drosophila melanogaster* and *Drosophila simulans* during vintage and non-vintage periods. *Oecologia 14*:1–16.

McKenzie, J. A. 1975. The influence of low temperature on survival and reproduction in populations of *Drosophila melanogaster. Aust. J. Zool. 23*:237–247.

McKenzie, J. A. 1978. The effect of developmental temperature on population flexibility in *Drosophila melanogaster* and *D. simulans. Aust. J. Zool. 26*:105–112.

McKenzie, J. A., and S. W. McKechnie. 1979. A comparative study of resource utilization in natural populations of *Drosophila melanogaster* and *D. simulans. Oecologia 40*:299–309.

McKenzie, J. A., and P. A. Parsons. 1971. Variations in mating propensities in strains of

Drosophila melanogaster with different scutellar chaeta numbers. *Heredity* 26:313–322.

McKenzie, J. A., and P. A. Parsons. 1972. Alcohol tolerance: an ecological parameter in the relative success of *Drosophila melanogaster* and *Drosophila simulans*. *Oecologia* 10:373–388.

McKenzie, J. A., and P. A. Parsons. 1974. Microdifferentiation in a natural population of *Drosophila melanogaster* to alcohol in the environment. *Genetics* 77:385–394.

McKusick, V. A. 1978. *Mendelian Inheritance in Man*, 5th ed. Baltimore: Johns Hopkins University Press.

McLaren, A., and D. Michie. 1956. Studies on the transfer of fertilized mouse eggs to uterine foster mothers. 1. Factors affecting the implantation and survival of native and transferred eggs. *J. Exp. Biol.* 33:394–416.

McLaren, A., and D. Michie. 1959. Studies on the transfer of fertilized mouse eggs to uterine foster mothers. 2. The effect of transferring large numbers of eggs. *J. Exp. Biol.* 36:40–50.

McNeil, E. B. 1970. *The Psychoses*. Englewood Cliffs, N. J.: Prentice-Hall.

Mainardi, D. 1963. Eliminazione della barriera etologica all'isolamento riproduttivo tra *Mus musculus domesticus* e *M.m. bactrianus* mediante azione sull'apprendiento infantile. Istituto Lombardo. (Rend. Sc.) B97:291–299.

Mainardi, D., M. Marsan, and A. Pasquali. 1965. Causation of sexual preferences in the house mouse: the behaviour of mice raised by parents whose odour was artificially altered. *Atti. Soc. Ital. Sci. Nat. Museo Civ. Milano* 104:325–338.

Malagolowkin-Cohen, C., A. S. Simmons, and H. Levene. 1965. A study of sexual isolation between certain strains of *Drosophila paulistorum*. *Evolution* 19:95–103.

Mandel, P., J. L. Nussbaum, N. Neskovic, L. Sarlière, E. Farkas, and O. Robain. 1973. The use of neurological mutants as experimental models. In *The Biochemistry of Gene Expression in Higher Organisms*. J. K. Pollak and J. W. Lee (eds.). Hingham, Mass.: D. Reidel, pp. 410–422.

Mangum, C. 1978. Nonhuman models of hereditary porphyrias. *Science* 201:1043.

Manning, A. 1959. The sexual behavior of two sibling *Drosophila* species. *Behaviour* 15:123–145.

Manning, A. 1961. The effects of artificial selection for mating speed in *Drosophila melanogaster*. *Anim. Behav.* 9:82–92.

Manning, A. 1963. Selection for mating speed in *Drosophila melanogaster* based on the behavior of one sex. *Anim. Behav.* 11:116–120.

Manning, A. 1966. Corpus allatum and sexual receptivity in female *Drosophila melanogaster*. *Nature* (London) 211:1321–1322.

Manning, A. 1967a. The control of sexual receptivity in female *Drosophila*. *Anim. Behav.* 15:239–250.

Manning, A. 1967b. Pre-imaginal conditioning in *Drosophila*. *Nature* (London) 216:338–340.

Manning, A. 1968. The effects of artificial selection for slow mating in *D. simulans*. *Anim. Behav.* 16:108–113.

Manosevitz, M., G. Lindzey, and D. D. Thiessen. 1969. *Behavioral Genetics: Method and Research*. New York: Appleton.

Martin, N. G., and H. J. Eysenck. 1976. Genetic factors in sexual behaviour. In *Sex and Personality*. H. J. Eysenck, (ed.). London: Open Books, pp. 192–219.

Mather, K. 1942. The balance of polygenic combinations. *J. Genet.* 43:309–336.

Mather, K. 1949. *Biometrical Genetics*. London: Methuen.

Mather, K. 1966. Variability and selection. *Proc. R. Soc. Lond. (Biol.) 164*:328–340.

Mather, K., and B. J. Harrison. 1949. The manifold effect of selection. *Heredity 3*:1–52, 131–162.

Mather, K., and J. L. Jinks. 1977. *Introduction to Biometrical Genetics.* London: Chapman & Hall.

Matthysee, S. W., and K. K. Kidd. 1976. Estimating the genetic contribution to schizophrenia. *Am. J. Psychiat. 133*:185–191.

Maxson, S. C., J. S. Cowen, and P. Y. Sze. 1977. Pharmacogenetic differences in audiogenic seizure priming of C57BL/6Bg and DBA/1 Bg-asrmice. *Biochem. Behav. 7*:221–226.

Maxwell, J. 1969. Intelligence, education and fertility. A comparison between the 1932 and 1947 Scottish surveys. *J. Biosoc. Sci. 1*:217–247.

May, R. M. 1977. Population genetics and cultural inheritance. *Nature* (London) *268*:11–13.

Maynard Smith, J. 1966. Sympatric speciation. *Am. Nat. 100*:637–650.

Maynard Smith, J. 1976. Group selection. *Quart. Rev. Biol. 51*:277–283.

Maynard Smith, J. 1978. The evolution of behavior. *Sci. Am. 239*:136–145.

Mayr, E. 1942. *Systematics and the Origin of Species.* New York: Columbia University Press.

Mayr, E. 1963. *Animal Species and Evolution.* Cambridge, Mass.: Harvard University Press.

Mayr, E. 1970. *Populations, Species, and Evolution.* Cambridge, Mass.: Harvard University Press.

Mayr, E. 1974. Behavior programs and evolutionary strategies. *Am. Sci. 62*:650–659.

Mayr, E., and T. Dobzhansky. 1945. Experiments on sexual isolation in *Drosophila.* 4. Modification of the degree of isolation between *Drosophila pseudoobscura* and *Drosophila persimilis* and of sexual preferences in *Drosophila prosaltans. PNAS 31*:75–82.

Médioni, J. 1962. "Contribution à l'etude psycho-physiologique et génétique du phototrophisme d'un insecte *Drosophila melanogaster* Meigen." *These Fac. Sci.* Strasbourg.

Meier, H. 1963. *Experimental Pharmacogenetics, Physiopathology of Heredity, and Pharmacologic Responses.* New York: Academic.

Melnyk, J., F. Vanasek, H. Thompson, and A. Rucci. 1969. Failure of transmission of supernumerary Y chromosomes in man. *Abstr. Am. Soc. Hum. Genet.* San Francisco (Oct. 1–4)42.

Mendel, G. 1865. Versuche über pflanzenhybriden. *Verh. Naturf. Verein Brünn 4*:3–47. Reprint. Pennsauken, N.J.: Stechert, 1960. Translation: *Experiments in Plant-Hybridisation.* Cambridge, Mass.: Harvard University Press, 1965.

Mendlewicz, S., J. Fleiss, and R. Fieve. 1972. Evidence for X-linkage in the transmission of manic-depressive illness. *JAMA 222*:1624–1627.

Menne, D., and H. Spatz. 1977. Colour vision in *Drosophila melanogaster. J. Comp. Physiol. 114*:301–312.

Merrell, D. J. 1953. Selective mating as a cause of gene frequency changes in laboratory populations of *Drosophila melanogaster. Evolution 7*:287–296.

Mesibov, R., and J. Adler. 1972. Chemotaxis toward amino acids in *Escherichia coli. J. Bacteriol. 112*:315–326.

Metrakos, J. D., and K. Metrakos. 1969. Genetic studies in clinical epilepsy. In *Basic*

Mechanisms of the Epilepsies. H. Jasper, A. A. Ward, Jr., and A. Pope (eds.). Boston: Little, Brown, pp. 700–708.

Michener, C. D. 1975. The Brazilian bee problem. *Ann. Rev. Entomol. 20*:399–416.

Mitchell, T. B. 1929. Sex anomalies in the genus *Megachile* with descriptions of new species (Hymenoptera: Megachilidae). *Trans. Am. Entomol. Soc. 54*:333.

Mittler, P. 1971. *The Study of Twins.* Gloucester, Mass.: Peter Smith.

Money, J. 1970. Behavior genetics: principles, methods and examples for XO, XXY, and XYY syndromes. *Sem. Psychiatry 2*:11–29.

Money, J., and S. Mittenthal. 1970. Lack of personality pathology in Turner's syndrome: relation to cytogenetics, hormones, and physique. *Behav. Genet. 1*:43–56.

Moor, L. 1967. Niveau intellectuel et polygonosomie: confrontation du caryotype et du niveau mental de 374 malades dont le caryotype comporte un exces de chromosomes X ou Y. *Rev. Neuropsychiatr. Infant. 15*:325–348.

Moore, J. A. 1952. Competition between *Drosophila melanogaster* and *Drosophila simulans*. 1. Population cage experiments. *Evolution 6*:407–420.

Moore, J. A. 1975. *Rana pipiens*—The changing paradigm. *Amer. Zool. 15*:837–849.

Moran, P. A. P. 1973. A note on heritability and the correlation between relatives. *Ann. Hum. Genet. 37*:217.

Moray, N., and Connolly, K. 1963. A possible case of genetic assimilation of behaviour. *Nature 199*:358–360.

Morgan, T. H., and C. B. Bridges. 1919. Contributions to the genetics of *Drosophila melanogaster*. In *The Origin of Gynandromorphs*. Carnegie Institution Pub. No. 278. Washington, D. C.: Carnegie Institution, pp. 1–122.

Morrill, R., and N. B. Todd, 1978. Mutant allele frequencies in domestic cats of Denver, Colorado. *J. Heredity 69*:131–134.

Morris, J. M. 1953. Testicular feminization. *Am. J. Obstet. Gynecol. 65*:1192–1211.

Morton, N. E. 1972. Human behavioral genetics. In *Genetics, Environment, and Behavior: Implications for Educational Policy*. L. Ehrman, G. S. Omenn, and E. Caspari (eds.). New York: Academic, pp. 247–264.

Morton, N. E., C. S. Chung, and M. P. Mi. 1967. *Genetics of Interracial Crosses in Hawaii*. New York: Karger.

Mourant, A. 1954. *The Distribution of the Human Blood Groups*. Springfield, Ill.: Charles C Thomas.

Muller, H. 1942. Isolating mechanisms, evolution, and temperature. *Biol. Symp. 6*:71–125.

Muskavitch, M. A., E. N. Kost, M. S. Springer, M. F. Goy, and J. Adler. 1978. Attraction by repellents: an error in sensory information processing by bacterial mutants. *Science 201*:63–65.

Myers, J. H., and C. J. Krebs. 1971. Genetic, behavioral and reproductive attributes of dispersing field voles *Microtus pennsylvanicus* and *Microtus ochrogaster*. *Ecol. Monogr. 41*:53–78.

Myers, R. H., and D. A. Shafer. 1979. Hybrid ape offspring of a mating of gibbon and siamang. *Science 205*:308–310.

Nachman, M., C. Larue, and J. LeMagnen. 1971. The role of olfactory and orosensory factors in the alcohol preference of inbred strains of mice. *Physiol. Behav. 6*:53–59.

Neel, J. V. 1962. Diabetes mellitus: a "thrifty" genotype rendered detrimental by "progress"? *Am. J. Hum. Genet. 14*:353–362.

Neel, J. V., S. Fajans, J. Conn, and R. Davison. 1965. Diabetes mellitus. In *Genetics and Epidemiology of Chronic Diseases,* Pub. No. 1163. Washington D. C.: U. S. Dept. of Health, Education and Welfare, pp. 105–132.

Neel, J. V., M. Layrisse, and F. M. Salzano. 1977. Man in the tropics: the Yanomama Indians. In *Population Structure and Human Variation.* G. A. Harrison (ed.). Cambridge: Cambridge University Press, pp. 109–142.

Neel, J. V., and R. H. Post. 1963. Transitory "positive" selection for color blindness. *Eugen. Q. 10*:33–35.

Neel, J. V., and W. J. Schull. 1968. On some trends in understanding the genetics of men. *Perspect. Biol. Med. 11*:565–602.

Nelson, M. C. 1971. Classical conditioning in the blowfly (*Phormia regina*): associative and excitatory factors. *J. Comp. Physiol. Psychol. 77*:353–368.

Newman, H. H., F. N. Freeman, and K. J. Holzinger. 1937. *Twins: A Study of Heredity and Environment.* Chicago: University of Chicago Press.

Nicholls, J. R., and S. Hsiao. 1967. Addiction liability of albino rats: breeding for quantitative differences in morphine drinking. *Science 157*:561–563.

Nielson, J., and F. Henriksen. 1972. The incidence of chromosome aberrations among males in a Danish youth prison. *Acta Psychiatr. Scand. 48*:87–102.

Notkins, A. L. 1979. The causes of diabetes. *Scientific American,* Nov. *241*(5):62.

Oakeshott, J. G. and J. B. Gibson. 1980. *The Genetics of Alcohol Metabolism in Man.* (forthcoming.)

Ödegaard, Ö. 1963. The psychiatric disease entities in the light of genetic investigation. *Acta Psychiatr. Scand. 39*:169–194.

O'Donald, P. 1976. Territory size, breeding time and mating preference in the Arctic skua. *Nature* (London) *260*:774–775.

O'Donald, P. 1977. Mating preferences and sexual selection in the Arctic skua II. Behavioural mechanisms of the mating preferences. *Heredity 39*:111–119.

O'Ferrall, G. J. M., and E. P. Cunningham, 1974. Heritability of racing performance in thoroughbred horses. *Livest. Prod. Sci. 1*:87–97.

Ogilvie, D. M., and R. H. Stinson. 1966. Temperature selection in *Peromyscus* and laboratory mice, *Mus musculus. J. Mammal. 47*:655–660.

O'Hara, E., A. Pruzan, and L. Ehrman. 1976. Ethological isolation and mating experience in *Drosophila paulistorum. PNAS 73*:975–976.

Oliverio, A. 1974. Genetic and biochemical analysis of behavior in mice. *Prog. Neurobiol. 3*:193–215.

Oliverio, A. 1975. Genetic factors in the control of drug effects on the behaviour of mice. In *The Genetics of Behavior.* J. H. F. Van Abeelen (ed.). New York: North-Holland, pp. 375–395.

Olsen, H. H., and W. E. Peterson. 1951. Uniformity of semen production and behavior in monozygous triplet bulls. *J. Dairy Sci. 34*:489–490.

Olsen, H. H., and W. E. Peterson. 1952. *Uniformity and Nutritional Studies with Monozygotic Bulls.* Presented at 47th annual meeting, American Dairy Science Assoc., Davis, Calif.

Omenn, G. S., and A. G. Motulsky. 1972. Biochemical genetics and the evolution of human behavior. In *Genetics, Environment, and Behavior: Implications for Educational Policy.* L. Ehrman, G. S. Omenn, and E. Caspari (eds.). New York: Academic, pp. 129–171.

Osmond, H., and A. Hoffer. 1966. A comprehensive theory of schizophrenia. *Int. J. Neuropsychiatry 2*:302–309.

Parnell, R. W. 1958. *Behavior and Physique: An Introduction to Practical and Applied Somatometry*. London: Arnold.

Parsons, P. A. 1961. Fly size, emergence time and sternopleural chaeta number in *Drosophila*. *Heredity 16*:455–473.

Parsons, P. A. 1964. A diallel cross for mating speeds in *Drosophila melanogaster*. *Genetica 35*:141–151.

Parsons, P. A. 1965a. Assortative mating for a metrical characteristic in *Drosophila*. *Heredity 20*:161–167.

Parsons, P. A. 1965b. The determination of mating speeds in *Drosophila melanogaster* for various combinations of inbred lines. *Experientia 21*:478.

Parsons, P. A. 1967a. *The Genetic Analysis of Behaviour*. London: Methuen.

Parsons, P. A. 1967b. Behavioural homeostasis in mice. *Genetica 38*:138–142.

Parsons, P. A. 1971. Extreme-environment heterosis and genetic loads. *Heredity 26*:579–583.

Parsons, P. A. 1972a. Genetic determination of behavior (mice and men). In *Genetics, Environment and Behavior: Implications for Educational Policy*. L. Ehrman, G. S. Omenn, and E. Caspari (eds.). New York: Academic, pp. 75–98.

Parsons, P. A. 1972b. Variations between strains of *Drosophila melanogaster* and *D. simulans* in giving offspring in interspecific crosses. *Can. J. Genet. Cytol. 14*: 77–80.

Parsons, P. A. 1973. *Behavioural and Ecological Genetics: A Study in Drosophila*. Oxford: Oxford University Press.

Parsons, P. A. 1974a. Male mating speed as a component of fitness in *Drosophila*. *Behav. Genet. 4*:395–404.

Parsons, P. A. 1974b. The behavioral phenotype in mice. *Am. Nat. 108*:377–385.

Parsons, P. A. 1975. Quantitative variation in natural populations. *Genetics 79*: 127–136.

Parsons, P. A. 1977a. Isofemale strains and quantitative traits in natural populations of *Drosophila*. *Am. Natur. 111*:613–621.

Parsons, P. A. 1977b. Larval reaction to alcohol as an indicator of resource utilization differences between *Drosophila melanogaster* and *D. simulans*. *Oecologia 30*:141–146.

Parsons, P. A. 1977c. Lek behavior in *Drosophila* (*Hirtodrosophila*) *polypori* Malloch—an Australian rainforest species. *Evolution 31*:223–225.

Parsons, P. A. 1977d. Genes, behavior and evolutionary processes: the genus *Drosophila*. *Adv. Genet. 19*:1–32.

Parsons, P. A. 1978a. Boundary conditions for *Drosophila* resource utilization in temperate regions especially at low temperatures. *Am. Nat. 112*:1063–1074.

Parsons, P. A. 1978b. Habitat selection and evolutionary strategies in *Drosophila*. *Behav. Genet. 8*:511–526.

Parsons, P. A. 1979a. Larval reactions to possible resources in three *Drosophila* species as indicators of ecological divergence. *Aust. J. Zool. 27*:413–419.

Parsons, P. A. 1979b. Polygenic variation in natural populations of *Drosophila*. In *Quantitative Genetic Variation*. J. N. Thompson, Jr., and J. M. Thoday (eds.). New York: Academic, pp. 61–79.

Parsons, P. A. 1980. Habitat selection and speciation in *Drosophila*. In *Essays on Evolution and Speciation in Honor of M. J. D. White*. W. R. Atchley and D. S. Woodruff (eds.). Cambridge: Cambridge University Press (forthcoming).

Parsons, P. A., and I. R. Bock. 1979. The population biology of Australian *Drosophila*. *Ann. Rev. Ecol. Syst. 10*:229–245.

Parsons, P. A., and M. M. Green. 1959. Pleiotropy and competition at the vermilion locus in *Drosophila melanogaster. PNAS 45*:993–996.

Parsons, P. A., S. M. W. Hosgood, and B. T. O. Lee. 1967. Polygenes and polymorphism. *Mol. Gen. Genet. 99*:165–170.

Parsons, P. A., and D. Kaul. 1966. Mating speed and duration of copulation in *Drosophila pseudoobscura. Heredity 21*:219–225.

Parsons, P. A., and S. M. Stanley. 1980. Special ecological studies—Domesticated and widespread species. In *Genetics and Biology of Drosophila,* Vol. 3a. M. Ashburner, H. L. Carson, and J. N. Thompson (eds.). New York: Academic (forthcoming).

Partanen, J., M. Bruun, and T. Markkanen. 1966. *Inheritance of Drinking Behavior. A Study on Intelligence, Personality, and Use of Alcohol of Adult Twins.* Helsinki: The Finnish Foundation for Alcohol Studies.

Partridge, L. 1978. Habitat selection. In *Behavioural Ecology: An Evolutionary Approach.* J. R. Krebs, and N. B. Davies (eds.). Oxford: Blackwell, pp. 351–376.

Patterson, J. 1942. Isolating mechanisms in the genus *Drosophila. Biol. Symp. 6*:271–287.

Paul, T. D., I. K. Brandt, L. J. Elsas, C. E. Jackson, P. Mamunes, C. S. Nance, and W. E. Nance. 1978. Phenylketonuria heterozygote detection in families with affected children. *Am. J. Hum. Genet. 30*:293–301.

Payne, W. J. A., and J. Hancock. 1957. The direct effect of tropical climate on the performance of European-type cattle. *Emp. J. Exp. Agric. 25*:321–338.

Pearson, K., and A. Lee. 1903. On the laws of inheritance in man. 1. Inheritance of physical characters. *Biometrika 2*:357–462.

Penrose, L. S. 1954. Distance, size and shape. *Ann. Eugen. 18*:337–343.

Penrose, L. S. 1961. Genetics of growth and development of the fetus. In *Recent Advances in Human Genetics.* L. S. Penrose and H. L. Brown (eds.). London: Churchill, pp. 56–75.

Penrose, L. S. 1963. *The Biology of Mental Defect,* 3d ed. London: Sidgwick & Jackson.

Pérez-Miravete, A. (ed). 1973. *Behavior of Micro-organisms.* New York: Plenum.

Perttunen, V. 1963. Effect of desiccation on the light reactions of some terrestrial arthropods. *Ergeb. Biol. 26*:90–97.

Petit, C. 1958. Le determinisme génétique et psycho-physiologique de la competition sexuelle chez *Drosophila melanogaster. Bull. Biol. Fr. Belg. 92*:1–329.

Petit, C., and L. Ehrman. 1969. Sexual selection in *Drosophila. Evol. Biol. 3*:177–223.

Petras, M. L. 1967. Studies of natural populations of *Mus.* 1. Biochemical polymorphisms and their bearing on breeding structure. *Evolution 21*:259–274.

Petters, R. M., D. S. Grosch, and C. S. Olson, 1978. A flightless mutation in the wasp *Habrobracon juglandis. J. Heredity 69*:113–116.

Pettigrew, T. F. 1971. Race, mental illness and intelligence: a social psychological view. In *The Biological and Social Meaning of Race.* R. H. Osborne (ed.). San Francisco: Freeman, pp. 87–124.

Philippart, M. 1979. Gaucher disease. In *Birth Defects Compendium.* 2d ed. D. Bergsma (ed.). New York: Liss, pp. 458–459.

Pickford, R. W. 1972. Colour-defective art students in four art schools. *Br. J. Physiol. Opt. 27*:102–114.

Pittendrigh, C. S. 1958. Adaptation, natural selection, and behavior. In *Behavior and Evolution.* A. Roe and G. Simpson (eds.). New Haven: Yale University Press, pp. 390–416.

Pollack, G., and R. Hoy. 1979. Temporal pattern as a cue for species-specific calling song recognition in crickets. *Science 204*:429–432.

Pollin, W. 1971. A possible genetic factor related to psychosis. *Am. J. Psychiat.* *128*:311–317.

Pollitzer, W. S. 1972. Discussion. In *Genetics, Environment, and Behavior: Implications for Educational Policy*. L. Ehrman, G. S. Omenn, and E. Caspari (eds.). New York: Academic, pp. 123–127.

Porter, I. H. 1968. *Heredity and Disease*. New York: McGraw-Hill.

Potegal, M. 1971. A note on spatial-motor deficits in patients with Huntington's disease: a test of a hypothesis. *Neuropsychologia 9*:233–235.

Powell, J. R. 1975. Protein variation in natural populations of animals. In *Evolutionary Biology*. T. Dobzhansky, M. K. Hecht, and W. C. Steere (eds.). New York: Plenum, vol. 8, pp. 79–120.

Price, W., and P. Whatmore. 1967. Criminal behavior and the XYY male. *Nature* (London) *213*:815–816.

Prince, G. J., and P. A. Parsons, 1980. Resource utilization specificity in three cosmopolitan *Drosophila* species. *J. Nat. Hist. 14*:559–563.

Propping, P. 1977. Genetic control of ethanol action on the central nervous system. *Hum. Genet. 35*:309–334.

Prout, T. 1971a. The relation between fitness components and population prediction in *Drosophila*. 1. The estimation of fitness components. *Genetics 68*:127–149.

Prout, T. 1971b. The relation between fitness components and population prediction in *Drosophila*. 2. Population prediction. *Genetics 68*:151–167.

Provine, W. B. 1971. *The Origins of Theoretical Population Genetics*. Chicago: University of Chicago Press.

Pruzan, A. 1976. Effects of age, rearing, and mating experiences on frequency dependent sexual selection in *Drosophila pseudoobscura*. *Evol. 30*:130–145.

Pruzan, A., P. Applewhite, and M. Bucci. 1977. Protein synthesis inhibition alters *Drosophila* mating behavior. *Pharmac. Biochem. Behav. 6*:355–357.

Pruzan, A., and G. Bush. 1977. Genotypic differences in larval olfactory discrimination in two *Drosophila melanogaster* strains. *Behav. Genet. 7*:457–464.

Pruzan, A., and L. Ehrman. 1974. Age, experience, and rare-male mating advantages in *Drosophila pseudoobscura*. *Behav. Genet. 4*:159–164.

Pruzan, A., L. Ehrman, I. Perelle, and J. Probber. 1979. Sexual selection, *Drosophila* age and experience. *Experientia 35*:1023–1024.

Quinn, W. G., and Y. Dudai. 1976. Memory phases in *Drosophila*. *Nature* (London) *262*:576–577.

Quinn, W. G., and J. L. Gould. 1979. Nerves and genes. *Nature* (London) *278*:19–23.

Quinn, W. G., W. A. Harris, and S. Benzer. 1974. Conditioned behavior in *Drosophila melanogaster*. *PNAS 71*:708–712.

Ramirez, I., and Fuller, J. L. 1976. Genetic influence on water and sweetened water consumption in mice. *Physiol. Behav. 16*:163–168.

Reed, E. W., and S. C. Reed. 1965. *Mental Retardation: A Family Study*. Philadelphia: Saunders.

Regen, J. 1914. Über die Anlockung des Weibchens von *Gryllus campestris* L. durch telephonisch übertrangene Stridulationslaute des Mannchens. *Pflüger's Arch. 155*:193–200.

Reimer, J. D., and M. L. Petras. 1967. Breeding structure of the house mouse, *Mus musculus,* in a population cage. *J. Mammal. 48*:88–89.

Richmond, R., and L. Ehrman. 1974. The incidence of repeated mating in the superspecies *D. paulistorum*. *Experientia 30*:489–490.

Riddle, D. L. 1977. A genetic pathway for dauer larva formation in *Caenorhabditis elegans*. In *Stadler Genetics Symposium,* Columbia: University of Missouri, vol. 9, pp. 101–120.

Riddle, D. L. 1978. The genetics of development and behavior in *Caenorhabditis elegans*. *J. Nematol. 10*:1–15.

Riekhof, P. L., W. A. Horton, D. J. Harris, and R. N. Schimke. 1972. Monozygotic twins with the Turner syndrome. *Am. J. Obstet. Gynecol. 112*:59–61.

Rife, D. C. 1938. Genetic studies of monozygotic twins. *J. Hered. 29*:83–90.

Rimoin, D. L., and R. N. Schimke. 1971. *Genetic Disorders of the Endocrine Glands*. St. Louis: Mosby.

Roberts, J. A. F. 1952. The genetics of mental deficiency. *Eugen. Rev. 44*:71–83.

Robinson, A., H. A. Lubs, and D. Bergsma (eds.). 1979. *Sex Chromosome Aneuploidy: Prospective Studies on Children*. The National Foundation–March of Dimes. *Birth Defects: Original Article Series, 15*(1). New York: Alan R. Liss.

Rockwell, R. F., F. Cooke, and R. Harmsen. 1975. Photobehavioral differentiation in natural populations of *Drosophila pseudoobscura* and *D. persimilis. Behav. Genet. 5*:189–202.

Rockwell, R. F., and M. B. Seiger. 1973a. Phototaxis in *Drosophila:* a critical evaluation. *Am. Sci. 61*:339–345.

Rockwell, R. F., and M. B. Seiger. 1973b. A comparative study of photoresponse in *Drosophila pseudoobscura* and *Drosophila persimilis. Behav. Genet. 3:* 163–174.

Roderick, G. W. 1968. *Man and Heredity*. New York: Macmillan.

Rodgers, D. A., and G. E. McClearn. 1962. Mouse strain differences in preference for various concentrations of alcohol. *Q. J. Stud. Alcohol 23*:26–33.

Rogoff, W. M., G. H. Gretz, M. Jacobson, and M. Beroza. 1973. Confirmation of (Z)-9-tricosene as a sex pheromone of the house fly. *Ann. Entomol. Soc. Am. 66*:739–741.

Rose, A., and P. A. Parsons. 1970. Behavioural studies in different strains of mice and the problems of heterosis. *Genetica 41*:65–87.

Rose, R. J., J. Miller, M. Dumont-Driscoll, and M. Evans. 1979. Twin-family studies of perceptual speed ability. *Beh. Genet. 9*:71–86.

Rosenthal, D. 1970. *Genetic Theory and Abnormal Behavior*. New York: McGraw-Hill.

Rosenthal, D. 1971. *Genetics of Psychopathology*. New York: McGraw-Hill.

Rothenbuhler, N. 1964. Behavior genetics of nest cleaning in honey bees. 4. Responses of F_1 and backcross generations to disease-killed brood. *Am. Zool. 4*:111–123.

Rundquist, E. A. 1933. Inheritance of spontaneous activity in rats. *J. Comp. Psychol. 16*:415–438.

Russell, L. 1961. Genetics of mammalian sex chromosomes. *Science 133*:1795–1803.

Salzano, F. M. 1972. Genetic aspects of the demography of American Indians and Eskimos. In *The Structure of Human Populations*. G. A. Harrison and A. J. Boyce (eds.). Oxford: Clarendon Press, pp. 234–251.

Sank, D. 1963. Genetic aspects of early total deafness. In *Family and Mental Health Problems in a Deaf Population*. J. D. Rainer, K. Z. Altschuler, and F. J. Kallmann (eds.). New York: Columbia University Press, pp. 28–81.

Satinder, K. P. 1971. Genotype-dependent effects of D-amphetamine sulphate and caffeine on escape-avoidance behavior of rats. *J. Comp. Physiol. Psychol. 76*:359–364.

Satow, Y., and C. Kung. 1974. Genetic dissection of the active electrogenesis in *Paramecium aurelia*. *Nature* (London) *247*:69–71.

Saul, S. H., M. J. Sinsko, P. R. Grimstad, and G. B. Craig, Jr. 1978. Population genetics of the mosquito *Aedes triseriatus:* genetic-ecological correlation at an esterase locus. *Am. Nat. 112*:333–339.

Savage, T. F., and W. M. Collins. 1972. Inheritance of star-gazing in Japanese quail. *J. Hered. 63*:88.

Say, B., N. J. Carpenter, P. R. Lanier, M. C. Cora Banez, K. Jones, and J. G. Coldwell. 1977. Chromosome variants in children with psychiatric disorders. *Am. J. Psychiat. 134*:424–426.

Scarr, S., and R. A. Weinberg. 1976. IQ test performance of black children adopted by white families. *Am. Psychol. 31*:726–739.

Schaffer, J. 1962. A specific cognition deficit observed in gonadal aplasia (Turner's syndrome). *J. Clin. Psychol. 18*:403.

Scharloo, W. 1971. Reproductive isolation by disruptive selection: Did it occur? *Am. Nat. 105*:83–96.

Schlesinger, K., R. C. Elston, and W. Boggan. 1966. The genetics of sound-induced seizure in inbred mice. *Genetics 54*:95–103.

Schlesinger, K., and B. J. Griek. 1970. The genetics and biochemistry of audiogenic seizures. In *Contributions to Behavior-Genetic Analysis: The Mouse as a Prototype.* G. Lindzey and D. D. Thiessen (eds.). New York: Appleton, pp. 219–257.

Schoen, A. M. S., E. M. Banks, and S. E. Curtis. 1976. Behavior of young Shetland and Welsh ponies (*Equus caballus*). *Biol. Behav. 1*:192:216.

Schroder, J. H. 1978. Differential response to irradiation of offspring of freshwater and seawater substrains of *Poecilia* (*Lebistes*) *reticulata* Peters in the "guppy male courtship activity test." *Theor. Appl. Genet. 51*:223–232.

Schuckit, M., D. W. Goodwin, and G. A. Winokur. 1972a. The half-sibling approach in a genetic study of alcoholism. In *Life History Research in Psychopathology,* Vol. 2. M. Roff, L. N. Robins, and M. Pollack (eds.). Minneapolis: University of Minnesota Press, pp. 120–127.

Schuckit, M., D. W. Goodwin, and G. A. Winokur. 1972b. A study of alcoholism in half-siblings. *Am. J. Psychiat. 128*:1132–1136.

Schuckit, M. A. and V. Rayes. 1979. Ethanol ingestion: differences in blood acetaldehyde concentrations in relatives of alcoholics and controls. *Science 203*:54–55.

Scott, J. P. 1943. Effects of single genes on the behavior of *Drosophila*. *Am. Nat. 77*:184–190.

Scott, J. P. 1977. Social genetics. *Behav. Genet. 7*:327–346.

Scott, J. P., and J. L. Fuller. 1965. *Genetics and the Social Behavior of the Dog.* Chicago: University of Chicago Press.

Scott, J. P., J. Stewart, and V. DeGhett. 1973. Separation in infant dogs. In *Separation and Depression*. J. P. Scott and E. C. Senay (eds.). Washington, D.C.: American Association for the Advancement of Science, pp. 3–32.

Selander, R. K. 1970. Behavior and genetic variation in natural populations. *Am. Zool. 10*:53–66.

Selander, R. K., and D. W. Kaufman. 1973. Genic variability and strategies of adaptation in animals. PNAS 70:1875–1877.

Sewell, D., B. Burnet, and K. Connolly. 1975. Genetic analysis of larval feeding behaviour in *Drosophila melanogaster. Genet. Res. 24*:163–173.

Sharpe, R. S., and P. A. Johnsgard. 1966. Inheritance of behavioral characters in F_2 mallard × pintail (*Anas platyrhynchos* L. × *Anas acuta* L.) hybrids. *Behaviour 27*:259–272.

Sheldon, W. H., and S. S. Stevens. 1942. *The Varieties of Temperament: A Psychology of Constitutional Differences.* New York: Harper & Row.

Sheldon, W. H., S. S. Stevens, and W. B. Tucker. 1940. *The Varieties of Human Physique: An Introduction to Constitutional Psychology.* New York: Harper & Row.

Shepard, T. H. 1961. Increased incidence of nontasters of phenylthiocarbamide among congenital athyrotic cretins. *Science 131*:929.

Shields, J. 1962. *Monozygotic Twins Brought up Apart and Brought up Together.* London: Oxford University Press.

Shorey, H., and R. J. Bartell. 1970. Role of a volatile female sex pheromone in stimulating male courtship behavior in *Drosophila melanogaster. Anim. Behav. 18*:159–164.

Sidman, R. L., S. H. Appel, and J. L. Fuller. 1965. Neurological mutants of the mouse. *Science 150*:513–516.

Sidman, R. L., and M. C. Green. 1965. Retinal degeneration in the mouse: Location of the *rd* locus in linkage group XVII. *J. Hered. 56*:23–29.

Siegel, I. M. 1967. Heritability and threshold determinations of the optomotor response in *Drosophila melanogaster. Anim. Behav. 15*:299–306.

Siegel, P. B. 1972. Genetic analysis of male mating behavior in chickens. 1. Artificial selection. *Anim. Behav. 20*:564–570.

Siegel, P. B. 1979. Behavior-genetics in chickens: a review. *World's Poult. Sci. J. 35*:9–19.

Silcock, M., and P. A. Parsons. 1973. Temperature preference differences between strains of *Mus musculus,* associated variables, and ecological implications. *Oecologia 12*:147–160.

Silverstein, R. M. 1977. Complexity, diversity and specificity of behavior modifying chemicals. In *Chemical Control of Insect Behavior.* H. Shorey and J. McKelvey, Jr. (eds.). New York: Wiley, pp. 231–251.

Simoons, F. J. 1970. Primary adult lactose intolerance and the milking habit: a problem in biologic and cultural interrelations. 2. A culture historical hypothesis. *Am. J. Digest. Dis. 15*:695–710.

Simoons, F. J., J. D. Johnson, and N. Kretchmer. 1977. Perspective on milk-drinking and malabsorption of lactose. *Amer. Acad. Pediat. 59*:98–109.

Simpson, G. G. 1969. *Biology and Man.* New York: Harcourt, Brace & World.

Sinnock, P. 1970. Frequency dependence and mating behavior in *Tribolium castaneum. Am. Nat. 104*:469–476.

Skodak, M., and H. M Skeels. 1949. A final follow-up study of one hundred adopted children. *J. Genet. Psychol. 75*:85–125.

Slater, E., and V. Cowie. 1971. *The Genetics of Mental Disorders.* London: Oxford University Press.

Snyder, L. H., and P. R. David. 1957. *The Principles of Heredity,* 5th ed. Boston: Heath.

Snyder, L., and D. F. Davidson. 1937. Studies of human inheritance. 18. The inheritance of taste deficiency to diphenylguanidine. *Eugen. News 22*:1–2.

Sonneborn, T. M. 1970. Methods in *Paramecium* research. In *Methods of Cell Physiology.* Vol. 4. D. M. Prescott (ed.). New York: Academic, pp. 241–339.

Southern, H. N., and E. M. O. Laurie. 1946. The house-mouse (*Mus musculus*) in corn ricks. *J. Anim. Ecol.* *15*:134–149.

Spassky, B., and T. Dobzhansky. 1967. Responses of various strains of *Drosophila pseudoobscura* and *Drosophila persimilis* to light and gravity. *Am. Nat.* *101*:59–63.

Spatz, H. C., A. Emanns, and H. Reichart. 1974. Associative learning in *Drosophila melanogaster*. *Nature* (London) *248*:359–361.

Spielman, R. S., E. C. Migliazza, and J. V. Neel. 1974. Regional linguistic and genetic differences among Yanomama Indians. *Science* *184*:637–644.

Spiess, E. B. 1962. *Papers on Animal Population Genetics*. Boston: Little Brown.

Spiess, E. B. 1968. Low frequency advantage in mating of *Drosophila pseudoobscura* karyotype. *Am. Nat.* *102*:363–379.

Spiess, E. B. 1977. *Genes in Populations*. New York: Wiley.

Spiess, E. B., and B. Langer. 1961. Chromosomal adaptive polymorphism in *Drosophila persimilis*. 3. Mating propensity of homokaryotypes. *Evolution* *15*:535–544.

Spiess, E. B., and B. Langer. 1964a. Mating speed control by gene arrangements in *Drosophila pseudoobscura* homokaryotypes. *PNAS* *51*:1015–1018.

Spiess, E. B., and B. Langer. 1964b. Mating speed control by gene arrangements in *Drosophila persimilis*. *Evolution* *18*:430–444.

Spiess, E. B., B. Langer, and L. Spiess. 1966. Mating control by gene arrangements in *Drosophila pseudoobscura*. *Genetics* *54*:1139–1149.

Spiess, E. B., and L. Spiess. 1967. Mating propensity, chromosomal polymorphism, and dependent conditions in *Drosophila persimilis*. *Evolution* *21*:672–678.

Spiess, L. D., and E. B. Spiess. 1969. Minority advantage in interpopulational matings of *Drosophila persimilis*. *Am. Nat.* *103*:155–172.

Spieth, H. T. 1952. Mating behavior within the genus *Drosophila* (Diptera). *Bull. Am. Mus. Nat. Hist.* *99*:105–145.

Spieth, H. T. 1958. Behavior and isolating mechanisms. In *Behavior and Evolution*. A. Roe and G. G. Simpson (eds.). New Haven: Yale University Press, pp. 363–389.

Spieth, H. T. 1974a. Courtship behavior in *Drosophila*. *Ann. Rev. Entomol.* *19*:385–405.

Spieth, H. T. 1974b. Mating behavior and evolution of the Hawaiian *Drosophila*. In *Genetic Mechanisms of Speciation in Insects*. M. J. D. White (ed.). Sydney: ANZ Book Company, pp. 94–101.

Spuhler, J. N. 1962. *Empirical Studies on Quantitative Human Genetics*. WHO Seminar on the Use of Vital and Health Statistics for Genetic and Radiation Studies. New York: U. N. Publ., pp. 241–252.

Spuhler, J. N. (ed.). 1967. *Genetic Diversity and Human Behavior*. Chicago: Aldine.

Spuhler, J. N. 1968. Assortative mating with respect to physical characteristics. *Eugen. Q.* *15*:128–140.

Spuhler, J. N., and G. Lindzey. 1967. Racial differences in behavior. In *Behavior-Genetic Analysis*. J. Hirsch (ed.). New York: McGraw-Hill, pp. 366–414.

Spuhler, K. P., D. W. Crumpacker, J. S. Williams, and B. P. Bradley. 1978. Response to selection for mating speed and changes in gene arrangement frequencies in descendants from a single population of *Drosophila pseudoobscura*. *Genetics* *89*:729–749.

Staats, J. 1966. The laboratory mouse. In *Biology of the Laboratory Mouse*. E. L. Green (ed.). New York: McGraw-Hill, pp. 1–9.

Stalker, H. D. 1942. Sexual isolation studies in the species complex *Drosophila virilis*. *Genetics* *27*:238–257.

Stalker, H. D. 1976. Chromosome studies in wild populations of *D. melanogaster*. *Genetics* *82*:323–347.

Stebbins, G. L. 1950. *Variation and Evolution in Plants.* New York: Columbia University Press.

Stein, Z., M. Susser, G. Saenger, and F. Marolla. 1972. Nutrition and mental performance. *Science 178*:708–713.

Stern, C. 1973. *Principles of Human Genetics,* 3d. ed. San Francisco: Freeman.

Stevenson, A. C., B. C. C. Davidson, and M. W. Oakes. 1970. *Genetic Counseling.* Philadelphia: Lippincott.

Strickberger, M. W. 1976. *Genetics,* 2d ed. New York: Macmillan.

Sturtevant, A. H. 1915. Experiments on sex recognition and the problem of sexual selection in *Drosophila. J. Anim. Behav. 5*:351–366.

Sutton, H. E. 1975. *An Introduction to Human Genetics,* 2d ed. New York: Holt.

Szebenyi, A. 1969. Cleaning behavior in *Drosophila melanogaster. Anim. Behav. 17*:641–651.

Takahashi, M. 1979. Behavioral mutants in *Paramecium caudatum. Genetics 91*:393–408.

Tan, C. C. 1946. Genetics of sexual isolation between *Drosophila pseudoobscura* and *Drosophila persimilis. Genetics 31*:558–573.

Taylor, C. E., and Powell, J. R. 1977. Microgeographic differentiation of chromosomal and enzyme polymorphism in *Drosophila persimilis. Genetics 85*:681–695.

Taylor, W. O. 1971. Effects on employment of defects in colour vision. *Br. J. Ophthalmol. 55*:753–760.

Thiessen, D. D. 1972. *Gene Organization and Behavior.* New York: Random House.

Thiessen, D. D., K. Owen, and M. Whitsett. 1970. Chromosome mapping of behavioral activities. In *Contributions to Behavior Genetic Analysis: The Mouse as a Prototype.* G. Lindzey and D. D. Thiessen (eds.). New York: Appleton, pp. 161–204.

Thoday, J. M. 1961. Location of polygenes. *Nature* (London) *191*:368–370.

Thoday, J. M., and J. B. Gibson, 1970. Environmental and genetical contributions to class difference: a model experiment. *Science 167*:990–992.

Thompson, J., and M. Thompson. 1973. *Genetics in Medicine,* 2d ed. Philadelphia: Saunders.

Thompson, J. N., Jr., and J. M. Thoday (eds.). 1979. *Quantitative Genetic Variation.* New York: Academic.

Thompson, W. R. 1953. The inheritance of behaviour: behavioural differences in fifteen mouse strains. *Can. J. Psychol. 7*:145–155.

Thompson, W. R. 1956. The inheritance of behavior: activity differences in five inbred mouse strains. *J. Hered. 47*:147–148.

Thuline, H., and D. Norby. 1961. Spontaneous occurrence of chromosomal abnormality in cats. *Science 134*:554–555.

Thurstone, L. L., and T. G. Thurstone. 1941. *The Primary Mental Abilities Tests.* Chicago: Science Research.

Tindale, N. B. 1953. Tribal and intertribal marriage among the Australian aborigines. *Hum. Biol. 25*:169–190.

Tobach, E., J. S. Bellin, and D. K. Das. 1974. Differences in bitter taste perception in three strains of rats. *Behav. Genetics 4*:405–410.

Tobach, E., and B. Rosoff (eds.). 1978. *Genes and Gender.* New York: Gordian Press.

Todd, N. B. 1962. Behavior and genetics of the domestic cat. *Cornell Veterinarian 53*:99.

Todd, N. B. 1978. An ecological, behavioral genetic model for the domestication of the cat. *Carnivore 1*:52–60.

Tolman, E. C. 1924. The inheritance of maze-learning ability in rats. *J. Comp. Psychol.* *4*:1–18.

Trivers, R. L. 1972. Parental investment and sexual selection. In *Sexual Selection and the Descent of Man.* B. Campbell (ed.). Chicago: Aldine, pp. 136–179.

Tryon, R. C. 1942. Individual differences. In *Comparative Psychology,* 2d ed. F. A. Moss (ed.). Englewood Cliffs, N. J.: Prentice-Hall.

Tschudy, D. P. 1979. Porphyria. In *Birth Defects Compendium,* 2d ed. D. Bergsma (ed.). New York: Liss, pp. 880–883.

Tsuboi, T., and S. Endo. 1977. Incidence of seizures and EEG abnormalities among offspring of epileptic patients. *Hum. Genet. 36*:173–189.

Vale, J. R., C. A. Vale, and J. P. Harley. 1971. Interaction of genotype and population number with regard to aggressive behavior, social grooming, and adrenal and gonadal weight in male mice. *Commun. Behav. Biol. 6*:209–221.

Valenstein, E. S., W. Riss, and W. C. Young. 1955. Experimental and genetic factors in the organization of sexual behavior in male guinea pigs. *J. Comp. Physiol. Psychol. 48*:397–403.

Van Abeelen, J. H. F. (ed.). 1975. *The Genetics of Behavior.* New York: North-Holland.

Vandenberg, S. 1967. Hereditary factors in psychological variables in man, with special emphasis on cognition. In *Genetic Diversity and Human Behavior.* J. N. Spuhler (ed.). Chicago: Aldine, pp. 99–133.

Vandenberg, S. 1972. The future of human behavior genetics. In *Genetics, Environment, and Behavior: Implications for Educational Policy.* L. Ehrman, G. S. Omenn, and E. Caspari (eds.). New York: Academic, pp. 273–289.

Van Riper, C. 1971. *The Nature of Stuttering.* Englewood Cliffs, N.J.: Prentice-Hall.

Van Valen, L., L. Levine, and J. A. Beardmore. 1962. Temperature sensitivity of chromosomal polymorphism in *Drosophila pseudoobscura. Genetica 33*: 113–127.

Vessie, P. R. 1932. On the transmission of Huntington's chorea for 300 years: the Bures family group. *J. Nerv. Ment. Dis. 76*:553–573.

Voaden, D. J., M. Jacobson, W. M. Rogoff, and G. H. Gretz. 1972. Chemical investigation of the sex pheromone of the house fly. *J. Econ. Entomol. 65*:359–385.

Vogel, F., and G. Röhrborn. 1970. *Chemical Mutagenesis in Mammals and Man.* New York: Springer-Verlag.

Vogel, F., and E. Schalt. 1979. The electroencephalogram (EEG) as a research tool in human behavior genetics: psychological examinations in healthy males with various inherited EEG variants. III. Interpretation of the results. *Hum. Genet. 47*:81–111.

Vogel, F., E. Schalt, and J. Krüger. 1979. The electroencephalogram (EEG) as a research tool in human behavior genetics: psychological examination in healthy males with various inherited EEG variants. II. Results. *Hum. Genet. 47*: 47–80.

Vogel, F., E. Schalt, J. Krüger, P. Propping, and K. Lehnert. 1979. The electroencephalogram (EEG) as a research tool in human behavior genetics: psychological examinations in healthy males with various inherited EEG variants. I. Rationale of the study, materials, methods. Heritability of test parameters. *Hum. Genet. 47*:1–45.

Waddington, C. H. 1957. *The Strategy of the Genes*. London: Allen & Unwin.

Wagner, R. P., B. H. Judd, B. G. Sanders, and R. H. Richardson. 1980. *Introduction to Modern Genetics*. New York: Wiley.

Wahlsten, D. 1972. Genetic experiments with animal learning: a critical review. *Behavioral Biology* 7:143–182.

Wahlund, S. 1928. Zusammensetzung von Populationen und Korrelationserscheinungen von Standpunkt der Vererbungslehre aus betrachet. *Hereditas* 11:65–106.

Walker, R. N. 1962. Body build and behavior in young children. 1. Body build and nursery school teachers' ratings. *Monogr. Soc. Res. Child Dev.* 27(3):1–94.

Wallace, B. 1954. Genetic divergence of isolated populations of *Drosophila melanogaster*. *Caryologia* 6 (Suppl.):761–764.

Waller, J. H. 1971. Achievement and social mobility: relationships among IQ score, education, and occupation in two generations. *Soc. Biol.* 18:252–259.

Ward, S. 1973. Chemotaxis by the nematode *Caenorhabditis elegans:* identification of attractants and analysis of the response by use of mutants. *PNAS* 70:817–821.

Ward, S. 1977. Invertebrate neurogenetics. *Ann. Rev. Genet.* 11:415–450.

Washburn, S. L. 1978. The evolution of man. *Sci. Am.* 239:146–169.

Washburn, S. L., and R. Moore. 1974. *Ape Into Man*. Boston: Little, Brown.

Watson, J. D., and F. H. Crick. 1953. Molecular structure of nucleic acids: a structure for deoxyribose nucleic acid. *Nature* (London) 171:737–738.

Weber, P. G., and S. P. Weber. 1975. The effect of female color, size, dominance and early experience upon mate selection in male convict cichlids, *Cichlasoma nigrofasciatum* Gunther (Pisces: Cichlidae). *Behaviour* 16:116–135.

Wecker, S. C. 1964. Habitat selection. *Sci. Am.* 211(4):109–116.

White, N. G. 1979. "Tribes, Genes and Habitats: Genetic Diversity among Aboriginal Populations in the Northern Territory of Australia." PhD thesis, La Trobe University, Bundoora, Australia.

White, N. G., and P. A. Parsons. 1973. Genetic and socio-cultural differentiation in the aborigines of Arnhem Land, Australia. *Am. J. Phys. Anthropol.* 38:5–14.

White, N. G., and P. A. Parsons. 1976. Population genetic, social, linguistic and topographical relationships in north-eastern Arnhem Land, Australia. *Nature* (London) 261:223–225.

Whiting, P. W. 1932. Reproductive reactions of sex mosaics of a parasitic wasp. *J. Comp. Psychol.* 14:345–363.

Whiting, P. W. 1934. Eye colors in the parasitic wasp *Habrobracon* and their behavior in multiple recessives and mosaics. *J. Genet.* 29:99–107.

Whiting, P. W. 1939. Mutant body colors in the parasitic wasp *Habrobracon juglandis* and their behavior in multiple recessives and mosaics. *Proc. Am. Phil. Soc.* 80:65–85.

Wienckowski, L. A. 1972. *Schizophrenia: Is There an Answer?* Washington, D.C.: U.S. Department of Health, Education and Welfare.

Wiersma, L. (ed.). 1967. *Invertebrate Nervous Systems: Their Significance for Mammalian Neurophysiology*. (Conference on Invertebrate Nervous Systems. California Institute of Technology.) Chicago: University of Chicago Press.

Wilcock, J. 1969. Gene action and behavior: an evaluation of major gene pleiotropism. *Psychol. Bull.* 72:1–29.

Williams, G. C. 1966. *Adaptation and Natural Selection*. Princeton: Princeton University Press.

Williamson, R., W. D. Kaplan, and D. Dagan. 1974. A fly's leap from paralysis. *Nature* (London) 252:224–226.

Wilson, E. O. 1973. Group selection and its significance for ecology. *Biol. Sci. 23*:631–638.

Wilson, E. O. 1975. *Sociobiology: The New Synthesis*. Harvard: Belknap Press.

Wilson, R. S. 1972. Twins: early mental development. *Science 175*:914–917.

Wilson, R. S. 1975. Twins: patterns of cognitive development as measured on the Wechsler preschool and Primary Scale of Intelligence. *Dev. Psychol. 11*:126–134.

Wilson, R. S. 1977. Twins and siblings: concordance for school-age mental development. *Child Dev. 48*:211–216.

Winokur, G. 1973. Genetic aspects of depression. In *Separation and Depression*. J. P. Scott and E. C. Senay (eds.). Washington, D.C.: American Association for the Advancement of Science.

Wisniewski, L., T. Hassold, J. Heffelfinger, and J. V. Higgins. 1979. Clinical studies in five cases of Inv Dup (15). *Hum. Genet. 50*:259–270.

Witkin, H., S. Mednick, F. Schulsinger, E. Bakkestrøm, K. Christiansen, D. Goodenough, K. Hirschorn, C. Lundsteen, D. Owen, J. Philip, D. Rubin, and M. Stocking. 1976. Criminality and XYY and XXY men. *Science 193*:547–555.

Wolken, J. J., A. D. Mellom, and G. Contis. 1957. Photoreceptor structure. 2. *Drosophila melanogaster. J. Exp. Zool. 134*:383–406.

Wolkin, J. R. 1977. "Structural and Behavioral Characteristics of a Young Hybrid Ape (*Hylobates lar moloch* × *Symphalangus syndactylus*)." Master's thesis, Georgia State University.

Wood-Gush, D. G. M. 1972. Strain differences in response to sub-optimal stimuli in the fowl. *Anim. Behav. 20*:72–76.

Woolf, C. M. 1971. Congenital cleft lip: a genetic study of 496 propositi. *J. Med. Genet. 8*:65–84.

Wright, J., and R. Pal. (eds.). 1967. *Genetics of Insect Vectors of Disease*. New York: Elsevier.

Wright, S. 1934. The results of crosses between inbred strains of guinea pig, differing in number of digits. *Genetics 19*:537–551.

Wright, S. 1955. Population genetics: the nature and cause of genetic variability in populations. *Cold Spring Harbor Symp. Quant. Biol. 7*:16–24.

Wright, S. 1960. Physiological genetics, ecology of populations and natural selection. In *Evolution After Darwin*, Vol. 1. S. Tax (ed.). Chicago: University of Chicago Press, pp. 429–475.

Wright, S., and T. Dobzhansky. 1946. Genetics of natural populations. 12. Experimental reproduction of some of the changes caused by natural selection in certain populations of *Drosophila pseudoobscura. Genetics 31*:125–156.

Wynne, L. C. (ed.). 1978. *The Nature of Schizophrenia*. New York: Wiley.

Yamazaki, K., E. A. Boyse, V. Mike, H. T. Thaler, B. J. Mathieson, J. Abbott, J. Boyse, Z. A. Zayas, and L. Thomas. 1976. Control of mating preferences in mice by genes in the major histocompatibility complex. *J. Exp. Med. 144*:1324–1335.

Yanai, J., P. Y. Sze, and B. E. Ginsburg. 1975. Effects of aminergic drugs and glutamic acid on audiogenic seizures induced by early exposure to ethanol. *Epilepsia 16*:67–71.

Zerbin-Rudin, E. 1969. Zur Genetik der Depressiven Erkrankungen. In *Das Depressive Syndrom*. H. Hippius and H. Selbach (eds.). Munich: Urban & Schwarzenberg.

Name Index

Page numbers in *italic* indicate illustrations or tables.

Subject Index

Page numbers in *italic* indicate illustrations or tables.

438